T3-BUV-203

PHYSICS FOR RADIOLOGISTS

Physics for Radiologists

P.P. Dendy, BA PhD
Chief Physicist,
Addenbrooke's Hospital, Cambridge

B. Heaton, MSc PhD
Lecturer in Biomedical Physics,
University of Aberdeen

Distributed in the USA and Canada by
BLACKWELL / YEAR BOOK MEDICAL PUBLISHERS • INC.

BLACKWELL SCIENTIFIC PUBLICATIONS

OXFORD LONDON EDINBURGH

BOSTON PALO ALTO MELBOURNE

Editorial offices:
Osney Mead, Oxford, OX2 0EL
(*Orders*: Tel. 0865 240201)
8 John Street, London, WC1N 2ES
23 Ainslie Place, Edinburgh, EH3 6AJ
52 Beacon Street, Boston
Massachusetts 02108, USA
667 Lytton Avenue, Palo Alto
California 94301, USA
107 Barry Street, Carlton
Victoria 3053, Australia

First published 1987

Set by Katerprint Typesetting Services Ltd, Oxford
Printed in Great Britain
at the University Press, Cambridge

DISTRIBUTORS

USA
Year Book Medical Publishers
35 East Wacker Driver
Chicago, Illinois 60601
(*Orders*: Tel. 312 726-9733)

Canada
The C. V. Mosby Company
5240 Finch Avenue East,
Scarborough, Ontario
(*Orders*: Tel. 416-298-1588)

Australia
Blackwell Scientific Publications
(Australia) Pty Ltd
107 Barry Street
Carlton, Victoria 3053
(*Orders*: Tel. (03) 347 0300)

British Library
Cataloguing in Publication Data

Dendy, P.P.
Physics for radiologists.
1. Radiology, Medical 2. Medical physics
I. Title II. Heaton, B.
530′.024616 R895

ISBN 0-632-01351-6

Contents

Introduction

The past 20 years have seen a rapid development in the range of imaging techniques available to the diagnostic radiologist and over the same period a marked increase in the level of sophistication of imaging equipment. These changes have inevitably led to changes in the background knowledge of physics that a radiologist is expected to acquire. For example, the concept of linking an image intensifier to a television camera in order to produce digitized images introduces many new ideas; many radiology departments now provide a nuclear medicine service, and of course the technique of magnetic resonance imaging is quite new. All of these subjects, and several others, require a knowledge and understanding of additional physical principles.

Fortunately, as equipment has become more sophisticated, it has also become more reliable. Furthermore the base-line knowledge of physics possessed by radiologists in training has risen. Thus the requirement to teach fundamental current electricity and electronics at this level has diminished considerably.

For all these reasons the content of a physics course for radiologists has changed quite appreciably and this book has been written to move with the times and to cover that physics now thought to be most relevant as reflected in the current syllabus of the Royal College of Radiologists (London).

The book has been designed as a comprehensive, compact primer, for aspiring radiologists, but to achieve this a number of hard decisions have had to be taken. First, it has been assumed that the reader will have studied physics to the standard required for university entrance (with some allowance for the knowledge being a little rusty), and is familiar with the terminology of such a physics syllabus. SI units have been adopted, with alternatives given where the reader might be confused.

Second, the subject matter is very interactive and this can lead to a tendency to repetition. In the interests of economy of space, most concepts are only considered once. Occasionally this requires a simple knowledge of material discussed later in the book. For example, the idea of filtration, which is dealt with in Chapter 2, requires a knowledge of beam attenuation, not discussed in detail until Chapter 3. Therefore it has been assumed that the reader will already possess a rudimentary knowledge of the subject or will be prepared to return to more difficult aspects at a second reading or during revision. The text is extensively cross-referenced and analogies have been drawn wherever possible.

Finally, for reasons both of ease of learning and compactness, lateral thinking has been encouraged where possible. For example, the ideas of exponential decay developed in Chapter 1 when discussing radioactivity are later applied to attenuation of X-ray beams and the measurement of relaxation times in magnetic resonance imaging with little further explanation.

The material is presented in a logical order. After a review of the fundamental principles of radiation physics and radioactivity in Chapter 1, Chapter 2 deals with the production of X-rays. All aspects of this topic have been considered here including X-ray tube rating and quality control measurements. Chapters 3 and 4, dealing respectively with the physics of the interaction of radiation with matter and the image receptor, provide the necessary basic information for consideration in Chapter 5 of the formation of the radiological image and the factors controlling image quality, discussed in qualitative terms. A number of more specialized radiographic techniques have been grouped together in Chapter 6.

Particular emphasis has been placed on those aspects of the subject covered in less detail in some other books used by radiologists. Thus, in Chapter 7 the difficult subject of radiation dosimetry has been presented with the requirements of radiologists particularly in mind. Similarly in Chapter 8 the physics of nuclear medicine is considered in some detail. This reflects the increasing involvement of radiologists in such work. The idea of digitized information is also introduced here for the first time. Quantification, a relatively new concept in diagnostic radiology, is the theme of Chapter 9, where both quantification of image quality via the production of digitized images, and quantification of imager performance are discussed. This chapter also deals with an increasingly important concept, the need to assess, in objective terms, the diagnostic value of a particular imaging investigation. All aspects of tomographic imaging are considered in Chapter 10.

Chapter 11 is seen as one of the most important chapters in the book since it brings together, from a wide variety of sources, most of the information a radiologist should know in radiobiology and radiological protection. The material presented here is completely up-to-date in respect of the UK Ionising Radiations Regulations 1985.

Chapters 12 and 13 scarcely do justice to ultrasound imaging and magnetic resonance imaging, but are thought to reflect fairly the amount of relevant physics a radiologist can reasonably be expected to learn in the time allocated to the subjects.

The detail given in the book is thought to be appropriate to a teaching course some 50 hours in length. A list of relevant references, both textbooks and review articles, has been provided for further reading at the end of each chapter, where a set of questions is also given to exercise the mind of the reader.

It is hoped that this text will provide both a starting point and a firm basis for revision for any modern course of physics designed to prepare radiologists for the Fellowship examinations.

Acknowledgements and thanks are due to many persons, but in particular to Dr Linda Eastwood, Miss Karen Palmer, Mr R.W. Barber, Dr R.H. Harrison, Dr W.I. Keyes, Dr C.J. Martin, Mr J. Sephton Wright and Dr P.F. Sharp for reading and commenting on early drafts of the manuscript.

Thanks are also due to Mrs S. Ellwood, Mrs J. Taylor and Mrs A. Dendy for their patience in typing and retyping several drafts of the manuscript.

The front cover is based on material published by Picker International, with permission.

1

Fundamentals of Radiation Physics and Radioactivity

1.1 Structure of the atom

All matter is made up of atoms, each of which has an essentially similar structure. All atoms are formed of small, dense, positively charged nuclei, typically some 10^{-15} m in diameter, orbited at much larger distances (about 10^{-10} m) by negatively charged, very light particles. The atom as a whole is electrically neutral.

The positive charge in the nucleus consists of a number of **protons** each of which has a charge of 1.6×10^{-19} coulombs (C) and a mass of 1.7×10^{-27} kilograms (kg). The negative charges are **electrons**. An electron carries the same numerical charge as the proton, but of opposite sign. However, an electron has only about 1/2000th the mass of the proton (9×10^{-31} kg). Each element is characterized by a specific number of protons, and an equal number of orbital electrons. This is called the **atomic number** and is normally denoted by the symbol Z. For example, Z for aluminium is 13, whereas for lead $Z = 82$.

Electrons can only exist at fairly well defined distances from the nucleus and are described as being in 'shells' around the nucleus (Fig. 1.1). More important than the distance of the electron from the nucleus is the electrostatic force binding the electron to the nucleus, or the amount of energy the electron would have to be given to escape from the field of the nucleus. This is equal to the amount of energy a free electron will lose when it is captured by the electrostatic field of a nucleus. It is possible to think in terms of an energy 'well' that gets deeper as the electron is trapped in shells closer and closer to the nucleus.

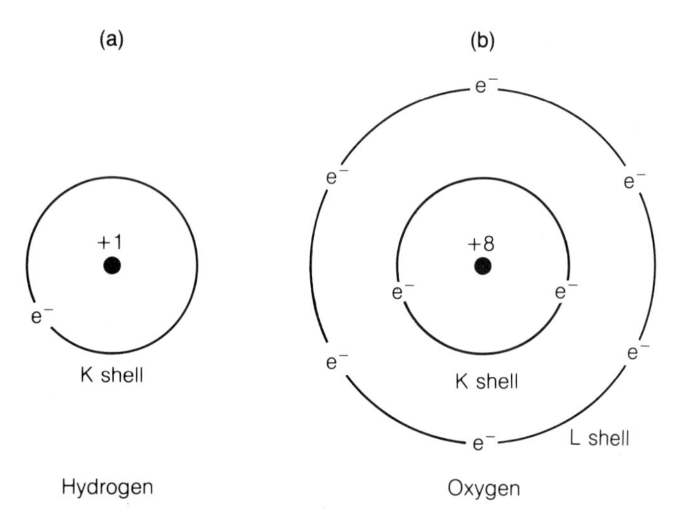

Fig. 1.1. Examples of atomic structure. (a) Hydrogen with one K shell electron. (b) Oxygen with two K shell electrons and six L shell electrons.

The unit in which electron energies are measured is the **electron volt** (eV)—this is the energy one electron would gain if it were accelerated through 1 volt of potential difference. One thousand electron volts is a kilo electron volt (keV) and one million electron volts is a mega electron volt (MeV). Some typical electron shell energies are shown in Fig. 1.2. Note that

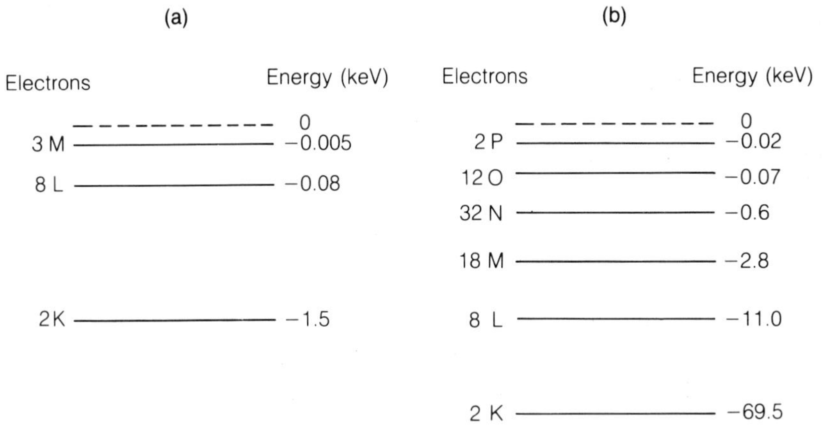

(a)

Electrons	Energy (keV)
	0
3 M	−0.005
8 L	−0.08
2K	−1.5

(b)

Electrons	Energy (keV)
	0
2 P	−0.02
12 O	−0.07
32 N	−0.6
18 M	−2.8
8 L	−11.0
2 K	−69.5

Fig. 1.2. Typical electron energy levels. (a) Aluminium (Z=13). (b) Tungsten (Z=74).

1 if a free electron is assumed to have zero energy, all electrons within atoms have negative energy—i.e. they are bound to the nucleus and must be given energy to escape;
2 the energy levels are not equally spaced and the difference between the K shell and the L shell is much bigger than any of the other differences.

1.2 Nuclear stability and instability

If a large number of protons were forced together in a nucleus they would immediately explode owing to electrostatic repulsion. Very short-range attractive forces are therefore required within the nucleus for stability, and these are provided by **neutrons**, uncharged particles with a mass almost identical to that of the proton.

The total number of protons and neutrons within the nucleus is called the **mass number**, usually given the symbol A. Each particular combination of Z and A defines a **nuclide**. One notation used to describe a nuclide is $^{A}_{Z}X$. The number of protons Z defines the element X, so for hydrogen $Z = 1$, for oxygen $Z = 8$ etc, but the number of neutrons is variable. Therefore an alternative notation that carries all necessary information is X-A.

Nuclides that have the same number of protons but different numbers of neutrons are known as **isotopes**. Thus O-16, the most abundant isotope of oxygen, has 8 protons (by definition) and 8 neutrons. O-17 is the isotope of oxygen which has 8 protons and 9 neutrons.

The number of neutrons required to stabilize a given number of protons lies within fairly narrow limits and Fig. 1.3a shows a plot of these numbers. Note that for many elements of biological importance the number of neutrons is equal to the number of protons, but the most abundant form of hydrogen, which has one proton but no neutrons, is an important exception. At higher atomic numbers the number of neutrons begins to increase faster than the number of protons—lead, for example, has 126 neutrons but only 82 protons.

An alternative way to display the data is to plot the sum of neutrons and protons against the number of protons (Fig. 1.3b). This is essentially a plot of nuclear mass against nuclear charge (or the total charge on the orbiting electrons). This concept will be useful when considering the interaction of ionizing radiation with matter, and in section 3.4 the near-constancy of mass/charge (A/Z is close to 2) for most of the biological range of elements will be considered in more detail.

If the ratio of neutrons to protons is outside narrow limits, the nuclide is

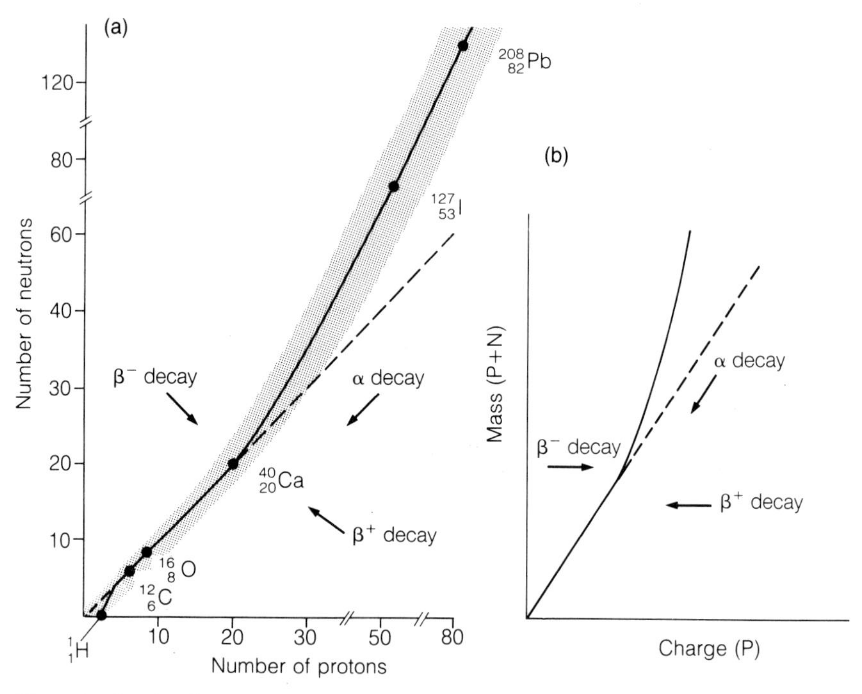

Fig. 1.3. Graphs showing the relationship between number of neutrons and number of protons for the most abundant stable elements. (a) Number of neutrons plotted against number of protons. The *dashed line* is at 45°. The *cross-hatched area* shows the range of values for which the nucleus is likely to be stable. (b) Total number of nucleons (neutrons and protons) plotted against number of protons.

On each graph the changes associated with β^+, β^- and α decay are shown.

radioactive or a **radionuclide**. For example H-1 (normal hydrogen) is stable, H-2 (deuterium) is also stable, but H-3 (tritium) is radioactive. A nuclide may be radioactive because it has too many or too few neutrons.

A simple way to make radioactive nuclei nowadays is to bombard a stable element with a flux of neutrons in a reactor. For example radioactive phosphorus may be made by the reaction shown below

$$^{31}_{15}\text{P} + {}^{1}_{0}\text{n} = {}^{32}_{15}\text{P} + {}^{0}_{0}\gamma$$

(the emission of a gamma ray as part of this reaction will be discussed later). However, this method of production results in a radionuclide that is mixed with the stable isotope since the number of protons in the nucleus has not changed and not all the P-31 is converted to P-32. Radionuclides that are 'carrier free' can be produced by bombarding with charged particles such as protons or deuterons, in a cyclotron; for example, if sulphur is bombarded with protons,

$$^{34}_{16}\text{S} + {}^{1}_{1}\text{P} = {}^{34}_{17}\text{Cl} + {}^{1}_{0}\text{n}$$

The radioactive product is now a different element and thus may be separated by chemical methods.

The **activity** of a source is a measure of its rate of decay or the number of disintegrations per second. It is measured in becquerels (Bq) where 1 Bq is equal to one disintegration per second. The becquerel has replaced the older unit of the curie (Ci), but since the latter is still encountered frequently in textbooks and published papers, it is important to know the conversion factor

$$1 \text{ Ci} = 3.7 \times 10^{10} \text{ Bq}$$

Hence

$$1 \text{ mCi (millicurie)} = 3.7 \times 10^{7} \text{ Bq (37 megabecquerels or MBq)}$$
$$1 \text{ }\mu\text{Ci (microcurie)} = 3.7 \times 10^{4} \text{ Bq (37 kilobecquerels or kBq)}$$

1.3 Radioactive concentration and specific activity

These two concepts are frequently confused.

Radioactive concentration

This relates to the amount of radioactivity per unit volume. Hence it will be expressed in Bq ml^{-1}. It is important to consider the radioactive concentration when giving a bolus injection. If one wishes to inject a large activity of technetium-99m (Tc-99m) in a small volume, perhaps for a dynamic nuclear medicine investigation, it is preferable to elute a 'new' molybdenum-technetium generator when the yield might be 8 GBq (200 mCi) in a 10 ml eluate [0.8 GBq ml^{-1} (20 mCi ml^{-1})] rather than an old generator when the yield might be only about 2 GBq (50 mCi) [0.2 GBq ml^{-1} (5 mCi ml^{-1})]. For a fuller discussion of the production of Tc-99m and its use in nuclear medicine see section 1.7 and Chapter 8.

Specific activity
This relates to the proportion of nuclei of the element of interest that are actually labelled. Non-radioactive material, for example iodine-127 (I-127) in a sample of I-125 may be present as a result of the preparation procedure or may have been added as carrier. The unit for the total number of atoms or nuclei present is the mole so the proportion that are radioactive or the **specific activity** can be expressed in Bq mol^{-1} or Bq kg^{-1}. The specific activity of a preparation should always be checked since it determines the total amount of the element being administered. Modern radiopharmaceuticals generally have a very high specific activity so the total amount of the element administered is very small, and problems such as iodine sensitivity do not normally arise in diagnostic nuclear medicine.

1.4 Radioactive decay processes

Three types of radioactive decay that result in the emission of charged particles will be considered at this stage.

β^- decay
A negative β particle is an electron. Its emission is actually a very complex process but it will suffice here to think of a change **in the nucleus** in which a neutron is converted into a proton. The particles are emitted with a range of energies. Note that although the process results in emission of electrons, it is a **nuclear** process and has nothing to do with the orbiting electrons.

The mass of the nucleus remains unchanged but its charge increases by one, thus this change is favoured by nuclides which have too many neutrons.

β^+ decay
A positive β particle, or positron, is the complementary particle to an electron, having the same mass and equal but opposite charge.

It is released when a proton in the nucleus is converted to a neutron.

The mass of the nucleus again remains unchanged but its charge decreases by one, thus this change is favoured by nuclides which have too many neutrons.

α decay
An α particle is a helium nucleus, thus it comprises two protons and two neutrons. After α emission, the charge is reduced by two units and the mass by four units.

The effects of β^-, β^+ and α decay are shown in Fig. 1.3a and Fig. 1.3b. Note that emission of α particles only occurs for the higher atomic number nuclides.

1.5 Exponential decay

Although it is possible to predict, from the number of protons and neutrons in the nucleus, which type of decay might occur, it is not possible to predict how fast decay will occur. One might imagine that nuclides that were furthest from the stability line would decay fastest. This is not so and the factors which determine the rate of decay are beyond the scope of this book.

However, all radioactive decay processes do obey a very important rule. This states that the only variables affecting the number of nuclei ΔN decaying in a short interval of time Δt are the number of unstable nuclei present N and the time interval Δt. Hence

$$\Delta N \propto N \Delta t$$

If the time interval is very short, the equation becomes

$$\mathrm{d}N = -k\, N \mathrm{d}t$$

where the constant of proportionality k is characteristic of the radionuclide, known as its decay constant, and the negative sign has been introduced to show that, mathematically, the number of radioactive nuclei actually decreases with elapsed time.

This equation may be integrated to give the well-known exponential relationship

$$N = N_0 \exp(-kt)$$

where N_0 is the number of unstable nuclei present at $t = 0$. Since the activity of a source, A, is equal to the number of disintegrations per second,

$$A = \frac{\mathrm{d}N}{\mathrm{d}t} = -kN = kN_0 \exp(-kt)$$

when $t = 0$,

$$\left(\frac{\mathrm{d}N}{\mathrm{d}t}\right)_{t=0} = kN_0 = A_0, \text{ so}$$

$$A = A_0 \exp(-kt) \qquad \text{(Equation 1.1)}$$

Thus the activity also decreases exponentially.

1.6 Half-life

An important concept is the **half-life** or the time ($T_{\frac{1}{2}}$) after which the activity has decayed to half its original value.

If A is set equal to $A_0/2$ in Equation 1.1,

$$\tfrac{1}{2} = \exp(-kT_{\frac{1}{2}}) \text{ or } kT_{\frac{1}{2}} = \ln 2$$

Hence

$$A = A_0 \exp\left(\frac{-\ln 2 \cdot t}{T_{\frac{1}{2}}}\right) = \exp\left(\frac{-0.693t}{T_{\frac{1}{2}}}\right) \qquad \text{(Equation 1.2)}$$

since $\ln 2 = 0.693$. Equally,

$$N = N_0 \exp\left(\frac{-0.693t}{T_{\frac{1}{2}}}\right)$$

Two extremely important properties of exponential decay must be remembered.

1 The idea of half-life may be applied from any starting point in time. Whatever the activity at a given time, after one half-life the activity will have been halved.

2 The activity never becomes zero, since there are many millions of radioactive nuclei present, so their number can always be halved to give a residue of radioactivity.

Clearly, if the value of $T_{\frac{1}{2}}$ is known, and the rate of decay is known at one time, the rate of decay may be found at any later time by solving Equation 1.2 above. However, the activity may also be found, with sufficient accuracy, by a simple graphical method. Proceed as follows:

1 Use the y-axis to represent activity and the x-axis to represent time.

2 Mark the x-axis in equal units of half-lives.

3 Assume the activity at $t = 0$ is 1. Hence the first point on the graph is (0,1).

4 Now apply the half-life rule. After one half-life, the activity is $\frac{1}{2}$, so the next point on the graph is $(1,\frac{1}{2})$.
5 Apply the half-life rule again to obtain the point $(2,\frac{1}{4})$ and successively $(3,\frac{1}{8})$ $(4,\frac{1}{16})$ $(5,\frac{1}{32})$. See Fig. 1.4a.
Note that, so far, the graph is quite general without consideration of any particular nuclide, half-life, or activity. To answer a specific problem, it is now only necessary to relabel the axes with the given data, e.g. 'The activity of an oral dose of I-131 is 90 MBq at 12 noon on Tuesday, 4 October. If the half-life of I-131 is 8 days, when will the activity be 36 MBq?' Figure 1.4b shows the same axes as Fig. 1.4a relabelled to answer this specific problem. This quickly yields the answer of $10\frac{1}{2}$ days, i.e. at 12 midnight on 14 October.*

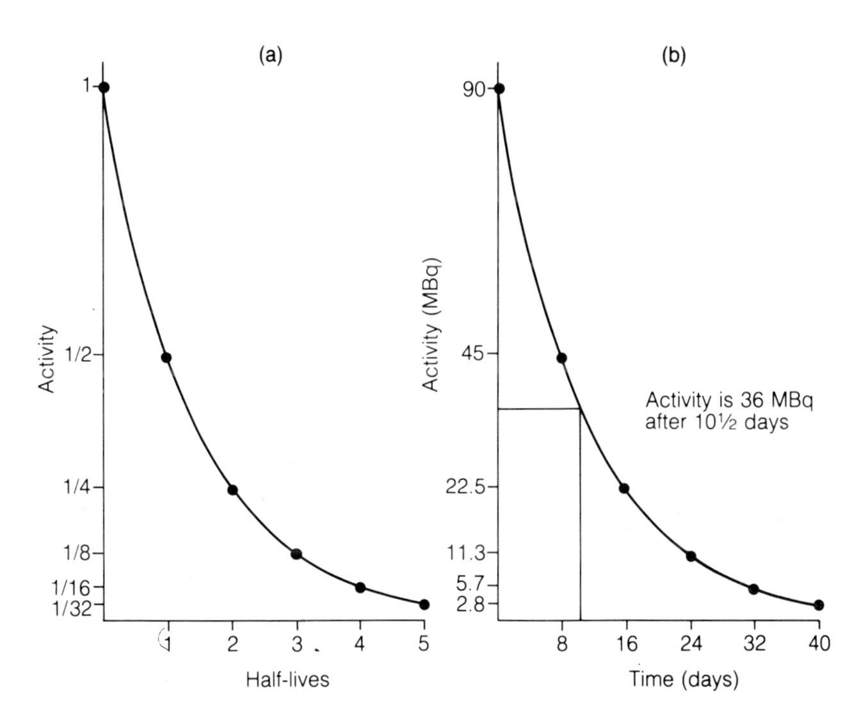

Fig. 1.4. Simple graphical method for solving any problem where the behaviour is exponential. (a) A basic curve that may be used to describe any exponential process. (b) The same curve used to solve the specific problem on radioactive decay set in the text.

*To solve this problem using Equation 1.2:
$36 = 90 \exp(-\ln 2 \cdot t/8)$ where t is the required time in days.

$$\ln\left(\frac{90}{60}\right) = \frac{(\ln 2)\cdot t}{8} \text{ from which } t = 10.6 \text{ days.}$$

This graphical approach may be applied to any problem that can be described in terms of simple exponential decay.

1.7 Secular and transient equilibrium

As already explained, radioactive decay is a process by which the nucleus attempts to achieve stability. It is not always successful at the first attempt and further decay processes may be necessary. For example, four decay schemes occur in nature each of which involves a long sequence of decay processes, terminating finally in stable lead-208 (Pb-208).

In such a sequence the nuclide which decays is frequently called the 'parent' and its decay product the 'daughter'. If both the parent and daughter nuclides are radioactive, and the parent has a longer half-life than the daughter, the rate of decay of the daughter is determined not only by its own half-life but also by the rate at which it is produced. As a first approximation, assume that the activity of the parent remains constant, or is constantly replenished so that the rate of production of the daughter remains constant. If none of the daughter is present initially, its rate of production will at first exceed its rate of decay and equilibrium will be reached when the rate of production is just equal to the rate of decay (Fig. 1.5a).

The curve is of the exponential type so the activity never actually reaches equilibrium. The rate of approach to equilibrium depends on the half-life of the daughter and after 10 half-lives the activity will be within 0.1% of equilibrium.* The equilibrium activity is governed by the activity of the parent.

Two practical situations should be distinguished.

1 The half-life of the parent is much longer than the half-life of the daughter; for example, radium-226, which has a half-life of 1620 years, decays to radon gas which has a half-life of 3.82 days. For most practical purposes the activity of the radon gas reaches a constant value, only changing very slowly as the radium decays. This is known as **secular equilibrium**.

*Mathematically, the shape of Fig. 1.5a is given by

$$N = N_{max}[1 - \exp(-\ln 2 \cdot t)/T_{\frac{1}{2}}]$$

where $T_{\frac{1}{2}}$ is now the half-life of the daughter radionuclide. Thus after n half-lives

$$N = N_{max}[1 - \exp(-n \ln 2)] = N_{max}[1 - (\tfrac{1}{2})^n]$$

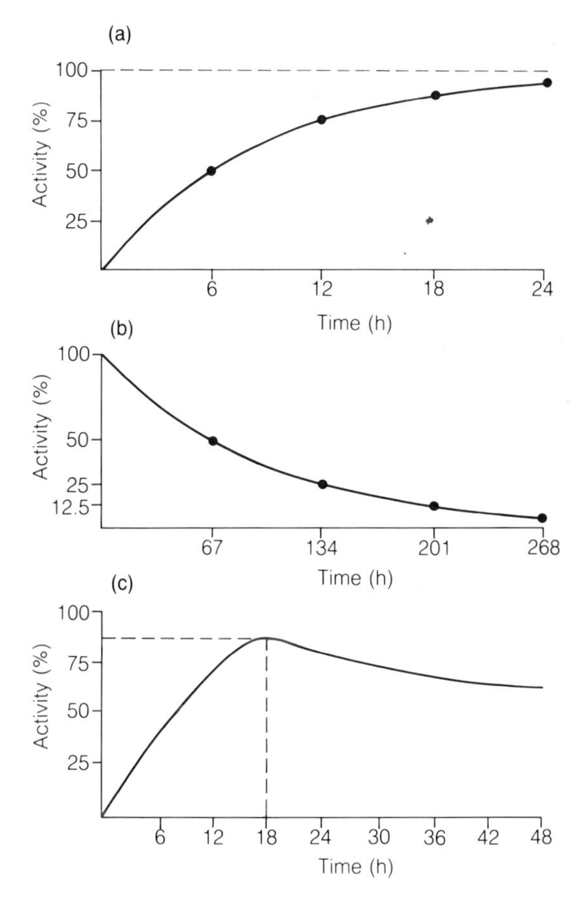

Fig. 1.5. (a) Increase in activity of a daughter when the activity of the parent is assumed to be constant. Generation of Tc-99m from Mo-99 has been taken as a specific example, but with the assumption that the supply of Mo-99 is constantly replenished.

(b) The decay curve for Mo-99 which has a half-life of 67 h.

(c) Increase in activity of a daughter when the activity of the parent is decreasing. Generation of Tc-99m from Mo-99 has been taken as a specific example. Curves (a) and (b) are multiplied to give the resultant activity of Tc-99m.

2 The half-life of the parent is not much longer than that of the daughter. The most important example for radiologists arises in diagnostic nuclear medicine and is molybdenum-99 (Mo-99) which has a half-life of 67 h before decaying to technetium-99m (Tc-99m) which has a half-life of 6 h. Now the growth curve for Tc-99m when the Mo-99 activity is assumed constant (Fig. 1.5a) must be multiplied by the decay curve for Mo-99 (Fig. 1.5b). The resultant (Fig. 1.5c) shows that an actual maximum of Tc-99m activity is reached after about 18 h. By the time the 10 half-lives (60 h) required for Tc-99m to come to equilibrium with Mo-99 have elapsed, the activity of Mo-99 has fallen to half its original value.

This is known as **transient equilibrium** because although the Tc-99m is in equilibrium with the Mo-99, the activity of the Tc-99m is not constant. It explains why the amount of activity that can be eluted from a Mo-Tc generator (see section 1.3 and Chapter 8) is much higher when the generator is first delivered than it is a week later.

1.8 Biological and effective half-life

When a radionuclide is administered, either orally or by injection, in addition to the reduction of activity with time due to the physical process of decay, activity is also lost from the body as a result of biological processes. Generally speaking, these processes also show exponential behaviour so the concentration of substance remaining at time t after injection is given by

$$C = C_0 \exp(-\ln 2 \cdot t/T_{\frac{1}{2}\,\mathrm{biol}})$$

(cf. Equation 1.2), where $T_{\frac{1}{2}}$ is the biological half-life.

When physical and biological processes are combined, the overall loss is the product of two exponential terms and the activity at any time after injection is given by

$$\begin{aligned} A &= A_0 \exp(-\ln 2 \cdot t/T_{\frac{1}{2}\,\mathrm{phys}}) \cdot \exp(-\ln 2 \cdot t/T_{\frac{1}{2}\,\mathrm{biol}}) \\ &= A_0 \exp[-\ln 2 \cdot t\,(1/T_{\frac{1}{2}\,\mathrm{phys}} + 1/T_{\frac{1}{2}\,\mathrm{biol}})] \end{aligned}$$

To find the effective half-life $T_{\frac{1}{2}\,\mathrm{eff}}$, set

$$A = A_0 \exp(-0.693t)/T_{\frac{1}{2}\,\mathrm{eff}})$$

Hence, by inspection,

$$1/T_{\frac{1}{2}\,\mathrm{eff}} = 1/T_{\frac{1}{2}\,\mathrm{phys}} + 1/T_{\frac{1}{2}\,\mathrm{biol}}$$

Note that if $T_{\frac{1}{2}\,\mathrm{phys}}$ is much shorter than $T_{\frac{1}{2}\,\mathrm{biol}}$, the latter may be neglected, and vice versa. For example if $T_{\frac{1}{2}\,\mathrm{phys}} = 1$ h and $T_{\frac{1}{2}\,\mathrm{biol}} = 20$ h,

$$1/T_{\frac{1}{2}\,\mathrm{eff}} = 1 + 1/20 = 1.05$$

and $T_{\frac{1}{2}\,\mathrm{eff}} = 0.95$ h or almost the same as $T_{\frac{1}{2}\,\mathrm{phys}}$.

1.9 Gamma radiation

Some radionuclides emit radioactive particles to gain stability. Normally, in addition to the particle, the nucleus also has to emit some energy, which it does in the form of gamma radiation. Note that emission of gamma rays as a mechanism for losing energy is very general and, as shown in section 1.2, may also occur when radionuclides are produced.

Although the emission of the particle and the gamma ray are, strictly speaking, separate processes, they normally occur very close together in time. However, some nuclides enter a metastable state after emitting the particle and emit their gamma ray some time later. When the two processes are separated in time in this way, the second stage is known as an **isomeric transition**. An important example in nuclear medicine is Tc-99m (the 'm' stands for metastable) which has a half-life of 6 h. This is long enough for it to be separated from the parent Mo-99 and the decay is then by gamma ray emission only which is particularly suitable for *in vivo* investigations (see Chapter 8).

Just as electrons in shells around the nucleus occupy well-defined energy levels, there are also well-defined energy levels in the nucleus. Since the gamma rays represent transitions between these levels, they are monoenergetic. However gamma rays with more than one well-defined energy may be emitted by the same nuclide, for example indium-111 emits gamma rays at 163 keV and 247 keV.

1.10 X-rays and gamma rays as forms of electromagnetic radiation

The propagation of energy by simultaneous vibration of electric and magnetic fields is known as electromagnetic (EM) radiation. Unlike sound, which is produced by the vibration of molecules and therefore requires a medium for propagation (see Chapter 12), EM radiation can travel through a vacuum. However, like sound, EM radiation exhibits many wave-like properties such as reflection, refraction, diffraction and interference and is frequently characterized by its wavelength. EM waves can vary in wavelength from 10^{-13} m to 10^{3} m and different parts of the EM spectrum are recognized by different names (see Table 1.1).

X-rays and gamma rays are both part of the EM spectrum and an 80 keV X-ray is in fact identical to, and hence indistinguishable from, an 80 keV gamma ray. In order to appreciate the reason for the apparent

Table 1.1. The different parts of the electromagnetic spectrum classified in terms of wavelength, frequency and quantum energy

	Radio waves	Infra-red	Visible light	Ultra violet	X-rays and gamma rays
Wavelength (m)	10^{3}–10^{-2}	10^{-4}–10^{-6}	5×10^{-7}	5×10^{-8}	10^{-9}–10^{-13}
Frequency (Hz)	3×10^{5}–3×10^{10}	3×10^{12}–3×10^{14}	6×10^{14}	6×10^{15}	3×10^{17}–3×10^{21}
Quantum energy (eV)	10^{-9}–10^{-4}	10^{-2}–1	2	20	10^{3}–10^{7}

confusion, it is necessary to consider briefly the origin of the discoveries of X-rays and gamma rays. As already noted, gamma rays were discovered as a type of radiation emitted by radioactive materials. They were clearly different from alpha rays and from beta rays, so they were given the name gamma rays. X-rays were discovered in quite a different way as 'emission by high energy machines of radiations that caused certain materials, such as barium platino-cyanide to fluoresce'. It was some time before the identity of X-rays produced by machines and gamma rays produced by radioactive materials was confirmed.

For a number of years, X-rays produced by machines were of lower energy than gamma rays, but with the development of linear accelerators and other high energy machines, this distinction is no longer useful.

No distinction between X-rays and gamma rays is totally self-consistent, but it is reasonable to describe gamma rays as the radiation emitted as a result of nuclear interactions, and X-rays as the radiation emitted as a result of events outwith the nucleus. For example, one method by which nuclides with too few neutrons may approach stability is by K-electron capture. This mode of radioactive decay has not yet been discussed. The nucleus 'steals' an electron from the K shell to neutralize one of its protons. The K shell vacancy is filled by electrons from outer shells and the associated characteristic radiations are referred to as X-rays, even though they result from radioactive decay.

An important concept is the **intensity** of a beam or X or gamma rays. This is defined as the amount of energy crossing unit area placed normal to the beam in unit time. In Fig. 1.6, if the total amount of radiant energy passing through the aperture of area a in time t is E, the intensity $I = E/(a \cos \theta \cdot t)$ where $a \cos \theta$ is the cross-sectional area of the aperture normal to the beam. If a is in m^2, t in s and E in joules, the units of I will be $J\ m^{-2}\ s^{-1}$.

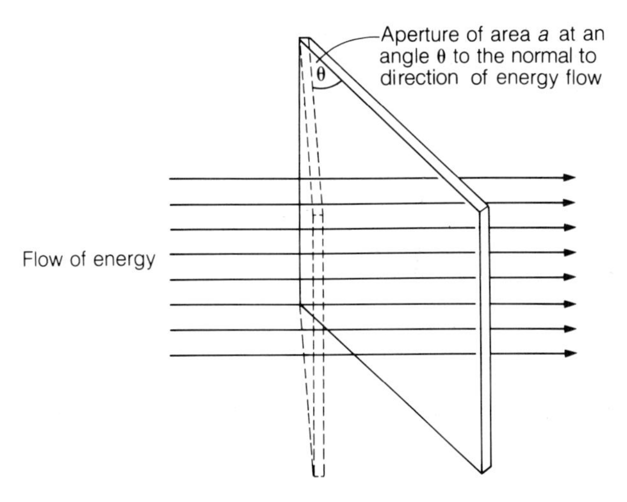

Fig. 1.6. A simple representation of the meaning of intensity.

1.11 Quantum properties of radiation

As well as showing the properties of waves, short wavelength EM radiation, such as X and gamma rays, can sometimes show particle-like properties. Each particle is in fact a small packet of energy and the size of the energy packet (ε) is related to frequency (f) and wavelength (λ) by the fundamental equations

$$\varepsilon = \mathrm{h}f = \mathrm{hc}/\lambda$$

where h is the Planck constant and c is the speed of electromagnetic waves.
Taking $c = 3 \times 10^8$ m s^{-1} and $h = 6.6 \times 10^{-34}$ J s

$$\varepsilon \text{ (in joules)} = \frac{2 \times 10^{-25}}{\lambda \text{ (in metres)}}$$

Thus the smaller the value of λ, the larger the value of the energy packet. For a typical X-ray wavelength of 10^{-12} m, the value of ε in joules for a single photon is inconveniently small, so the electron volt, a unit of energy that has already been introduced, is used where 1 eV = 1.6×10^{-19} J.

1.12 Inverse square law

Before considering the interaction of radiation with matter, one important law that all radiations obey under carefully defined conditions will be introduced. This is the **inverse square law** which states that for a point source, and in the absence of attenuation, the intensity of a beam of radiation will decrease as the inverse of the square of the distance from that source.

The law is essentially just a statement of conservation of energy, since if the rate at which energy is emitted as radiation is E, the energy will spread out in all directions and the amount crossing unit area per second at radius r, $I_r = E/4\pi r^2$ (Fig. 1.7). Similarly the intensity crossing unit area at radius R, $I_R = E/4\pi R^2$. Thus the intensity is decreasing as $1/(\text{radius})^2$.

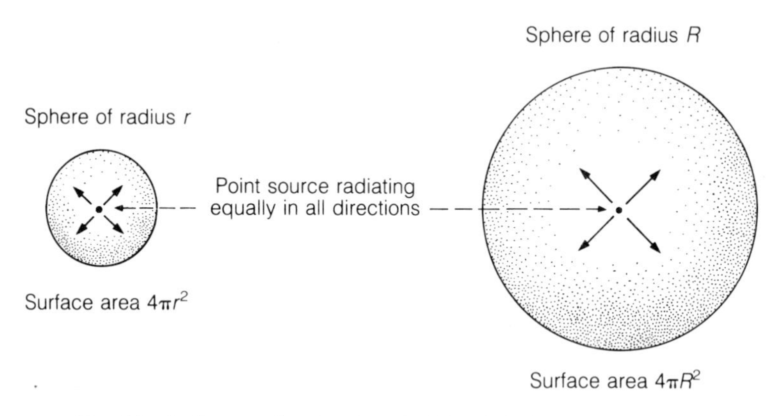

Fig. 1.7. A diagram showing the principle of the inverse square law.

1.13 Interaction of radiation with matter

As a simple model of the interaction of radiation with matter, consider the radiation as a stream of fast moving particles (alphas, betas or photons) and the medium as an array of nuclei each with a shell of electrons around it (Fig. 1.8). As the particle tries to penetrate the medium, it will collide with atoms. Sometimes it will transfer energy of excitation during a collision. This type of interaction will be considered in more detail in Chapter 3. The energy is quickly dissipated as heat. Occasionally, the interaction will be so violent that one of the outermost electrons will be torn away from the nucleus to which it was bound and become free. **Ionization** has occurred because an ion pair has been created. Sometimes, as in interaction C, the electron thus released has enough energy to cause further ionizations and a cluster of ions is produced.

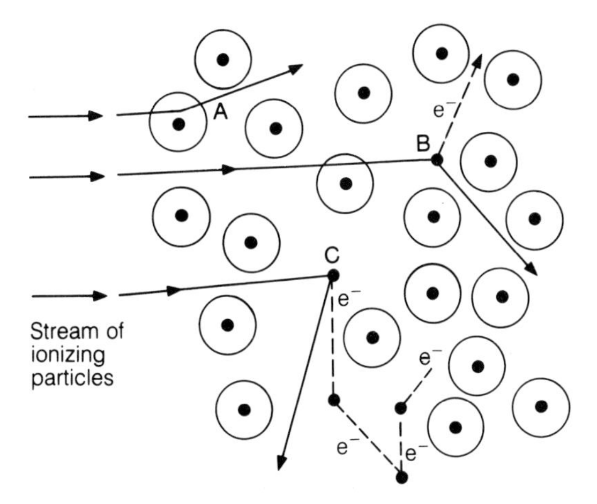

Fig. 1.8. Simple model of the interaction of radiation with matter. Interaction A causes excitation, interaction B causes ionization, and interaction C causes multiple ionizations. At each ionization an electron is released from the nucleus to which it was bound.

The amount of energy required to create an ion pair is about 35 eV. Charged particles of interest in medicine invariably possess this amount of energy. For EM radiation, a quantum of X or gamma rays always has more than 35 eV but a quantum of, say, ultraviolet or visible light does not. Hence the EM spectrum may be divided into ionizing and non-ionizing radiations.

The above, very simple model may also be used to predict how easily different types of radiation will be attenuated by different types of material. Clearly, as far as the stopping material is concerned, a high density of large nuclei (i.e. high atomic number) will be most effective for causing many collisions. Thus gases are poor stopping materials, but lead ($Z = 82$) is excellent and, if there is a special reason for compact shielding, even depleted uranium ($Z = 92$) is sometimes used.

With regard to the bombarding particles, size (or mass) is again important and since the particle is moving through a highly charged region, interaction is much more probable if the particle itself is charged and therefore likely to come under the influence of the strong electric fields associated with the electron and nucleus. Since X and gamma ray quanta are uncharged and have zero rest mass, they are difficult to stop and lead may be required to cause appreciable attenuation.

The β^- particle is more massive and is charged so it is stopped more easily—a few mm of low atomic number materials such as perspex will usually suffice. Since it will be shown in Chapter 3 that the mechanism of energy dissipation by X and gamma rays is via secondary electron formation, a table of electron ranges in soft tissue will be helpful (Table 1.2).

Protons and alpha particles are even more massive than β^- particles and are charged, so they are stopped easily. Alpha particles, for example, are so easily stopped, even by a sheet of paper, that great care must be taken when attempting to detect them to ensure that the detector has a thin enough window to allow them to enter the counting chamber. Neutrons are more penetrating because, although of comparable mass to the proton, they are uncharged.

One final remark should be made regarding the ranges of radiations. Charged particles eventually become trapped in the high electric fields around nuclei and have a finite range. Beams of X or gamma rays are stopped by random processes, and as shown in Chapter 3 are attenuated exponentially. This process has many features in common with radioactive decay. For example the rate of attenuation by a particular material is predictable but the radiation does not have a finite range.

Table 1.2. Approximate ranges of electrons in soft tissue

Electron energy (keV)	Approximate range (mm)
20	0.01
40	0.03
100	0.14
400	1.3

1.14 Linear energy transfer

Beams of ionizing radiation are frequently characterized in terms of their linear energy transfer (LET). This is a measure of the rate at which energy is transferred to the medium and hence of the density of ionization along the track of the radiation. Although a difficult concept to apply rigorously, it will suffice here to use a simple definition, namely that LET is the energy transferred to the medium per unit track length. It follows from this definition that radiations which are easily stopped will have a high LET, those which are more penetrating will have a low LET. Some examples are given in Table 1.3.

Table 1.3. Approximate values of linear energy transfer for different types of radiation

Radiation	LET ($keV\ \mu m^{-1}$)
1 MeV γ rays	0.5
100 kVp X-rays	6
20 keV β^- particles	10
5 MeV neutrons	20
5 MeV α particles	50

1.15 Summary of energy changes in radiological physics

Energy cannot be created or destroyed but can only be converted from one form to another. Therefore it is important to summarize the different forms in which energy may appear. Remember that **work** is really just another word for energy—stating that body A does work on body B means that energy is transferred from body A to body B.

Mechanical energy

This can take two well-known forms.

1 **Kinetic energy**, $\frac{1}{2}mv^2$, where m is the mass of the body and v its velocity.

2 **Potential energy**, mgh, where g is the gravitational acceleration and h the height of the body above the ground.

Kinetic energy is more relevant than potential energy in the physics of X-ray production and the behaviour of X-rays.

Electrical energy

When an electron, charge e, is accelerated through a potential difference V, it acquires energy eV. Thus if there are n electrons they acquire total energy neV. Note:

1 Current (i) is rate of flow of charge. Thus $i = ne/t$ where t is the time. Hence, rearranging, an alternative expression for the energy in a beam of electrons is Vit.

2 Just as Vit is the amount of energy gained by electrons as they accelerate through a potential difference V, it is also the amount of energy lost by electrons (usually as heat) when they fall through a potential difference of V, for example when travelling through a wire that has resistance R.

3 If the resistor is 'ohmic', that is to say it obeys **Ohms law**, then $V = iR$

and alternative expressions for the heat dissipated are V^2/R or i^2R. Note, however, that many of the resistors encountered in the technology of X-ray production are non-ohmic.

Heat energy

When working with X-rays, most forms of energy are eventually degraded to heat and when a body of mass m and specific heat capacity s receives energy E and converts it into heat, the rise in temperature ΔT will be given by

$$E = ms\Delta T$$

Excitation and ionization energy

Electrons are bound in energy levels around the nucleus of the atom. If they acquire energy of excitation they may jump into a higher energy level. Sometimes the energy may be enough for the electrons to escape completely (ionization). Note that if this occurs the electron may also acquire some kinetic energy in addition to the energy required to cause ionization.

Radiation energy

Radiation represents a flow of energy. This is usually expressed in terms of beam intensity I such that $I = E/(a \cdot t)$ where E is the total energy passing through an area a placed normal to the beam, in time t.

Quantum energy

X and gamma radiation frequently behave as exceedingly small energy packets. The energy of one quantum is $\mathrm{h}f$ where h is the Planck constant and f the frequency of the radiation. The energy of one quantum is so small that the joule is an inconveniently large unit so the electron volt is introduced where $1\ \mathrm{eV} = 1.6 \times 10^{-19}\ \mathrm{J}$.

Mass energy

As a result of Einstein's work on relativity, it has become apparent that mass is just an alternative form of energy. If a small amount of matter, mass m, is converted into energy, the energy released $E = mc^2$ where c is the speed of electromagnetic waves. This change is encountered most frequently in radioactive decay processes. Careful calculation, to about one part in a million, shows that the total mass of the products is slightly less than the total mass of the starting materials, the residual mass having been converted to energy according to the above equation.

Further reading

Chackett K. F. (1981) *Radionuclide Technology—an Introduction to Quantitaive Nuclear Medicine*. Van Nostrand Reinhold, New York.

Gifford D. (1984) *A Handbook of Physics for Radiologists and Radiographers*. Wiley, New York.

Johns H. E. and Cunningham J. R. (1983) *The Physics of Radiology*, 4th edn. Thomas, Springfield.

Meredith W. J. and Massey J. B. (1977) *Fundamental Physics of Radiology*, 3rd edn. Wright, Bristol.

Exercises

1 Describe in simple terms the structure of the atom and explain what is meant by atomic number, atomic weight and radionuclide.

2 What is meant by the binding energy of an atomic nucleus? Define the unit in which it is normally expressed and indicate the order of magnitude involved.

3 Describe the different ways in which radioactive disintegration can occur.

4 What is meant by the decay scheme of a radionuclide and radioactive equilibrium?

5 What is a radionuclide generator?

6 A radiopharmaceutical has a physical half-life of 6 h and a biological half-life of 20 h. How long will it take for the activity in the patient to fall to 25% of that injected?

7 The decay constant of iodine-123 is 1.34×10^{-5} s^{-1}. What is its half-life and how long will it take for the radionuclide to decay to one-tenth of its original activity?

8 Investigate whether the values of radiation intensity given below decrease exponentially with time:

Intensity (J m^{-2} s^{-1})	100	70	50	33	25	20	10	6.7	5.0	4.0
Time (s)	0	1.0	2.0	3.0	4.0	5.0	10.0	15.0	20.0	25.0

9 A radionuclide A decays into a nuclide B which has an atomic number one less than that of A. What types of radiation might be emitted either directly or indirectly in the disintegration process? Indicate briefly how they are produced.

10 Give typical values for the ranges of α particles and β^- particles in soft tissue. Why is the concept of range not applicable to gamma rays?

11 For an unknown sample of radioactive material explain how it would be possible to determine by simple experiment

(a) the types of radiation emitted,

(b) the half-life.

12 State the inverse square law for a beam of radiation and give the conditions under which it will apply exactly.

13 A surface is irradiated uniformly with a monochromatic beam of X-rays of wavelength 2×10^{-11} m. If 20 quanta fall on each square cm of the surface per second, what is the intensity of the radiation at the surface? (Use data given in section 1.11.)

14 Place the following components in order of power dissipation:

(a) a fluorescent light

(b) an X-ray tube

(c) an electric fire

(d) a pocket calculator

(e) an electric iron

2

Production of X-rays

2.1 Introduction

When electrons are accelerated to energies in excess of 5 keV and are then directed onto a surface, X-rays may be emitted. The X-rays originate principally from rapid deceleration of the electrons when they strike the target atoms. These X-rays are known as 'Bremsstrahlung' or braking radiation.

The essential features of a simple X-ray tube are shown in Fig. 2.1 and comprise:

1 a heated metal filament to provide a copious supply of electrons by thermionic emission and to act as cathode;

2 an evacuated chamber across which a high potential difference can be applied;

3 a metal anode (the target) with a high efficiency for conversion of electron energy into X-ray photons;

4 a thick-walled glass chamber with a thinner window that will be transparent to most of the X-rays.

In this chapter the mechanisms of X-ray production will be considered in detail and the main components of a modern X-ray tube will be described. Physical factors affecting the design and performance of X-ray sets will be considered along with implications for the taking of good radiographs.

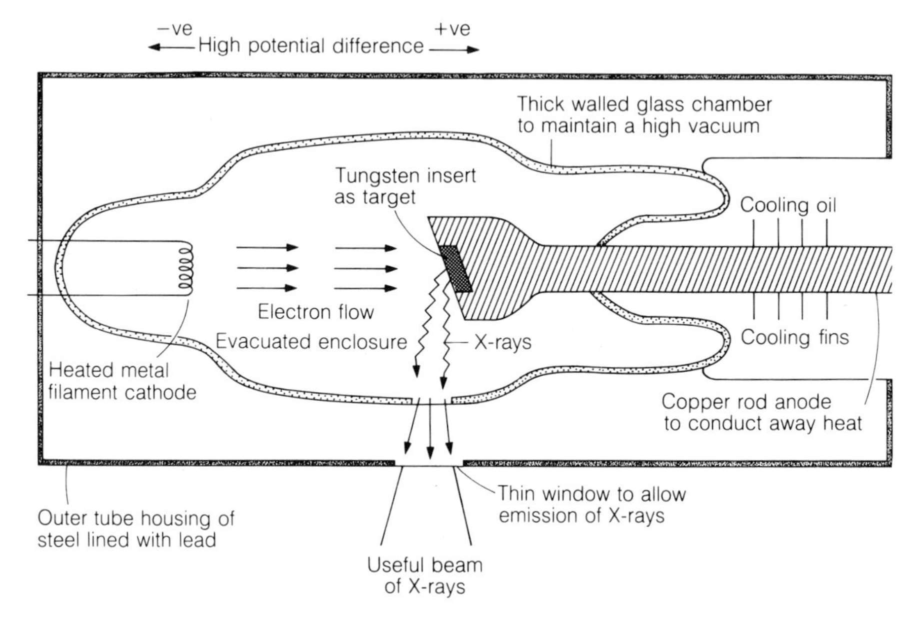

Fig. 2.1. The essential features of a simple, stationary anode X-ray tube.

2.2 The X-ray spectrum

If the accelerating voltage across the X-ray tube shown in Fig. 2.1 were about 100 kV, the spectrum of radiation that would be used for radiology might be something like that shown in Fig. 2.2. The various features of this spectrum will now be discussed.

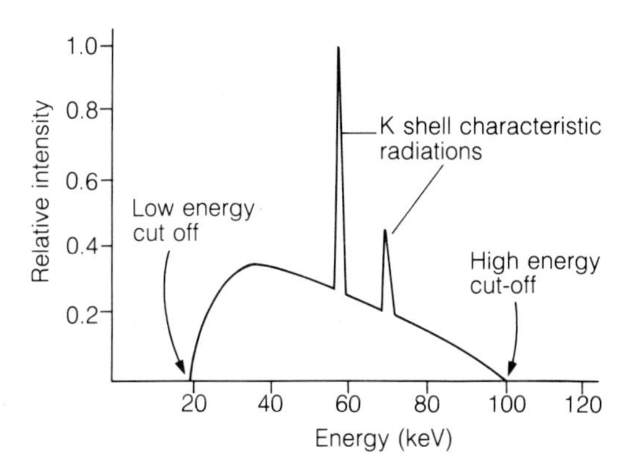

Fig. 2.2. Spectrum of radiation incident on the patient from an X-ray tube operating at 100 kVp using a tungsten target and 2.5 mm aluminium filtration.

2.2.1 The continuous spectrum

When a fast-moving electron strikes the anode, several things may happen. The most common is that the electron will suffer a minor interaction with an orbital electron as depicted at A in Fig. 2.3. This will result in the transfer to the target of a small amount of energy which will appear eventually as heat. At diagnostic energies, at least 99% of the electron energy is converted into heat and the dissipation of this heat is a major technical problem that will be considered in section 2.5.

Occasionally, an electron will come close to the nucleus of a target atom, where it will suffer a much more violent change of direction because the charge and mass of the nucleus are so much greater than those of an electron (example B). The electron does not penetrate the nucleus because the energy barrier presented by the positive charge is far in excess of the electron energy but is deviated around it. The interaction results in a change of energy of the electron and the emission of electromagnetic

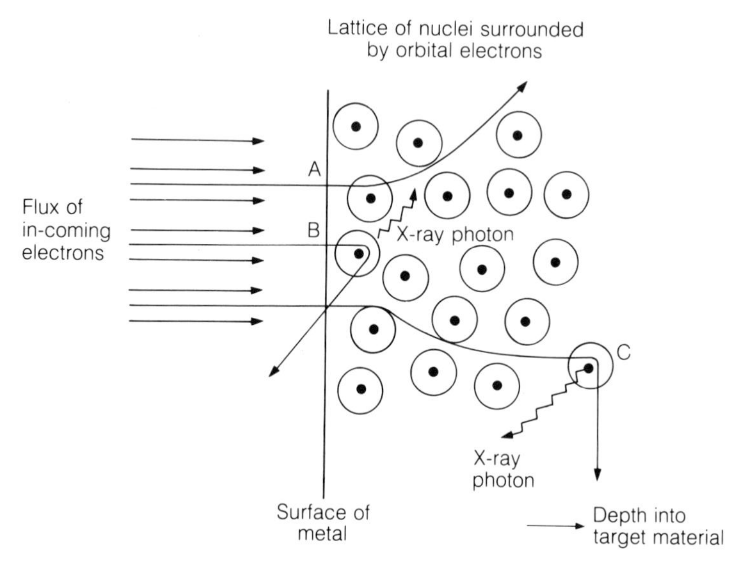

Fig. 2.3. Schematic representation of the interaction of electrons with matter. (A) Interaction resulting in the generation of low energy electromagnetic radiations (infra red, visible, ultraviolet and very soft X-rays). All these are rapidly converted into heat. (B) Interaction resulting in the production of an X-ray. (C) Production of an X-ray after previous interactions that resulted only in heat generation.

radiation that is in the X-ray range of the spectrum. The amount of energy lost by the electron in such a collision is very variable and hence the energy given to the X-ray can take a wide range of values. Note that X-ray emission may occur after two or three earlier slight deviations (example C). Therefore not all emissions occur from the surface of the anode. This factor is important when the spatial distribution of X-ray emission is considered.

2.2.2 The low and high energy cut-off

These parts of the spectrum are simply explained. Low energy electromagnetic radiations are easily attenuated and below a certain energy they are so heavily attenuated—by the materials of the anode, by the window of the X-ray tube and by any added filtration—that the intensity emerging is negligible. X-ray attenuation is discussed in detail in Chapter 3.

To explain the high energy cut-off, recall that an electron may, very occasionally, lose all its energy to X-ray production (section 2.2.1). Hence for any given electron energy, i.e. accelerating voltage across the X-ray tube, there is a well-defined maximum X-ray energy equal to the energy of a single electron. This corresponds to a minimum X-ray wavelength. Note that it is not possible, by quantum theory, for the energy of several electrons to be stored up in the anode to produce a jumbo-sized X-ray quantum.

It is useful to calculate the electron velocity, the maximum X-ray photon energy and the minimum X-ray photon wavelength associated with a given tube kilovoltage. To avoid complications associated with relativistic effects, a tube operating at only 30 kV is considered.

The energy of each electron is given by the product of its charge (e) and the accelerating voltage (V):

$$\begin{aligned} eV &= 1.6 \times 10^{-19} \times 3 \times 10^{4} \\ &= 4.8 \times 10^{-15}\ \text{J (or } 3 \times 10^{4} \text{ electron volts i.e. 30 keV)} \end{aligned}$$

Note the distinction between an accelerating voltage, measured in kV, and an electron energy measured in keV.

The electron velocity is obtained from the fact that its kinetic energy is $\frac{1}{2}m_e v_e^2$ where m_e is the mass of the electron and v_e its velocity. Hence

$$\tfrac{1}{2}m_e v_e^2 = 4.8 \times 10^{-15}\ \text{J}.$$

Since $m_e = 9 \times 10^{-31}$ kg, v_e is approximately 10^8 m s^{-1}. This is one third the speed of light which shows that relativistic effects are important even at quite low tube kilovoltages.

From above, the maximum X-ray photon energy ε is 4.8×10^{-15} J and the minimum wavelength is obtained by substitution in:

$$\varepsilon = hf = hc/\lambda$$

(h is the Planck constant, c the velocity of light and λ the wavelength of the resulting X-ray). Hence

$$\lambda = hc/\varepsilon = \frac{6.6 \times 10^{-34} \times 3 \times 10^{8}}{4.8 \times 10^{-15}} = 4.1 \times 10^{-11}\ \text{m (or 0.041 nm)}$$

Note that calculations giving the maximum X-ray photon energy and minimum wavelength are valid even when the electrons travel at relativistic speeds.

2.2.3 Shape of the continuous spectrum

A detailed treatment of the continuous spectrum is beyond the scope of this book, but the following approach is helpful since it involves some other important features of the X-ray production process. First, imagine a very thin anode, and consider the production of X-rays, not the X-rays that finally emerge. It may be shown by theoretical arguments that the intensity of X-rays produced will be constant up to a maximum X-ray energy determined by the energy of the electrons (see Fig. 2.4a).

A thick anode may now be thought of as composed of a large number of thin layers. Each will produce a similar distribution to that shown in Fig. 2.4a, but the maximum photon energy will gradually be reduced because the incident electrons lose energy as they penetrate the anode material. Thus the composite picture for X-ray production might be as shown in Fig. 2.4b.

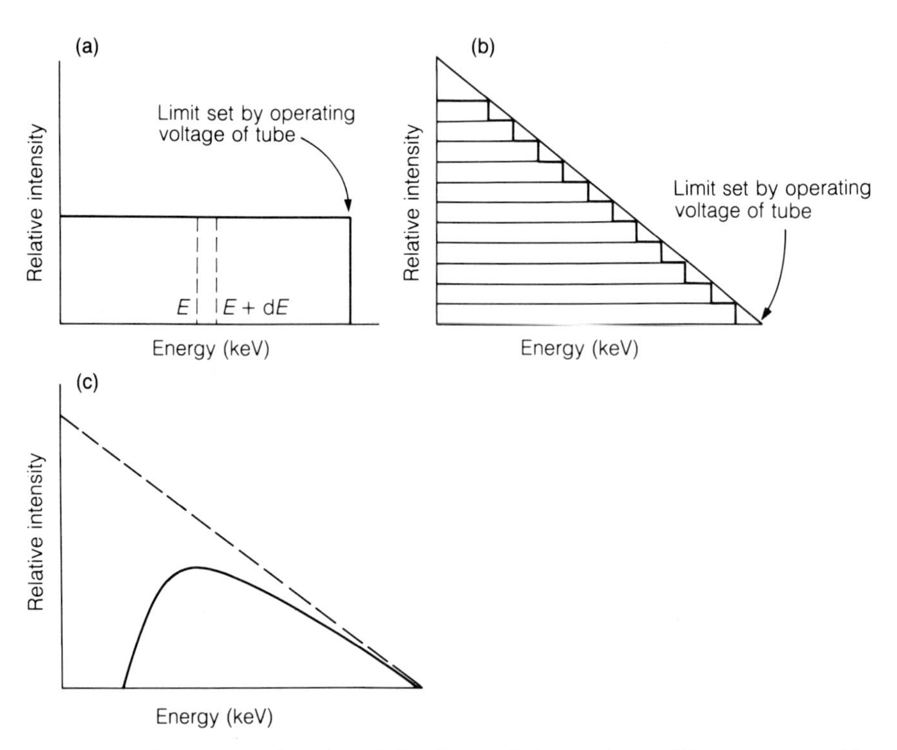

Fig. 2.4. A simplified explanation of the shape of the continuous X-ray spectrum. (a) Production of X-rays from a very thin anode. Note that the intensity of the beam in the small range E to $E + dE$ will be equal to the number of photons per square metre per second multiplied by the photon energy. Fewer high energy photons are produced but their energy is higher and the product is constant. (b) Production of X-rays from a thicker anode, treated as a series of thin anodes. (c) X-ray emission (*solid line*) compared with X-ray production (*dotted line*).

However, before the X-rays emerge, the intensity distribution will be modified in two ways. First, X-rays produced deep in the anode will be attenuated in reaching the surface of the anode and secondly all the X-rays will be attenuated in penetrating the window of the X-ray tube. Both processes reduce the intensity of the low energy radiation more than that of the higher energies so the resultant is the solid curve in Fig. 2.4c.

In the absence of further filtration (see section 3.8) the X-ray energy corresponding to maximum intensity will be about one third of the highest energy X-ray photons.

2.2.4 Line or characteristic spectra

Superimposed on the continuous spectrum there may be a set of line spectra which result from an incoming electron interacting with a bound orbital electron in the target. If the incoming electron has sufficient energy to overcome the binding energy, it can remove the bound electron creating a vacancy in the shell. The probability of this happening is greatest for the innermost shells. This vacant energy level is then filled by an electron from a higher energy level falling into it and the excess energy is emitted as an X-ray. Thus, if for example the vacancy is created in the K shell, it may be filled by an electron falling from either the L shell, the M shell or outer shells. Even a free electron may fill the vacancy but the most likely transition is from the L shell.

As discussed in section 1.1, orbital electrons must occupy well-defined energy levels and these energy levels are different for different elements. Thus the X-ray photon emitted when an electron moves from one energy level to another has an energy equal to the difference between the two energy levels in that atom and hence is characteristic of that element.

Reference to Fig. 1.2b shows that the K series of lines for tungsten ($Z = 74$) will range from 58.5 keV (for a transition from the L shell to the K shell) to 69.5 keV (if a free electron fills the K shell vacancy). Transitions to the L shell are of no practical importance in diagnostic radiology since the maximum energy change for tungsten is 11 keV.

Lower atomic number elements produce characteristic X-rays at lower energies. The K shell radiations from molybdenum ($Z = 42$) at circa 19 keV are important in mammography (see section 6.6). Note that characteristic radiation cannot be produced unless the operating kV to the X-ray tube is high enough to remove the relevant bound electrons from the anode target atoms.

2.2.5 Factors affecting the X-ray spectrum

If the spectrum changes in such a manner that its shape remains unaltered i.e. the intensity or number of photons at every photon energy changes by the same factor, there has been a change in radiation **quantity**. If on the other hand, the shape of the spectrum changes, there has also been a change in radiation **quality** (the penetrating power of the X-ray beam). A number of factors that affect the X-ray spectrum may be considered.

Tube current, I_c

This determines the number of electrons striking the anode. Thus the exposure E is proportional to tube current, but only the quantity of X-rays is affected ($E \propto I_c$).

Time of exposure

This again determines the number of electrons striking the anode so exposure is proportional to time but only the quantity of X-rays is affected ($E \propto t$).

Applied voltage

If other tube operating conditions are kept constant, the flux of X-rays produced, or exposure, increases approximately as the square of tube kilovoltage ($E \propto \text{kV}^2$). Two factors contribute to this increase. First the electrons have more energy to lose when they hit the target. Second, as shown in Table 2.1, the efficiency of conversion of electrons into X-rays rather than into heat also increases with tube kilovoltage. The change associated with a large increase in kV is shown to emphasize the effect.

Furthermore, increasing the tube kilovoltage also alters the radiation **quality** since the high energy cut-off has now increased. Note that the position of any characteristic lines will not change.

Table 2.1. Efficiency of conversion of electron energy into X-rays as a function of tube kilovoltage

Tube kilovoltage (kV)	Heat %	X-rays %
60	99.5	0.5
200	99	1.0
4000	60	40

Profile of applied voltage

So far it has been assumed that the X-ray tube is operating from direct current, whereas in practice it operates from alternating current (Fig. 2.5a). Since in Fig. 2.1 one end of the X-ray tube must act as a 'cathode' and the other end an 'anode', no current flows when an alternating potential is applied during the half cycle when the cathode is positive with respect to the anode. Half wave rectification (Fig. 2.5b) may be achieved by inserting a rectifier in the anode circuit (see section 2.3.4) but since X-rays are only emitted for half the cycle, output is poor. Improved output can be achieved by full wave rectification obtained by using a simple bridge circuit. However, the tube is still not emitting X-rays all the time (Fig. 2.5c). Furthermore, the majority of X-rays are emitted at a kilovoltage below the peak value (kVp).

A more constant voltage will improve the quality of the radiation and this can be achieved by using a three phase supply. The X-ray tube is now

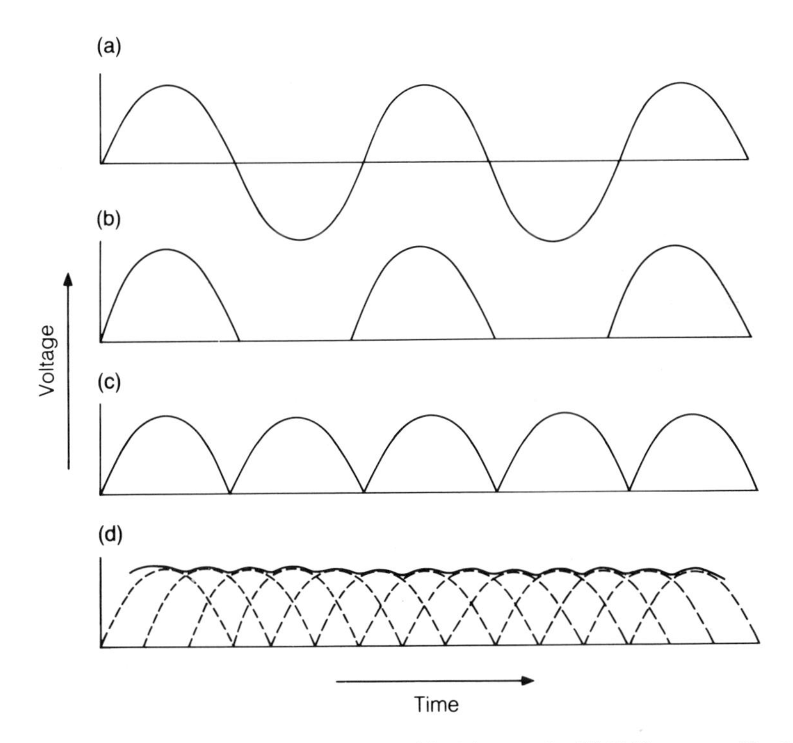

Fig. 2.5. Examples of different voltage profiles. (a) Mains supply. (b) Half-wave rectification. (c) Full wave rectification obtained using a bridge rectifier. (d) Three phase supply (with rectification).

driven by three separate voltage supplies, each of which has been fully rectified. The three supplies are 60° out of phase and switching circuits ensure that each supply only drives the X-ray tube when the voltage is near to peak value. The resultant voltage profile (Fig. 2.5d) shows only about 15% variation. If the cathode supply is also three phase and is arranged to be 30° out of phase with the anode supply, fluctuations can be reduced to about 5%.

This subject is considered in more detail in section 2.3.4. In future, in accordance with standard practice, operating voltages will be expressed in kVp to emphasize that the peak voltage with respect to time is being given.

Filtration

This also has a marked effect on both the quantity and quality of the X-ray beam, not only reducing the overall output but also reducing the proportion of low energy photons. The effect of beam filtration is considered in detail in section 3.8.

Anode material

Choice of anode material affects the efficiency of X-ray production (see section 2.3.2) and the characteristic spectrum.

2.3 Components of the X-ray tube

2.3.1 The cathode

The cathode is constructed as a coiled wire filament of reasonably high resistance R so that for a given filament heating current I_F (typically in the region of 5 A), effective ohmic heating (I_F^2R) and minimum heat losses will occur. A metal is chosen for the cathode that will give a copious supply of electrons by thermionic emission at temperatures where there is very little evaporation of metal atoms into the vacuum (e.g. tungsten).

Between exposures, the filament is kept warm on stand-by because although its resistance may be typically 5 Ω at 2000 K, at room temperature it falls to about 0.1 Ω. Thus a large current would be required to heat the filament rapidly from room temperature to its working temperature.

For reasons related primarily to geometrical unsharpness in the image, a small target for electron bombardment on the anode is essential. However, unless special steps are taken, the random thermally induced velocities of the electrons leaving the cathode will cause a broad beam to strike the anode. Therefore the filament is surrounded by a metal cup, normally maintained at the same potential as the filament. This cup provides an electric field which exercises a focusing action on the electrons to produce a spot on the anode of the required size.

Most diagnostic X-ray tubes have a dual filament assembly, each filament having its own focusing cup, so as to produce two spots of different sizes. Note that spot size does vary somewhat with tube current and tube kilovoltage since the focusing action cannot be readily adjusted to compensate for variations in the mutual electrostatic repulsion between electrons when either their density or energy changes.

2.3.2 The anode material

The material chosen for the anode should satisfy a number of requirements. It should have:

1 A high conversion efficiency for electrons into X-rays. High atomic numbers are favoured since the X-ray intensity is proportional to Z. At 100 keV, lead ($Z = 82$) converts 1% of the energy into X-rays but aluminium ($Z = 13$) converts only 0.1%.

2 A high melting point so that the large amount of heat released causes minimal damage to the anode.

3 A high conductivity so that the heat is removed rapidly.

4 A low vapour pressure, even at very high temperatures, so that atoms are not boiled off from the anode.

5 Suitable mechanical properties for anode construction.

In stationary anodes the target area is pure tungsten (W) ($Z = 74$, melting point 3370°C) set in a metal of higher conductivity such as copper. Originally rotating anodes were made of pure tungsten. However, at the high temperatures generated in the rotating anode (see section 2.3.3), deep cracks developed at the point of impact of the electrons. The deleterious effects of damaging the target in this way are discussed in sections 2.3.5 and 2.4. The addition of 5–10% rhenium (Rh) ($Z = 75$, melting point 3170°C) greatly reduced the cracking by increasing the ductility of tungsten at high temperatures. However, pure W/Rh anodes would be extremely expensive so molybdenum is now chosen as the base metal. Molybdenum ($Z = 42$, melting point 2620°C) stores twice as much heat, weight for weight, as tungsten, but the anode volume is now greater because molybdenum has a smaller density than tungsten. As shown in Fig. 2.6a only a thin layer of W/Rh is used, to prevent distortion that might arise from the differences in thermal expansion of the different metals.

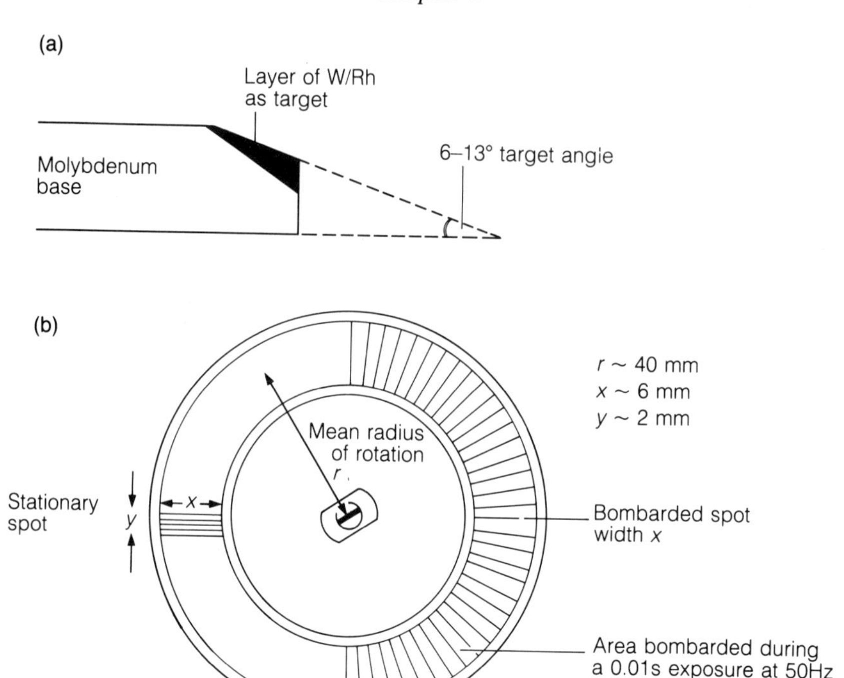

Fig. 2.6. (a) Detail of the target area on a modern rotating anode. (b) Principle of the rotating anode showing the area bombarded in a 0.01 s exposure at 50 Hz.

2.3.3 Anode design

The two principal requirements of anode design are first to make adequate arrangements for dissipation of the large quantity of heat generated and second to ensure a good spatial distribution of X-rays. Design features related primarily to heat dissipation are discussed below, the spatial distribution of X-rays is considered later.

Stationary anode

This form of anode is rarely used in a modern X-ray department but is still used in some 'low output' mobile X-ray systems used for fluoroscopy and in dental units. Tungsten in the form of a small disc about 1 mm thick and 1 cm in diameter is embedded in a large block of copper. As shown in Fig. 2.1 the copper protrudes through the tube envelope into the surrounding oil. Heat is transferred from the tungsten to copper by conduction and thence to the oil by convection. The cooling fins assist the convection process. The oil transfers this heat to the X-ray tube shield by conduction and it is eventually removed by air in the X-ray room by convection.

Rotating anode

This is the main form of anode used in diagnostic X-ray units. The principle of the rotating anode is very simple (Fig. 2.6b). If the required focal spot size on the target is, say 2 mm × 2 mm, for an anode angled to the beam at about 16° the dimensions of the area actually bombarded by electrons are about 6 mm × 2 mm (see section 2.4). The area over which heat is dissipated can be increased by arranging for the tungsten target to be an annulus of material which rotates rapidly. It may be shown that, if the exposure time is long enough for the anode to rotate at least once,

$$\frac{\text{Effective area for heat absorption with rotating anode}}{\text{Effective area for heat absorption with stationary anode}} = \frac{2\pi r \cdot x}{y \cdot x}$$

From the diagram this is an improvement of about 6 × 40/2 = 120 times and the heat input can be increased considerably (although not by a factor of 120).

Rotation rates range from 3000 rpm (the 50 Hz mains supply) to 10,000 rpm, ensuring that the anode rotates several times during even the shortest exposure. However, this does create some problems with respect to the type of mounting and cooling mechanism. Adequate electrical contact is maintained via bearings on which the anode rotates, but the area of contact is quite insufficient for adequate heat conduction. Since the anode is in an evacuated tube, there are no losses by convection. The initial mode of heat transfer from the anode to the cooling oil must therefore be radiation at a rate proportional to $(\text{anode temperature})^4 - (\text{oil temperature})^4$. With a rotating anode, heat loss by conduction is actually minimized since it might result in over-heating of the bearings. Thus the rotating anode is mounted on a long thin rod of low conductivity material such as molybdenum.

2.3.4 Electrical circuits

Only brief details will be given of the essential electrical components of an X-ray generator. For a much fuller explanation students are referred to Meredith & Massey (1977).

The transformer

This provides a method of converting high alternating currents of low potential difference to low alternating currents at high potential difference.

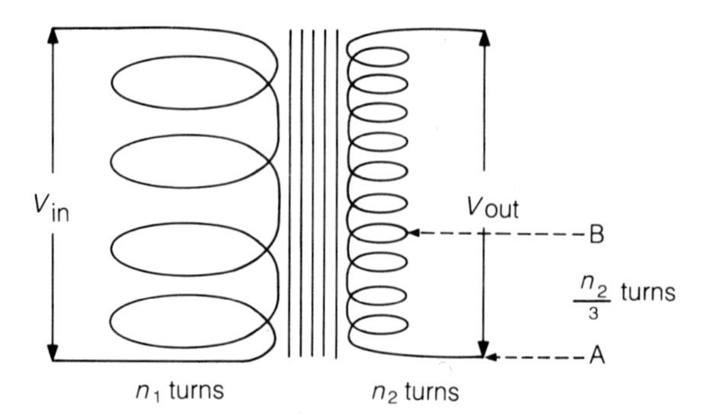

Fig. 2.7. Essential features of a simple transformer.

It consists of two coils of wire which are electrically insulated from one another but are wound on the same soft iron former (Fig. 2.7). If no energy is lost in the transformer

$$V_{out}/V_{in} = n_2/n_1$$

where n_1 and n_2 are the number of turns on the primary and secondary coils respectively.

Note:

1 By making a connection at different points, different output voltages may be obtained. For example the alternating potential difference across AB is given by

$$V_{AB} = 1/3V_{out}$$

2 Since, in the ideal case, all power is transferred from the input circuit to the output circuit

$$V_{in}I_{in} = V_{out}I_{out}$$

Hence if V_{out} and I_{out} to the X-ray set are 100 kVp and 50 mA respectively, since V_{in} will be the 240 V ac supply,

$$I_{in} = \frac{100 \times 10^3 \times 50 \times 10^{-3}}{240} = 20 \text{ A}$$

so input currents are very high. (Hence the requirement for special 30 A hospital circuits when using non-condenser discharge mobile X-ray units).

3 Power loss occurs in all transformers, and the amount depends on working conditions, especially I_{in}. Hence V_{out} and I_{out} also vary and auxiliary electrical circuits are required to stabilize them when operating with an X-ray set.

Figure 2.10 (see later) also shows, on the extreme left, an autotransformer. An autotransformer comprises one winding only and works on the principle of self-induction. Since the primary and secondary circuits are in contact, it cannot transform high voltages or step up from low to high voltages. However, it does give a variable secondary output on the low voltage side of the transformer and hence controls kV directly.

Generation of different voltage wave forms

As explained in section 2.2.5, the alternating potential must be rectified before it is applied to an X-ray set. The X-ray tube can act as its own rectifier (self-rectification) since it will only pass current when the anode is positive and the cathode is negative. However, this is a very inefficient method of X-ray production because if the anode gets hot, it will start to release electrons by thermionic emission. These electrons will be accelerated towards the cathode filament during the half cycle when the cathode is positive and will damage the tube. Thus the voltage supply is rectified independently.

If a gas-filled diode valve or a solid state p-n junction diode rectifier is placed in the anode circuit, half wave rectification (Fig. 2.5b) is obtained. The gas-filled diode is a simplified X-ray tube comprising a heated cathode filament and an anode in an evacuated enclosure. Electrons may only flow from cathode to anode but the diode differs from the X-ray set in that it is designed so that only a small proportion of the electrons boiled off the cathode travel to the anode. In terms of Fig. 2.17a (see later) the diode operates on the rapidly rising portion of the curve, whereas the X-ray tube operates on the near-saturation portion.

The design and mode of operation of a p-n junction diode will be considered in section 7.14 when its use as a radiation detector is discussed. It has many advantages over the gas-filled diode as a rectifier including its small size, long working lifetime, and robustness. It is also easy to manufacture in bulk, inexpensive, requires no filament heating circuit, has a low heat dissipation and a fast response time. For rectification silicon rectifiers have a number of advantages over selenium, including a negligible forward voltage drop and a very high reverse resistance resulting in negligible reverse current flow. They can also withstand high reverse bias voltages so only a few hundred silicon rectifiers are required rather than a few thousand if made of selenium, and they can work up to 200°C if required.

The essential features of full wave rectified and three phase supplies are shown in Fig. 2.8.

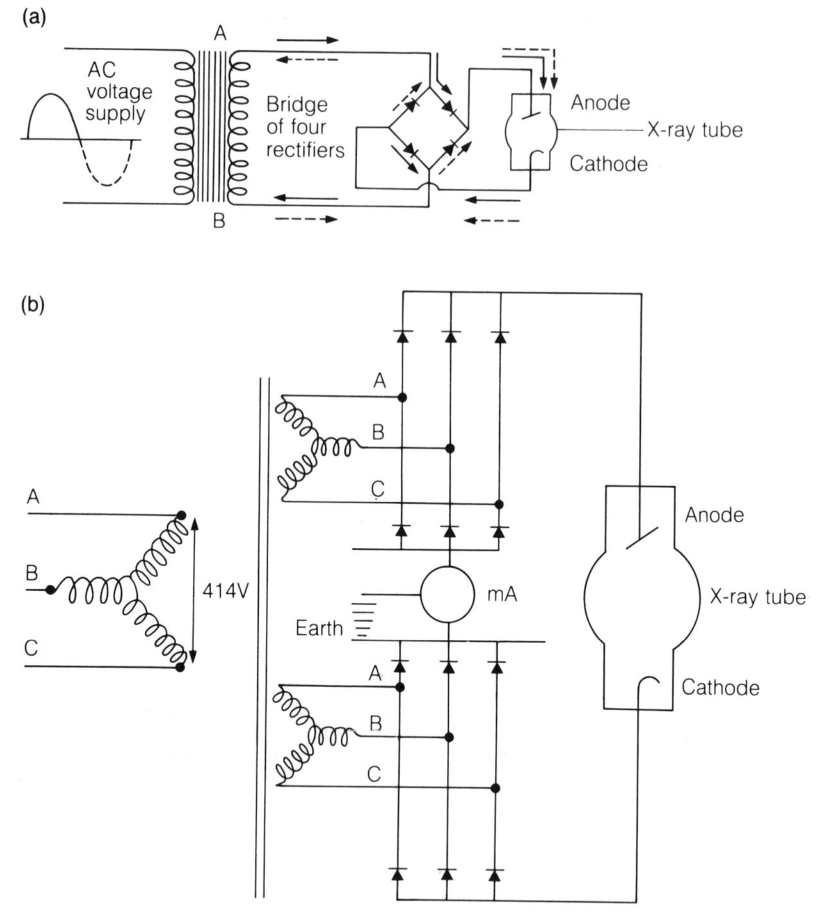

Fig. 2.8. Essential features of (a) full wave rectified, (b) three phase supplies. (a) *Solid* and *dotted arrows* show that irrespective of whether A or B is at a positive potential, the current always flows through the X-ray tube in the same direction. (b) Inductances connected in a star pattern to form a three phase transformer.

Action of smoothing capacitors

A capacitor in parallel with the X-ray unit will help to smooth out any variations in applied potential (Fig. 2.9).

Consider for example the full wave rectified supply shown in Fig. 2.5c. When electrons are flowing from the bridge circuit, some of them flow onto the capacitor plates and are stored there. When the potential across the bridge circuit falls to zero, electrons flow from the capacitor to maintain the current through the X-ray tube.

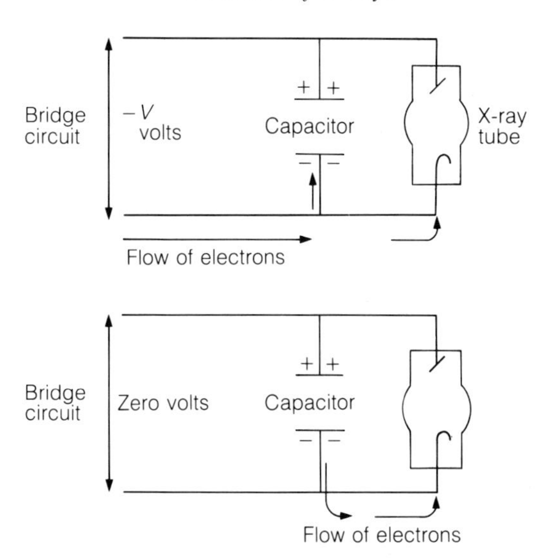

Fig. 2.9. Illustration of the use of a capacitor for voltage smoothing.

Tube kilovoltage and tube current meters

These are essential components of the circuit and are shown in relation to other components in Fig. 2.10. Note that the voltmeter is placed in the primary circuit so that a reading may be obtained before the exposure key is closed. There are two ammeters. A_F measures the filament supply current (I_F) which may be adjusted to give the required thermionic emission before expsoure starts. The actual tube current flowing during exposure (I_c) is measured by ammeter A_c.

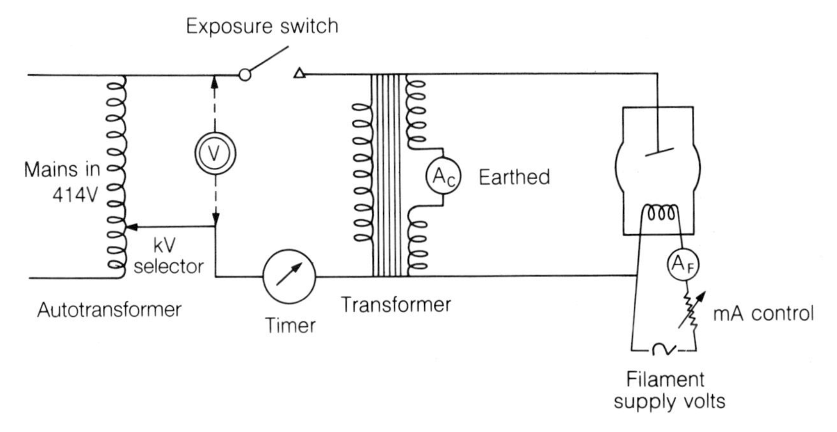

Fig. 2.10. The positions of kilovoltage and current meters in the electrical circuit.

2.3.5 The tube envelope and housing

The envelope

The envelope is of thick-walled glass and must be constructed under very clean conditions to a high precision so as to provide adequate insulation between the cathode and anode. It also provides a vacuum seal to the metallic components that protrude through it. Great care must be taken at the manufacturing stage to achieve a very high level of vacuum before the tube is finally sealed. If residual gas molecules are bombarded by electrons, the electrons may be scattered and strike the walls of the glass envelope, thereby causing reactions that result in release of gas from the glass and further reduction of the vacuum.

The presence of atoms or molecules of gas or vapour in the vacuum, whatever their origin, is likely to have a deleterious effect on the performance of the tube. For example, metal evaporation from the anode can cause a conducting film across the glass envelope, thereby distorting the pattern of charge across the tube. This can change the output characteristics since it is assumed that the flow of electrons from cathode to anode will be influenced by the repulsive effect of a static layer of charge on the tube envelope. If this charge is not static, the electrons in the beam are not repelled by the tube envelope and deviate to it. This diversion of current may significantly reduce tube output.

Both residual gas and anode evaporation cause a form of tube instability which may occasionally be detected during screening as a kick on the milliammeter as discharges take place. In the extreme case, the tube goes 'soft' and arcs over during an exposure.

The tube housing

This has various functions which may be summarized as follows:

1 shields against stray X-rays because it is lined with lead;
2 provides an X-ray window—which filters out some low energy X-rays;
3 contains the anode rotation power source;
4 provides high voltage terminals;
5 insulates the high voltage;
6 allows precise mounting of the X-ray tube envelope;
7 provides a means for mounting the X-ray tube;
8 provides a reference and attaching surface for X-ray beam collimation devices;
9 contains the cooling oil.

The advantages of filling this housing with oil are:

(a) high voltage insulation;
(b) effective conduction of heat from the X-ray insert tube;
(c) since the oil expands, an expansion diaphragm can be arranged to operate a switch when the oil reaches its maximum safe temperature.

2.3.6 Timing mechanisms

Although a variety of timing mechanisms has been used in the past, only the two that are most widely used will be discussed.

The electronic timer

If a capacitor C is charged to a fixed potential V_0, either positive or negative, and then placed in series with a resistor R, the rate of discharge of the capacitor depends on the values of C and R. A family of curves for fixed C and variable R is shown in Fig. 2.11. Note that a large resistance reduces the rate of flow of charge so the rate of fall of V is slower.

These curves may be used as the basis for a timer if a switching device is arranged to operate when the potential across C reaches say V_s. As shown in Fig. 2.11, if R has been preset to R_1 this will occur after t_1, but if R has been preset to R_2 it will occur after a longer time t_2.

Note the difference between timing circuits and switching circuits. The switching device can be a thyratron, which is simply a gas-filled triode, in

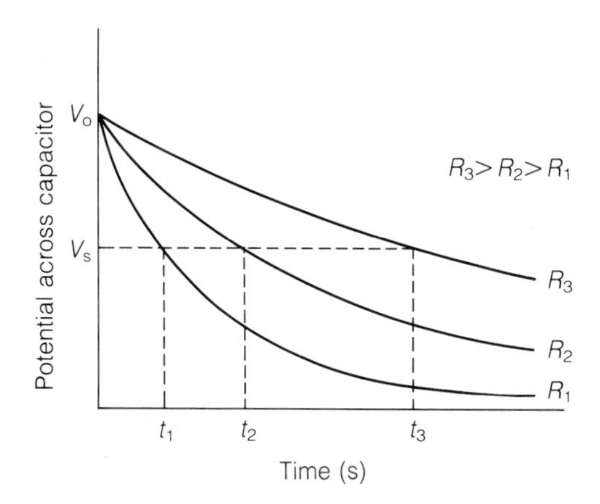

Fig. 2.11. Curves showing the rate of discharge of a capacitor through resistors of different resistance. The time taken to reach V_s, when the switching mechanism would operate, depends on the value of R.

series with a relay. Whilst the grid G of the thyratron is at a sufficiently high negative potential, electrons emitted from the triode filament F are repelled and do not reach the anode A. When the potential at G is less negative, a current flows through the thyratron circuit. One way in which the thyratron might be used as a timing device is illustrated in Fig. 2.12.

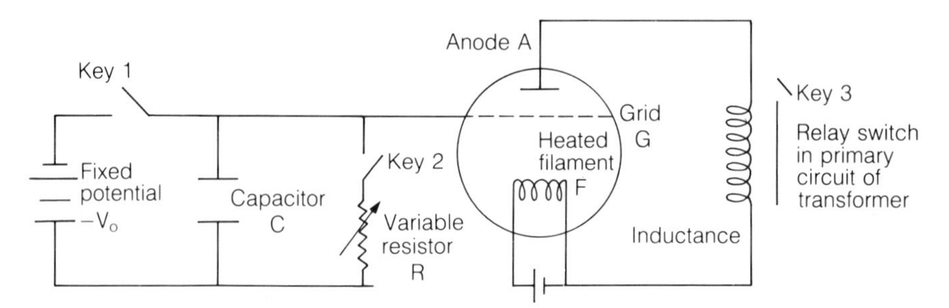

Fig. 2.12. Use of a thyratron to act as a switching mechanism.

The operation of this timer circuit is as follows:

1 Before exposure, key 1 is closed so that both capacitor and grid are charged to a negative potential, $-V_0$, that is sufficient to stop electrons flowing in the thyratron.

2 Exposure is started by closing key 3. The relay completes the circuit to the primary of the transformer provided that no current is flowing through the inductance coil.

3 As key 3 is closed, key 1 is opened and key 2 is closed. Hence the capacitor starts to discharge through R, which has been preset to a value such that after the requisite time interval the potential across the capacitor will fall to that numerically smaller negative potential $-V_s$ at which the thyratron starts to conduct.

4 When the thyratron starts to conduct, the relay opens key 3 and the exposure is terminated.

More recently solid state thyratrons (silicon controlled rectifiers) have almost completely replaced gas filled tubes. They have identical characteristics but are smaller and do not require a heater.

The photo timer

The weakness of the electronic timer, and other timers that predetermine the exposure, is that a change in any factor which affects the amount of radiation actually reaching the film, notably patient thickness, will alter the amount of film blackening. In the photo timer the exposure is linked more directly to the amount of radiation reaching the film. This is known as automatic exposure control.

One design places small ionization chamber monitors in the cassette tray system between the patient and the film–screen combination. The amount of radiation required to produce a given degree of film blackening with a given film–screen combination under standard development conditions is known, so when the ion chamber indicates that this amount of radiation has been received, the exposure is terminated. This type of exposure control does not need to be 'set' prior to each exposure, but some freedom of adjustment is provided to allow for minor variations in film blackening if required. Adjustment will also be required if screens of different sensitivity are used.

As an alternative to ion chambers, small photoelectric cells coupled to photomultiplier tubes (see Chapter 7) may be used. They have the disadvantage however of being X-ray opaque so they must be placed behind the cassette, where the X-ray intensity is low, and special radiolucent cassettes must be used. Photoelectric cells are not now so popular conventionally but are widely used in photofluorography.

A weakness with these types of phototimer is that the ion chamber or photomultiplier tube only monitors the radiation reaching a small part of the film and this may not be representative of the radiation reaching the rest of the film. This problem can be partially overcome by using several small ion chambers, usually three, and controlling the exposure with the one that is closest to the region of greatest interest on the resulting X-ray film.

For a fuller treatment of timing mechanisms the reader should consult references given at the end of the chapter.

2.3.7 Electrical safety features

A number of features of the design of X-ray sets are primarily for safety and should be summarized briefly.

The tube housing
As already indicated, this provides a totally enclosing metallic shield that can be firmly earthed thereby contributing to electrical safety.

High tension cables
High tension cables are constructed so that they can operate up to potentials of 150 kV. Since the outermost casing metal braid of the cable must be

at earth potential for safety, a construction of multiple coaxial layers of rubber and other insulators must be used to provide adequate resistance between the innermost conducting core and the outside to prevent current flow across the cable.

It is essential that high tension cables are not twisted or distorted in any other manner that might result in breakdown of the insulation. They must not be load-bearing.

Electrical circuits

These are designed in such a way that the control panel and all meters on the control panel are at earth potential. Nevertheless, it is important to appreciate that many parts of the equipment are at very high potential and the following simple precautions should be observed.

1 Ensure that equipment is installed and maintained regularly by competent technicians.

2 Record and report to the service engineer any evidence of excessive mechanical wear, especially to electrical cables, plugs and sockets.

3 Similarly, report any equipment malfunction.

4 Adopt all other safety procedures that are standard when working with electrical equipment.

2.4 Spatial distribution of X-rays

When 40 keV electrons strike a thin metal target, the directions in which X-rays are emitted are as shown in Fig. 2.13a. Most X-rays are emitted at angles between 45° and 90° to the direction of electron travel. The more energetic X-rays travel in a more forward direction (smaller value of θ). It follows that if the mean X-ray energy is increased by increasing the energy of the electrons, the lobes are tilted in the direction of the electron flow.

When electrons strike a thick metal target, the situation is more complicated because X-ray production may occur from the surface or it may occur at depth in the target. Also, the spatial distribution of X-rays will now depend on the angle presented by the anode to the incoming electron beam. Consider the anode shown in Fig. 2.13b and angled at 30° to the beam. X-rays produced in the direction B are much more heavily attenuated than those produced in the direction A because they travel further through anode material. This is clearly a disadvantage since a primary objective of good X-ray tube design is to ensure that the field of view is

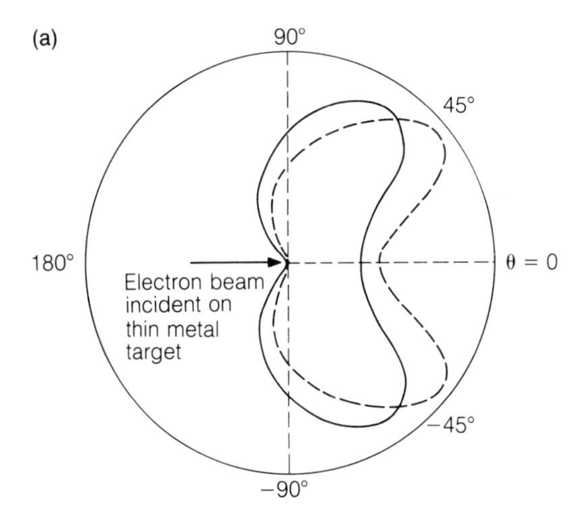

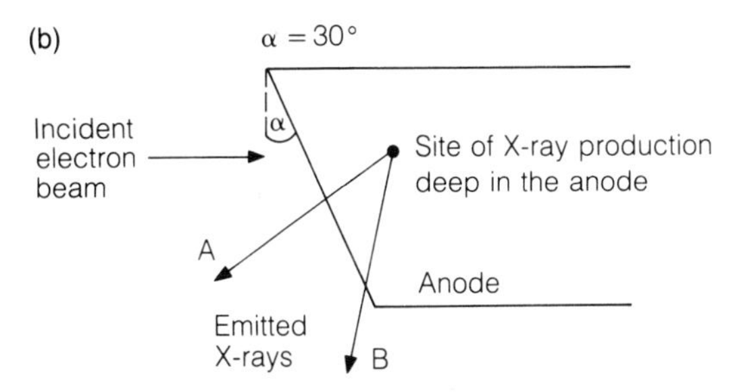

Fig. 2.13. (a) Approximate spatial distribution of X-rays generated from a thin metal target bombarded with 40 keV electrons. This figure is known as a polar diagram: the distance of the curve from the origin represents the relative intensity of X-rays emitted in that direction. The polar diagram that might be obtained with 100 keV electrons is shown *dotted*.
(b) The effect of self absorption within the target on X-ray production from a thick anode.

uniformly exposed to radiation. Only if this is achieved can variations in film blackening be attributed to variations in scatter and absorption within the patient. Variation in intensity across the field is minimized by carefully selecting the angle at which the anode surface is inclined to the vertical (Fig. 2.14).

Note the following additional points:

1 The radiation intensity reaching the film is still not quite uniform, being maximum near the centre of the field of view. This is due to

(a) an inverse square law effect—radiation reaching the edges of the field has to travel further, and

(b) a small obliquity effect—beams travelling through the patient at a

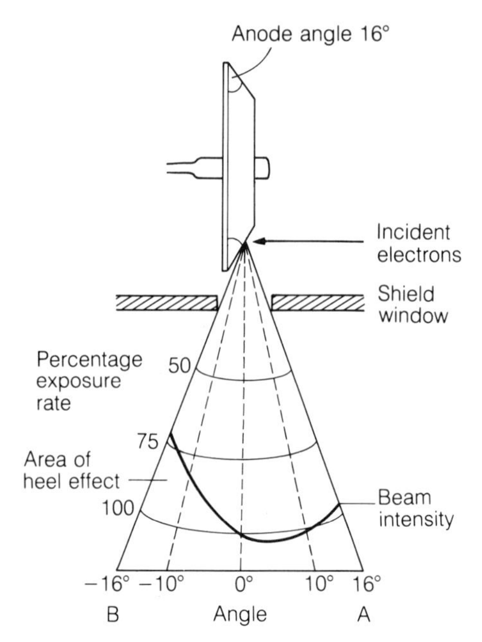

Fig. 2.14. Variation of X-ray intensity across the field of view for a typical anode target angle.

slight angle must traverse a greater thickness of the patient and are thus more attenuated.

Neither of these factors is normally of great practical importance.

2 The anode angle selected does not remove the asymmetry completely and this is known as the 'heel effect'. The effect of X-ray absorption in the target, which results in a bigger exposure at A than at B, is more important than asymmetry in X-ray production, which would favour a bigger exposure at B.

3 No such asymmetry exists in a direction normal to that of the incident electron beam so if careful comparison of the blackening on the two sides of the film is essential, the patient should be positioned accordingly.

4 The shape of the exposure profile is critically dependent on the quality of the anode surface. If the latter is pitted owing to overheating by bombarding electrons, much greater differences in exposure may ensue.

5 An angle of about 13–16° is frequently chosen and this has one further benefit. One linear dimension of the effective spot for the production of X-rays is less than the dimension of the irradiated area by a factor equal to $\sin \alpha$. Sin 13° is about 0.2, so angling the anode in this way allows the focusing requirement on the electron beam to be relaxed whilst ensuring a good focal spot for X-ray production (Fig. 2.15). If a very small focal spot

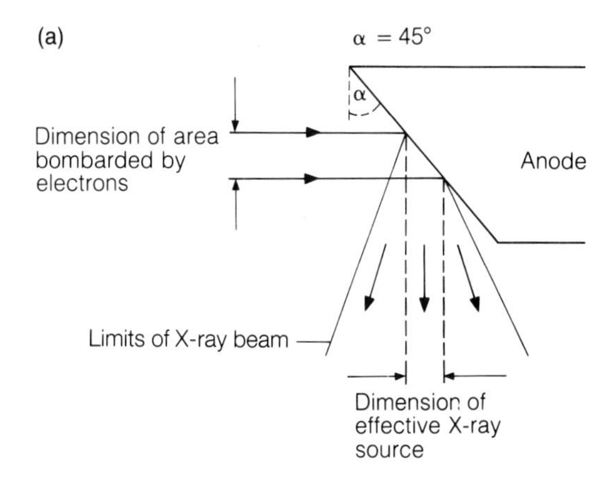

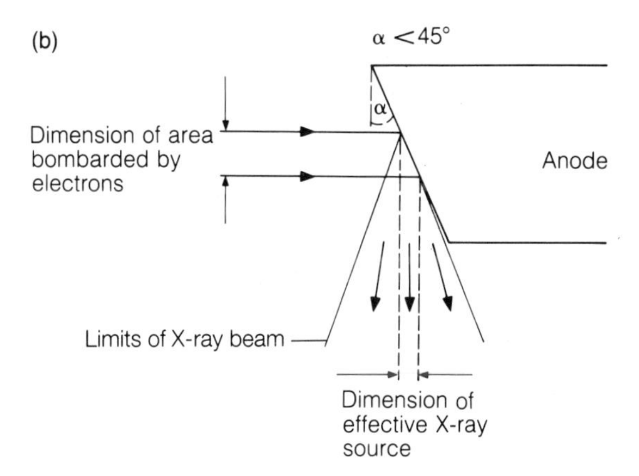

Fig. 2.15. The effect of the anode angle on the effective focal spot size for X-ray production. (a) If the anode is angled at 45° the effective spot for X-ray production is equal to the target bombarded by electrons. (b) If the anode is angled at less than 45° to the vertical the effective spot for X-ray production is less than the bombarded area. Note that in all cases these areas are measured normal to the beam. The actual area of impact on the anode will be greater for reasons explained in the text.

(~ 0.3 mm) is required, an angle of only 6° may be used. Note that with such a small anode angle, the heel effect greatly restricts the field size. This can only be compensated by increasing the focus–film distance with consequent loss of intensity at the film due to the inverse square law.

Even with a well-designed anode, a certain amount of extrafocal radiation arises from regions of the anode outwith the focal spot. These X-rays may be the result of poorly collimated electrons but are more usually the consequence of secondary electrons being scattered from the target and striking the anode elsewhere. Extrafocal X-rays may contribute as much as 15–20% of the total output dose of the tube but are of lower average energy. Many of them will fall outside the area defined by the light beam diaphragm and under extreme conditions may cast a shadow of the patient (Fig. 2.16).

Over the region of interest, the extrafocal radiation creates a uniform low-level X-ray intensity. This appears to have little effect on conventional X-rays but it can reduce the quality of 100 mm cine film and the effect of extrafocal radiation in digital radiology probably merits investigation.

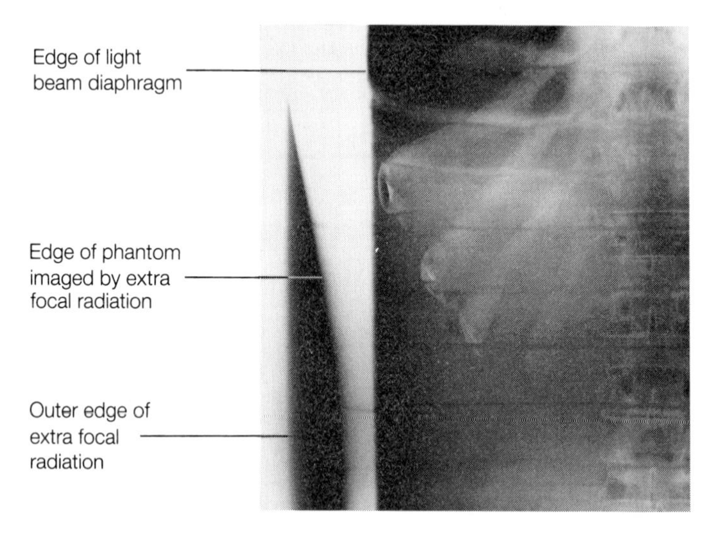

Fig. 2.16. Radiograph showing the effect of extra focal radiation. The field of view as defined by the light beam diaphragm is shown on the right but the outer edge of the 'patient' (a phantom in this instance) is also radiographed on the left by the extra focal radiation.

2.5 Rating of an X-ray tube

2.5.1 Introduction

The production of a good radiograph depends on the correct choice of tube kVp, current, exposure time and focal spot size. In many situations a theoretical optimum would be to use a point source of X-rays to minimize geometrical blurring (see section 5.9.1), and a very short exposure time, say 1 ms, to eliminate movement blurring (see section 5.9.2). However, these conditions would place impossible demands on the power requirement of the set. For example an exposure of 50 mA s would require a current of 50 A. Even if this current could be achieved, the amount of heat generated in such a small target area in such a short time would cause the anode to melt. This condition must be avoided by increasing the focal spot size or the exposure time, generally in practice the latter. Furthermore, during prolonged exposures, for example in fluoroscopy, a secondary limitation may be placed on the total amount of heat generated in the tube and shield.

Thus the design of an X-ray tube places both electrical and thermal constraints on its performance and these are frequently expressed in the form of rating charts, which recommend *maximum* operating conditions to ensure a reasonably long tube life when used in equipment that is properly designed, installed, calibrated and operated. Note that lower ratings should be used whenever possible to maximize tube life.

2.5.2 Electrical rating

Electrical limits are not normally a problem for a modern X-ray set but are summarized here for reference.

Maximum voltage

This will be determined by the design, especially the insulation, of the set and the cables. It is normally assumed that the high voltage transformer is centre grounded (see Fig. 2.9), i.e. that the voltages between each high voltage tube terminal and ground are equal. A realistic upper limit is 150 kVp.

Maximum tube current

This is determined primarily by the filament current. Very approximately, the tube current (I_C)will be about one tenth of the filament current (I_F). In other words only about one tenth of the electrons passing through the filament coil are 'boiled off' from it. A modern X-ray tube may be designed to operate up to 1000 mA but under normal conditions tube currents will be less than half this value. The lifetime of the tube can be significantly extended by a small reduction in current. The lifetime of a filament operating at 4.3 A is about 10 times that of one operating at 4.8 A.

If the voltage is increased at fixed filament current, the tube current will change as shown in Fig. 2.17a. At low voltages, the tube current increases as the kV is increased because more and more electrons from the space charge around the cathode are being attracted to the anode. In theory, the tube current should plateau when the voltage is large enough to attract all electrons to the anode. In practice there is always a cloud of electrons

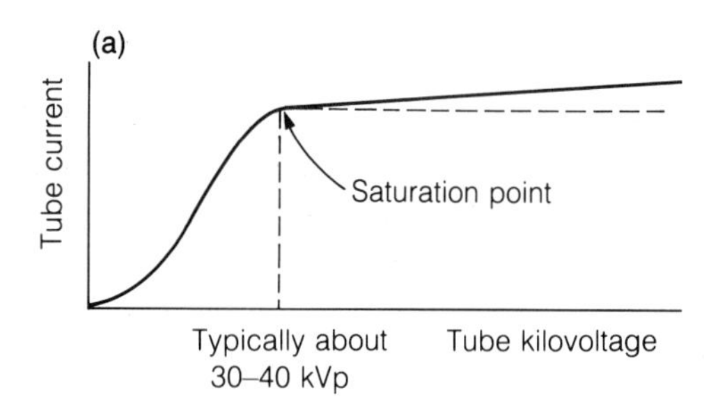

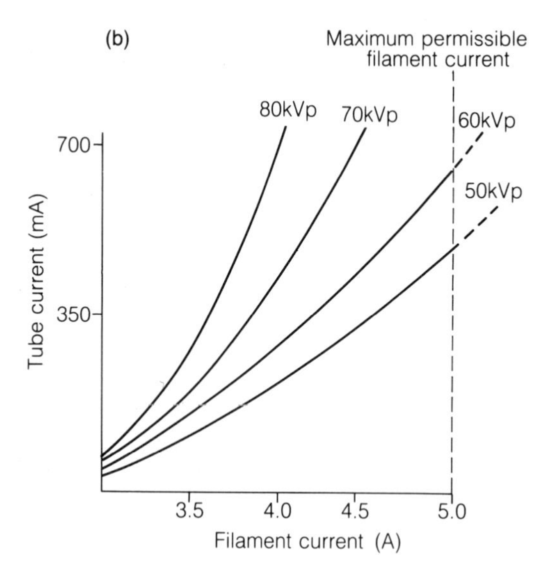

Fig. 2.17. (a) The effect of increasing tube kilovoltage on the tube current for a fixed filament current. (b) A family of curves relating tube current to filament current for different applied voltages.

(a space charge) around the cathode and as the potential difference is increased, a few more electrons are attracted to the anode. The result is that, as the tube kilovoltage is increased, the maximum tube current attainable also increases. Hence a typical family of curves relating tube current to filament current might be as in Fig. 2.17b. Modern X-ray tubes contain several compensating circuits one of which stabilizes the tube current against the effect of changes in voltage.

Maximum power

This is the product of tube current and voltage, but is not a practical limitation.

2.5.3 Thermal rating—considerations at short exposures

When electrons strike the anode of a diagnostic tube, 99% of their energy is converted into heat. If this heat cannot be adequately dissipated, the anode temperature may quickly rise to a value at which damage occurs due to excessive evaporation, or the anode may melt which is even worse. The amount of heat the anode can absorb before this happens is governed by its thermal rating.

For exposure times between 0.02 s and 10 s the primary thermal consideration is that the area over which the electrons strike the anode should not overheat. This is achieved by dissipating the heat over the anode surface as much as possible. The factors that determine heat dissipation will now be considered.

Effect of cooling
It is important to appreciate that when the maximum heat capacity of a system is reached, any attempt to achieve acceptable exposure factors by increasing the exposure time is dependent on the fact that during a protracted exposure some cooling of the anode occurs. Consider the extreme case of a tube operating at its anode thermal rating limit for a given exposure. If the exposure time is doubled in an attempt to increase film blackening then, in the absence of cooling, the tube current must be halved. This is because at a given kVp the energy deposited in the anode is directly proportional to the product of the current and the period of exposure. However, in the presence of cooling, longer exposure times do permit greater power dissipation as shown in Fig. 2.18.

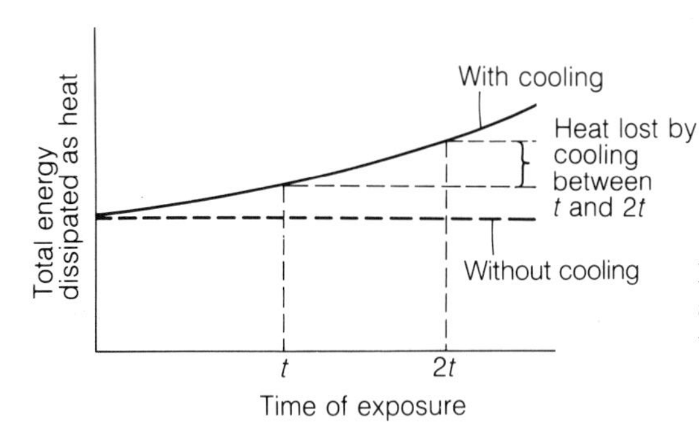

Fig. 2.18. Total energy dissipated as heat for different exposure times with and without cooling.

Target spot size
For fixed kVp and exposure time, the maximum permitted current increases with target spot size. For very small spots (~ 0.3 mm) the maximum

current is approximately proportional to the area of the spot since this determines the volume in which heat is generated. For larger spots (~ 2 mm) the maximum current is more nearly proportional to the perimeter of the spot since the rate at which heat is conducted away becomes the most important consideration.

Anode design

The main features that determine the instantaneous rating of a rotating anode are

1 its radius, which will determine the circumference of the circle on which the electrons fall,

2 its rate of rotation, and

3 for a fixed target spot size, the anode angle. Typical rating curves showing the maximum permissible tube current for different exposure times for anodes of different design are shown in Fig. 2.19. A small anode angle and rapid rotation give the highest rating but note that the differences between the curves become progressively less as the exposure time is extended.

Note also that since for the first complete rotation of the anode surface electrons are falling on unheated metal, the curve is initially almost horizontal. The maximum permissible tube current for a stationary anode operating under similar conditions would be much lower.

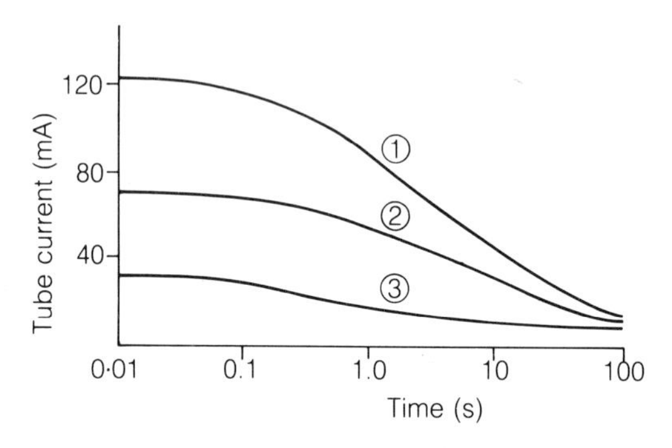

Fig. 2.19. Typical rating curves showing the maximum permissible tube current at different exposure times for anodes of different design. Each tube is operating at 100 kVp three phase with a 0.3 mm focal spot. (1) Type PX 410 4″ diameter anode with a 10° target angle and 150 Hz stator. (2) Type PX 410 4″ diameter anode with a 10° target angle and 50 Hz stator. (3) Type PX 410 4″ diameter anode with a 15° target angle and 50 Hz stator.

Curves (1) and (2) show the effect of increasing the speed of rotation of the anode. Curves (2) and (3) show the effect of changing the target angle. From Waters G. *J. Soc. X-ray Tech.* Winter 1968/69, 5.

Tube kilovoltage

As the kVp increases, the maximum permissible tube current for a fixed exposure time decreases (Fig. 2.20). This is self-evident if a given power dissipation is not to be exceeded.

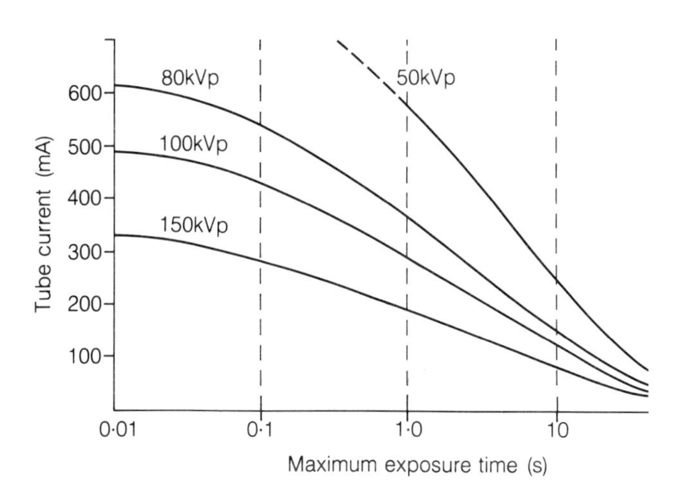

Fig. 2.20. Maximum permissible tube current as a function of exposure time for various tube kilovoltages for a typical rotating anode. Type PX 306 tube with a 3″ diameter anode, 15° target angle operating on single phase with a 60 Hz stator and a 2 mm focal spot.

Note: the tube current is higher than in Fig. 2.19 because a larger focal spot is being used; the *dotted line* indicates that the maximum permissible filament current would probably be exceeded under these conditions. Reproduced by permission of Picker International.

Such a rating chart may readily be used to determine if a given set of exposure conditions is admissible with a particular piece of equipment. For example is an exposure of 400 mA at 70 kVp for 0.2 s allowed? Reference to Fig. 2.20 shows that the maximum permissible exposure time for 400 mA at 70 kVp is about 1.0 s so the required conditions can be met. Note that for very long exposures the product kVp × mA × time is converging to the same value for all curves and the heat storage capacity of the anode then becomes the limiting factor (see section 2.5.5).

When full wave rectified and three phase supply rating charts are compared at the same kVp, all other features of anode design being kept constant, the curves actually cross (Fig. 2.21). For very short exposures higher currents can be used with a three phase than with a single phase supply, but the converse holds at longer exposures.

To understand why this is so, consider the voltage and current profiles for two tubes with the same kVp and mA settings (Fig. 2.22). Note:

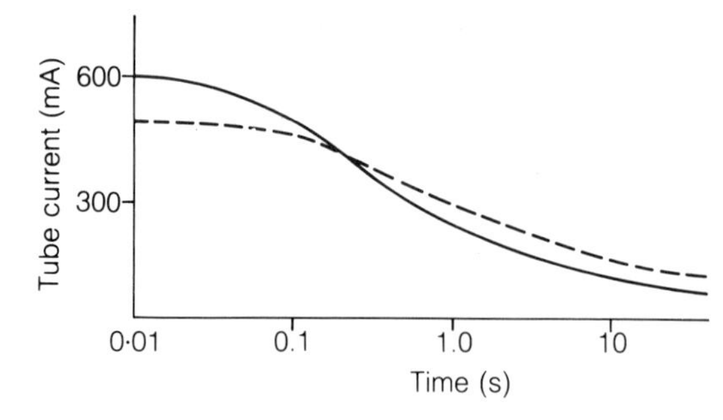

Fig. 2.21. Maximum permissible current as a function of exposure time for 80 kVp single phase, full wave rectified (*dotted line*), and 80 kVp three phase supplies (*solid line*).

1 The current does not follow the voltage in the full wave rectified tube. As soon as the potential difference is sufficient to attract all the thermionically emitted electrons to the anode, the current remains approximately constant.

2 The three phase current remains essentially constant throughout.

3 The peak value of the current must be higher for the full wave rectified tube than for the three phase tube, if the average values as shown on the meter are to be equal.

For very short exposures, instantaneous power is important. This is maximum at T_m and since the voltages are then equal, power is proportional to instantaneous current and is higher for the full wave rectified system. Inverting the argument, if power dissipation cannot exceed a predetermined maximum value, the average current limit must be lower for the full wave rectified tube.

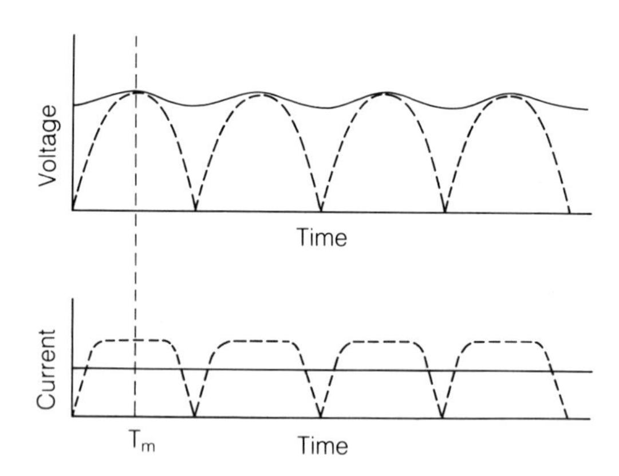

Fig. 2.22. Voltage and current profiles for two tubes with the same kVp and mA settings but with three phase (*solid line*) or full wave rectified (*dotted line*) supplies.

For longer exposures, average values of kV and mA are important. Average values of current have been made equal, so power is proportional to average voltage and this is seen to be higher for the three phase supply. Hence, again inverting the analysis, the current limit must be lower for the three phase tube as predicted by the rating curve graphs.

It is left as an exercise to the reader to explain, by similar reasoning, why the rating curve for a full wave rectified tube will always be above the curve for a half wave rectified tube.

2.5.4 Overcoming short-exposure rating limits

If a desired combination of kVp, mA, time and focal spot size is unattainable owing to rating limits, several things can be done, although all may degrade the image in some way. Increasing the focal spot size or the time of exposure have already been mentioned. The other possibility is to increase the kVp. At first inspection, this appears to give no benefit. Suppose the rating limit has been reached and the kVp is increased by 10%. The current will have to be reduced by 10% otherwise the total power dissipated as heat will increase. The gain from increasing the kVp appears to be negated by a loss due to reduced mA. However, although X-ray output will fall by 10% as a result of reducing mA, it will increase by about 20% as a result of the 10% increase in kVp (see section 2.2.5). Furthermore, X-ray transmission through the patient is better at the higher kVp and, in the diagnostic range, film sensitivity increases with kVp. Thus film blackening, which is ultimately the relevant criterion, is increased about 40% by the 10% increase in kVp and reduced by only about 10% due to reduction in mA, yielding a net positive gain. Some image degradation may occur as a result of loss of contrast at the higher kVp (see section 5.3).

2.5.5 Multiple or prolonged exposures

If too many exposures are taken in a limited period of time, the tube may overheat for three different reasons:

1 the surface of the target can be overheated by repeated exposures before the surface heat has time to dissipate into the body of the anode;

2 the entire anode can be overheated by repeating exposures before the heat in the anode has had time to radiate into the surrounding oil and tube housing;

3 the tube housing can be overheated by making too many exposures before the tube shield has had time to lose its heat to the surrounding air.

The heat capacity of the total system, or of parts of the system, is sometimes expressed in heat units (HU). By definition, 1.4 HU are generated when 1 J of energy is dissipated.

The basis of this definition can be understood for a full wave rectified supply:

$$\begin{aligned} \text{HU} &= 1.4 \times \text{energy} \\ &= 1.4 \times \text{root mean square (rms) kV} \times \text{average mA} \times \text{s} \end{aligned}$$

But

$$\text{rms kV} = 0.71 \times \text{kVp}$$

Hence

$$\text{HU} = \text{kVp} \times \text{mA} \times \text{s}$$

Thus the HUs generated in an exposure are just the product of (voltage) × (current) × (time) shown on the X-ray control panel, hence the introduction of the HU was very convenient for single phase generators.

Unfortunately, this simple logic does not hold for three phase supply. The mean kV is now much higher, perhaps 0.95 kVp, so:

$$\text{HU} = 1.4 \times 0.95 \text{ kVp} \times \text{mA} \times \text{s} = 1.35 \times \text{kVp} \times \text{mA} \times \text{s}$$

Hence for three phase supply the product of kVp and mAs as shown on the meters must be multiplied by 1.35 to obtain the heat units generated. With the increasing use of three phase generators, joules are becoming the preferred unit.

The rating charts already discussed may be used to check that the surface of the target will not overheat during repeat exposures. This cannot occur provided that the total heat units of a series of exposures made in rapid sequence does not exceed the heat units permissible, as deduced from the radiographic rating chart, for a single exposure of equivalent total exposure duration.

When the time interval between individual exposures exceeds 20 s there is no danger of focal track overheating. The number and frequency of exposures is now limited either by the anode or by the tube heat storage capacity. A typical set of anode thermal characteristic curves is shown in Fig. 2.23. Two types of curve are illustrated:

1 Input curves showing the heat stored in the anode after a specified, long period of exposure. Also shown, dotted, is the line for 470 watt input power in the absence of cooling. This line is a tangent to the curve at zero

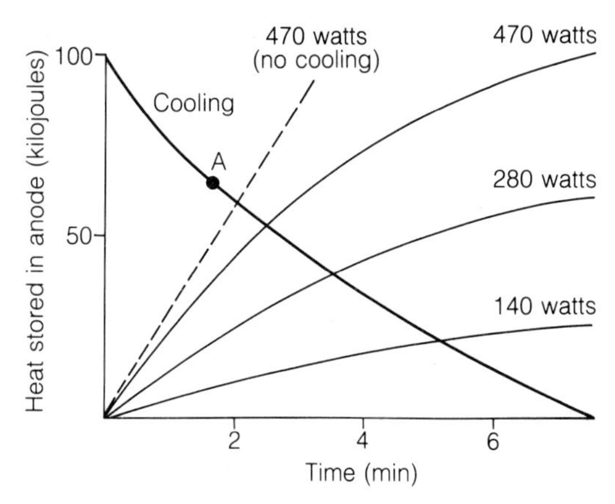

Fig. 2.23. Typical anode thermal characteristic curves, showing the heat stored in the anode as a function of time for different input powers. Reproduced by permission of Picker International.

time since the anode is initially cold and loses no heat. At constant kVp the initial slope is proportional to the current. As the anode temperature increases, the anode starts to lose heat and the curve is no longer linear.

2 A cooling curve showing the heat stored in the anode after a specified period of cooling. Note that if the heat stored in the anode after exposure is only 65×10^3 J, the same cooling curve may be used but the point A must be taken as $t = 0$.

Two other characteristics of the anode are important. First, the maximum anode heat storage capacity, which is 100×10^3 J here, must be known. For low screening currents, the heat stored in the anode is always well below its heat limit, but for higher input power the maximum heat capacity is reached and screening must stop.

The second characteristic is the maximum anode cooling rate. This is the rate at which the anode will dissipate heat when at its maximum temperature (360 watts) and gives a measure of the maximum current, for given kVp, at which the tube can operate continuously. Note that under typical modern screening conditions, say 2 mA at 80 kVp, the rate of heat production is only $2 \times 80 \times 10^3 \times 10^{-3} = 160$ W.

During screening, or a combination of short exposures and screening, the maximum anode heat storage capacity must not be exceeded. Exercises in the use of this rating chart are given at the end of the chapter.

When the total time for a series of exposures exceeds the time covered by the anode thermal characteristic chart, a tube shield cooling chart must be consulted. This is similar to the anode chart except that the cooling time will extend (typically) to 100 min and the maximum tube shield storage capacity may be as high as 10^6 J.

As a final comment on thermal rating, it is worth noting that a significant amount of power is required to set the anode rotating and this is also dissipated eventually as heat. In a busy accident department taking many short exposures in quick succession, three times as much heat may arise from this source as from the X-ray exposures themselves.

2.5.6 Falling load generators

Although thermal loading considerations indicate that a limit is imposed on the number and rate of multiple exposures that can be taken, in practice this limit is rarely reached with modern X-ray units. In recent years anode design has improved significantly and diagnostic tubes are currently available that are capable of operating at 1 ampere (1000 mA). However, before these improved anodes were available, falling load generators were introduced to enable rapid multiple exposures to be taken by running the tube as close to the maximum rating as possible. This method of operation uses the fact that the rate of heat loss from the anode is greatest when the anode is at its maximum working temperature, so the current through the tube is kept as high as possible without this maximum temperature being exceeded.

The anode temperature is monitored and if it reaches the maximum allowed, a motor driven rheostat, originally set at zero, introduces a resistance into the filament circuit thereby reducing the tube current in a step-wise manner. Because the transformer is not ideal, this lowering of tube current causes an increase in the kVp, and this has to be compensated for by increasing the resistance in the primary circuit in the transformer. In older falling load generators this was performed by introducing fixed resistors, but more modern units use a continuously variable resistor.

By maintaining the current at its maximum possible value, the minimum time will be required for a given exposure. To achieve this the exposure must be set and controlled using a meter calibrated in milliampere seconds or the exposure must be terminated using a phototimer for automatic exposure control (see section 2.3.6). Note however that the falling load generator will be of little value for short exposure times (say 0.4 s) because there will be insufficient time for the current to fall through many steps. Also there is wear on the tube at high current so lifetime is shortened by falling load operation. Thus a falling load generator might be a possibility for a busy orthopaedic clinic examining spines with heavy milliampsecond loadings and long exposure times. For chest work it would be useless.

2.5.7 Safety interlocks

These are provided to ensure that rating limits are not exceeded on short exposures. If a combination of kVp, mA, s and spot size is selected that would cause anode over-heating, a 'tube overload' warning light will appear and the tube cannot be energized.

During multiple exposures a photoelectric cell may be used to sample radiant heat from the anode and thereby determine when the temperature of the anode disc has reached a maximum safe value. A visual or audible warning is then triggered. In some modern systems the tube loading is under computer control. Anode temperature is continuously calculated from a knowledge of heat input and cooling characteristics. When the rating limit is reached, generator output is automatically reduced.

2.6 X-ray tube lifetime

The life of an X-ray tube can be extended by taking steps to avoid thermal stress and other problems associated with heating. For example, the anode is very brittle when cold and if a high current is used in this condition, deep cracks may develop. Thus at the start of operations several exposures at approximately 75 kVp and 400 mA s (200 mA for 2 s) should be made at 1 min intervals. Ideally, if the generator is idle for periods exceeding 30 min, the process should be repeated.

Keeping the 'prepare' time to a minimum will reduce filament evaporation onto the surface of the tube and also bearing wear in the rotating anode. The generator should be switched off when not in use.

The tube should be operated well below its rating limits whenever possible.

2.7 Quality assurance of performance for standard X-ray sets

It is sometimes difficult to identify the boundary between quality assurance and radiological protection in diagnostic radiology. This is because the primary purpose of good quality assurance is to obtain the best diagnostic image required at the first exposure. If this is achieved successfully, not only is the maximum information obtained for the radiation delivered, but also the radiation dose to the patient is minimized.

A number of performance checks should be carried out at regular intervals. These will, for example, confirm that:

1 the tube kVp is correctly set;
2 the milliammeter reading is accurate;
3 the exposure timer is accurate;
4 the radiation output is reproducible during repeat exposures.

All these factors affect the degree of film blackening and the level of contrast.

Tube kV may be checked by visual assessment of film blackening after the beam has passed through different thicknesses of different materials. A widely adopted method was suggested by Ardran and Crooks (1968) and made use of a specially adapted cassette. In the cassette there are two rows of holes, one of which has a strip of high sensitivity (fast) intensifying screen placed behind it, whilst the other has a strip of low sensitivity (slow) intensifying screen placed behind it. With this construction, the film backed by fast screen would be consistently blacker than the one backed by slow screen. However, if a copper step wedge is placed over the holes in front of the fast screen, there will be a certain thickness of copper for which attenuation in the copper will exactly compensate for the difference in sensitivity of the screens, and equal film blackening will result. Since beam attenuation in the copper is dependent on the kilovoltage of the radiation the step wedge can be calibrated so that the thickness of copper resulting in equal blackening can be converted to a kVp reading.

Exposure times for half wave and full wave rectified tubes are still occasionally checked using a spinning top. This consists of a metal disc with a radial line of holes in it. Each hole can be selected in turn. When placed over a film and spun during the exposure, an arc of dots is seen on the developed film. One dot is produced each time a pulse of X-rays is generated, the rate of X-ray pulses depending on the mains frequency and the type of rectification. As long as the top is spinning at the start and end of the exposure, and the top does not complete one revolution during the exposure, counting the number of dots allows the time to be calculated. (The multiple holes allow several times to be checked on one film.) This device will not work with three phase supplied tubes because there is insufficient fluctuation in the X-ray output. In these units exposures can be checked by placing a disc with a slit in it over a photographic film and rotating the disc at *known* angular velocity. The length of the blackened trace can be used to calculate exposure time.

Modern equipment is now available to simplify both these checks and nowadays, both kVp and time of exposure are usually measured more directly using a digital kV meter attached to a fast responding storage cathode ray oscilloscope (CRO). Several balanced photo detectors are used under filters of different materials and different thicknesses. By using internally programmed calibration curves, a range of kVps may be checked with the same filters. Accuracy to better than 5% should be achievable. This is particularly important for mammography (30 kVp ± 1.5 kV) because the absorption coefficient of soft tissue falls rapidly with increasing kV at these energies. Note that the CRO also displays the voltage profile so a fairly detailed analysis of the performance of the X-ray tube generator is possible.

Consistency of output may be checked by placing an ion chamber in the direct beam and making several repeat measurements. This arrangement can also be used to confirm a linear relationship between the tube output (measured in mGy in air for a fixed exposure time) and the preset mA. It is not normal to make an absolute calibration of tube current.

If an automatic exposure control is used, a check should be made that the three ion chambers are matched so that the same film density is achieved whichever chamber is selected.

There are recommended values for the total beam filtration (2.0 mm Al up to 100 kVp and 2.5 mm Al above 100 kVp) and these should be checked. To do so a ion chamber is placed in the direct beam and readings are obtained with different known thicknesses of aluminium in the beam. The half value layer (HVL) may be obtained by trial and error or graphically. Note that the HVL obtained in this way is *not* the beam filtration (although when expressed in mm of Al the values are sometimes very similar) and the filtration must be obtained from a look-up table (Table 2.2).

It is left as an exercise to the reader, after a careful study of Chapter 3, to explain why the look-up table will be different at other tube voltages.

Since all operators are urged to use the smallest possible field sizes, it is important to ensure that the optical beam, as defined by the light beam diaphragm, is in register with the X-ray beam. This may be done by placing an unexposed X-ray film on the table and using lead strips or a wire

Table 2.2. Relationship between beam filtration and half value layer for a full wave rectified X-ray tube operating at 70 kVp

Half value layer mm Al	1.0	1.5	2.0	2.5	3.0
Total filtration mm Al	0.6	1.0	1.5	2.2	3.0

rectangle to define the optical beam. An exposure is made, at very low mA s because there is no patient attenuation, and the film developed. The exposed area should correspond to the radiograph of the lead strips to better than about 1 cm at a focus–film distance of 1 m. When the alignment is poor there will be sufficient scattered radiation to provide a radiograph of the lead strips (Fig. 2.24). At the same time a check can be made that the axis of the X-ray beam is vertical by arranging two small (2 mm) X-ray opaque spheres vertically one above the other about 20 cm apart in the centre of the field of view. If their images are not superimposed on the developed radiograph, the X-ray beam axis is incorrectly aligned.

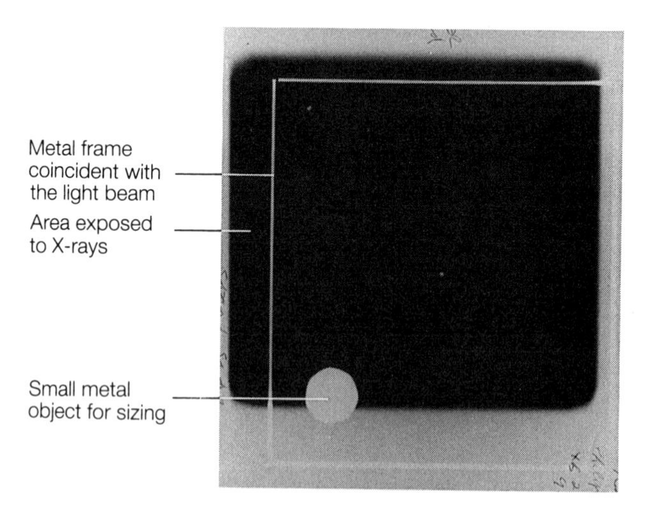

Fig. 2.24. Radiograph showing poor alignment of the X-ray beam and the light beam diaphragm.

Few centres check focal spot size regularly, perhaps because there is evidence from a range of routine X-ray examinations that in general quite large changes in spot size are not detectable in the quality of the final image. However, focal spot size is one of the factors affecting tube rating and significant errors in its value could affect the performance of the tube generator. A pin-hole technique, illustrated in Fig. 2.25, may be used to measure the size of the focal spot. The drawing is not to scale but typical dimensions for a 1 mm spot are shown.

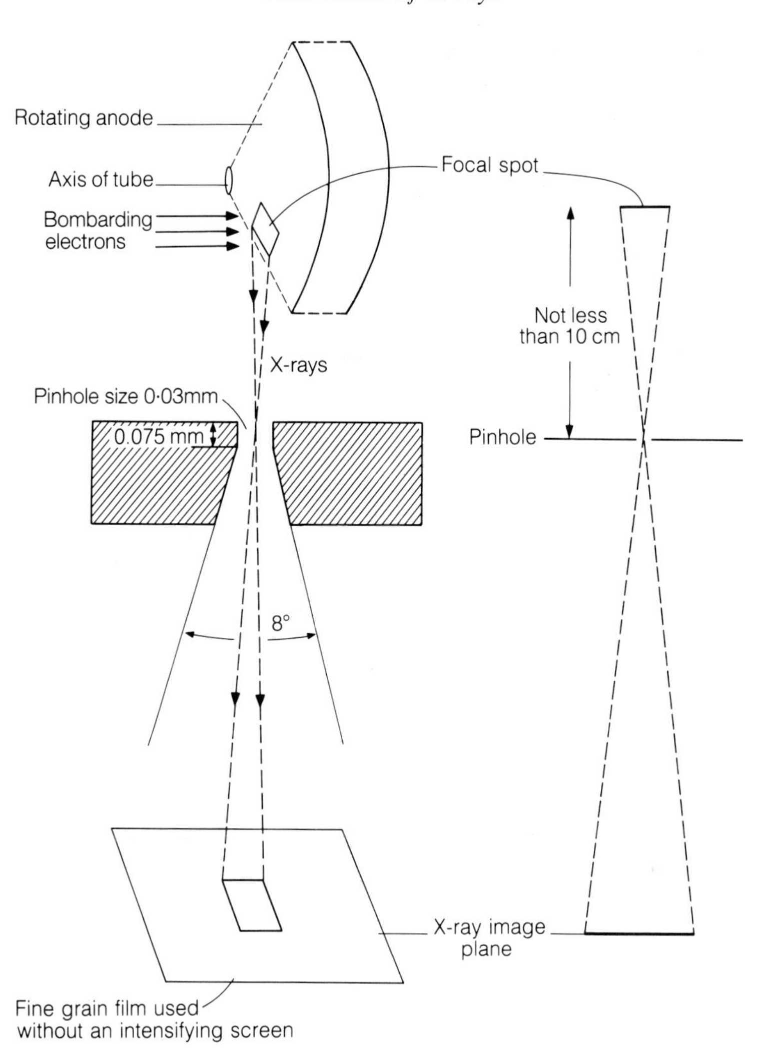

Fig. 2.25. Use of a pin-hole technique to check focal spot size.

By similar triangles

$$\frac{\text{Size of image}}{\text{Size of focal spot}} = \frac{\text{Pin-hole to film distance}}{\text{Focus to pin-hole distance}}$$

and for a pin-hole of this size this ratio is usually about 3. Note that the pin-hole must be small—its size affects the size of the image and hence the apparent focal spot size. The 'tunnel' in the pin-hole must be long enough for X-rays passing through the surrounding metal to be appreciably attenuated.

Focal spot sizes measured by the pin-hole technique are only accurate to about 25% but this is generally acceptable since image quality only changes slowly with focal spot size, and manufacturers' work to very generous tolerances on this design feature. Focal spot size can be measured more accurately, especially for smaller spots, and information may be obtained on uniformity of output within the spot, if required, by using a star test pattern. For fuller details see, for example, *Christensen's* (Curry *et al.* 1984). Such information would be important for example if one were attempting to image, say, a 0.4 mm blood vessel at 2 × magnification because image quality would then be very dependent on both spot size and shape.

For further information on quality control, the reader is referred to Conference Report Series 29 published by the Hospital Physicists' Association (1979).

2.8 Summary

In this chapter the basic principles of X-ray production have been discussed and the most important features may be summarized as follows:

1 An X-ray spectrum consists of a continuous component and, if the applied voltage is high enough, characteristic line spectra.

2 An important distinction must be made between radiation **quantity**, which is related to the overall intensity of X-rays produced, and radiation **quality** which requires a more detailed consideration of the distribution of X-ray intensities with photon energy. The former depends on a number of factors such as tube kilovoltage, time of exposure and atomic number of the target anode. Tube kilovoltage and beam filtration, which will be considered in detail in Chapter 3, affect radiation quality. Use of a three phase supply maintains the tube kilovoltage close to maximum throughout the exposure and both the quantity and quality of X-rays are thereby enhanced.

3 The high performance of modern X-ray equipment relies on careful design and construction of many components both in the X-ray tube itself and in the associated circuitry. Two features of anode design are particularly important. The first is a consequence of the fact that only about 0.5%

of the electron energy is converted into X-rays, whilst the remainder appears as heat which must be removed. The second is the requirement for the X-ray exposure to be as uniform as possible over the irradiated field. This is achieved by careful attention to the anode shape and particularly the angle at which it is presented to the electron flux.

4 Notwithstanding careful anode design, generation of heat imposes constraints on X-ray tube performance especially when very short exposures with small focal spot sizes are attempted. The limiting conditions are usually expressed in the form of rating curves. If a rating limit is exceeded, either the duration of exposure or the focal spot size must be increased. Occasionally the desired result may be achieved by increasing the tube kilovoltage.

5 To minimize the need for repeat X-rays, thereby increasing the overall radiation body burden to the population, careful quality control of the performance of X-ray sets at regular intervals is essential.

References and further reading

Ardran G. M. and Crooks H. E. (1968) Checking diagnostic X-ray beam quality. *Br. J. Radiol.* **41**, 193–198.

Chesney D. N. and Chesney M. O. (1984) *X-ray Equipment for Student Radiographers*, 3rd edn. Blackwell Scientific Publications.

Curry T. S., Dowdey J. E. and Murry R. C. Jr. (1984) *Christensen's Introduction to the Physics of Diagnostic Radiology*, 3rd edn. Lea & Febiger, Philadelphia.

Gifford D. (1984) *A Handbook of Physics for Radiologists and Radiographers*. Wiley, Chichester.

Harshbarger-Kelly M. E. (1985) Devices for measuring peak kilovoltage of diagnostic X-ray equipment, with emphasis on non-invasive electronic meters. *Health Care Instrum.* **1**, 27–33.

Hill D. R., ed. (1975) *Principles of Diagnostic X-ray Apparatus*. Macmillan, London.

Hospital Physicists' Association (1977) *The Physics of Radiodiagnosis*. (HPA Scientific Report Series 6, 2nd edn) The Hospital Physicists' Association, London.

Hospital Physicists' Association (1979) *Quality Assurance Measurements in Diagnostic Radiology*. (HPA Conference Report Series 29) The Hospital Physicists' Association, London.

Meredith W. J. and Massey J. B. (1977) *Fundamental Physics of Radiology*, 3rd end. Wright, Bristol.

Waggener R. G. and Wilson C. R., eds (1980) *Quality Assurance in Diagnostic Radiology*. (Medical Physics Monograph 4) American Association of Physicists in Medicine.

Wilks R. (1981) *Principles of Radiological Physics*. Churchill Livingstone, Edinburgh.

Exercises

1 Explain why the X-ray beam from a diagnostic set consists of photons with a range of energies rather than a monoenergetic beam.
2 What is meant by 'characteristic radiation'? Describe very briefly three processes in which characteristic radiation is produced.
3 Describe, with the aid of a diagram, the two physical processes that give rise to the production of X-rays from energetic electrons. How would the spectrum change if the target were made thin?
4 Explain why there is both an upper and a lower limit to the energy of the photons emitted by an X-ray tube.
5 What is the source of electrons in an X-ray tube and how is the number of electrons controlled?
6 The cathode of an X-ray tube is generally a small coil of tungsten wire.
(a) Why is it a small coil?
(b) Why is the material tungsten?
7 Draw a well-labelled diagram of the rotating anode X-ray tube as used in diagnosis. Explain the functions of the various parts and the advantages of the materials used.
8 Explain, with a diagram, the action of a timer for a 120 kV diagnostic X-ray set.
9 How would the output of an X-ray tube operating at 80 kVp change if the tungsten anode ($Z = 74$) were replaced by a tin anode ($Z = 50$)?
10 What is the effect on the output of an X-ray set of
(a) tube kilovoltage?
(b) the material of the anode?
11 It is required to take a radiograph with a very short exposure. Explain carefully why it may be advantageous to increase the tube kilovoltage.
12 What advantages does a rotating anode offer over a stationary anode in an X-ray tube?
13 Discuss the effect of the following on the rating of an X-ray tube:
(a) length of exposure,
(b) profile of the voltage supply as a function of time,
(c) previous use of the tube.
14 Discuss the factors that determine the upper limit of current at which a fixed anode X-ray tube can be used.
15 What do you understand by the thermal rating of an X-ray tube? Explain how suitable anode design may be used to increase the maximum permissible average beam current for
(a) short exposures,

(b) longer exposures.

16 For a fixed tube kilovoltage and focal spot size, explain why the maximum permissible average current for a three phase supply is sometimes higher and sometimes lower than for a single phase supply.

17 A technique calls for 550 mA, 0.05 s with the kV adjusted in accordance with patient thickness. If the rating chart of Fig. 2.20 applies, what is the maximum kVp that may be used safely?

18 A technique calls for 400 milliampseconds at 90 kVp. If the possible mA values are 500, 400, 300, 200, 100 and 50 and the rating chart in Fig. 2.20 applies, what is the shortest possible exposure time?

19 An exposure of 400 mA, 100 kVp, 0.1 s is to be repeated at the rate of six exposures per second for a total of 3 s. Is this technique safe if the rating chart of Fig. 2.20 applies?

20 A radiographic series consisting of six exposures of 280 mA, 75 kVp and 0.5 s has to be repeated. What is the minimum cooling time that must elapse before repeating the series if the rating chart of Fig. 2.23 applies?

21 If the series of example 20 is preceded by fluoroscopy at 100 kVp and 3 mA, for how long can fluoroscopy be performed prior to radiography?

22 Suggest reasons why radionuclides do not provide suitable sources of X-rays for medical radiography.

3

Interaction of X-rays and Gamma Rays with Matter

3.1 Introduction

The radiographic process depends on the fact that when a beam of X-rays passes through matter its intensity is reduced by an amount that is determined by the physical properties, notably thickness, density and atomic number, of the material through which the beam passes. Hence it is variations in these properties from one part of the patient to another that create detail in the final radiographic image. Since these variations are often quite small, a full understanding of the way in which they affect X-ray transmission under different circumstances, especially at different photon energies, is essential if image detail is to be optimized.

In this chapter an experimental approach to the problem of X-ray beam attenuation in matter will first be presented and then the results will be explained in terms of fundamental processes. Finally some implications of particular importance to radiology will be discussed.

3.2 Experimental approach to beam attenuation

X-rays and gamma rays are indirectly ionizing radiations. When they pass through matter they are absorbed by processes which set electrons in motion and these electrons produce ionization of other atoms or molecules in the medium. The electrons have short, finite ranges (see Table 1.2) and their kinetic energy is rapidly dissipated first as ionization and excitation, eventually as heat.

Conversely, the X-rays and gamma rays themselves do not have finite ranges, whatever their energy. If a fairly well collimated beam passes through different thicknesses of absorbing material, it will be found that equal thicknesses of stopping material reduce the beam intensity to the same fraction of its initial value, but the beam intensity is never reduced to zero. For simplicity a monoenergetic beam of gamma rays will be considered at this stage. The fact that an X-ray beam comprises photons with a range of energies introduces complications that will be considered towards the end of the chapter.

Referring to Fig. 3.1a, if a thickness x of material reduces the beam intensity by a fraction α, the beam intensity after crossing a further thickness x will be $\alpha x \cdot (\alpha I_0) = \alpha^2 I_0$.

If, as shown in Fig. 3.1b, $\alpha = \frac{1}{2}$ it may readily be observed that the variation of intensity with thickness is closely similar to the variation of radioactivity with time discussed in section 1.5. In other words the intensity decreases exponentially with distance, as expressed by the equation $I = I_0 e^{-\mu x}$ where I_0 is the initial intensity and I the intensity after passing through thickness x. μ is a property of the material and is known as the **linear attentuation coefficient**. If x is measured in mm, μ has units mm^{-1}.

A quantity analogous to the half-life of a radioactive material is frequently quoted. This is the **half value thickness** $H_{\frac{1}{2}}$ and is the thickness of material that will reduce the beam intensity to a half.

The analogy with radioactive decay is shown on the next page. Attenuation behaviour may be described in terms of either μ or $H_{\frac{1}{2}}$, since there is a simple relationship between them. If the value of $H_{\frac{1}{2}}$ is known, or is calculated from μ, then the graphical method described in section 1.6 may be used to determine the reduction in beam intensity caused by any

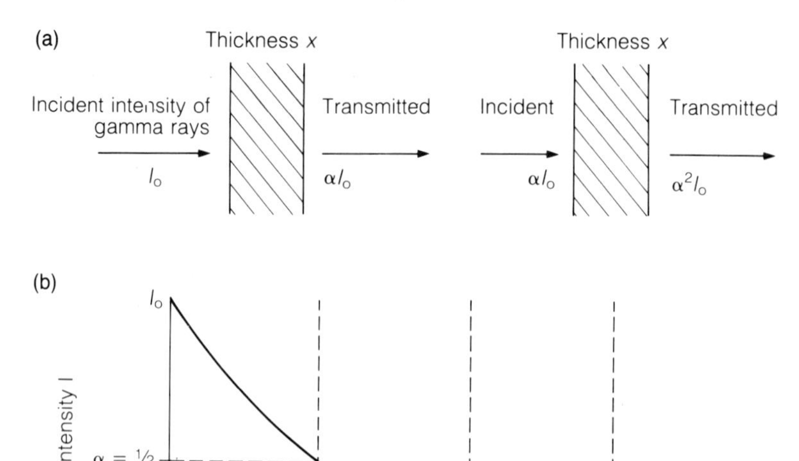

Fig. 3.1. (a) Transmission of a monoenergetic beam of gamma rays through layers of attenuating medium of different thickness. (b) Variation of intensity with thickness of attenuator.

Radioactive decay

$$A = A_0 e^{-\alpha t}$$

$$\alpha = \frac{\ln 2}{T_{\frac{1}{2}}} = \frac{0.693}{T_{\frac{1}{2}}}$$

$$A = A_0 e^{-0.693t/T_{\frac{1}{2}}}$$

Attenuation of a monoenergetic gamma ray beam

$$I = I_0 e^{-\mu x}$$

$$\mu = \frac{\ln 2}{H_{\frac{1}{2}}} = \frac{0.693}{H_{\frac{1}{2}}}$$

$$I = I_0 e^{-0.693x/H_{\frac{1}{2}}}$$

thickness of material. Conversely, the method may be used to find the thickness of material required to provide a given reduction in beam intensity. This is important when designing adequate shielding. The smaller the value of μ, the larger the value of $H_{\frac{1}{2}}$ and the more penetrating the radiation. Table 3.1 gives some typical values of μ and $H_{\frac{1}{2}}$ for monoenergetic radiations. The following are the main points to note:

1 In the diagnostic range μ decreases ($H_{\frac{1}{2}}$ increases) with increasing energy, i.e. the radiation becomes more penetrating.

Table 3.1. Typical values of μ and $H_{\frac{1}{2}}$ for monoenergetic radiations

Energy (keV)	Material	Atomic number	Density ($kg\ m^{-3}$)	μ (mm^{-1})	$H_{\frac{1}{2}}$ (mm)
30	Water	7.5	10^3	0.036	19
60				0.02	35
200				0.014	50
30	Bone	12.3	1.65×10^3	0.16	4.3
60				0.05	13.9
200				0.02	35
30	Lead	82	11.4×10^3	33	2×10^{-2}
60				5.5	0.13
200				1.1	0.6

2 μ increases ($H_{\frac{1}{2}}$ decreases) with increasing density. The radiation is less penetrating because there are more molecules per unit volume in the stopping material with which to collide.
3 Variation of μ with atomic number is complex although it clearly increases quite sharply with atomic number at very low energies. In Table 3.1 some of the trends are obscured by variations in density.
4 For water, which for the present purpose has properties very similar to those of soft tissue, $H_{\frac{1}{2}}$ in the diagnostic range is about 30 mm. Thus in passing through the body the intensity of an X-ray beam will be reduced by a factor of 2 for every 30 mm travelled. If a patient is 18 cm across this represents six half value thicknesses so the intensity is reduced by 2^6 or 64 times.
5 At similar energies, $H_{\frac{1}{2}}$ for lead is 0.1 mm or less so quite a thin layer of lead provides perfectly effective shielding for, say, the door of an X-ray room.

It is sometimes convenient to separate the effect of density ϱ from other factors. This is achieved by using a **mass attenuation coefficient**, μ/ϱ, and then the equation for beam intensity is rewritten

$$I = I_0 e^{-(\mu/\varrho)\varrho x}$$

When the equation is written in this form, it may be used to show that the stopping power of a fixed mass of material per unit area is constant, as one would expect since the gamma rays encounter a fixed number of atoms. Consider for example two containers, each filled with the same gas and each with the same area A, but of length $5l$ and l (Fig. 3.2). Let the densities of the two gases be ϱ_1 and ϱ_2. Furthermore let the mass of gas be the same in each container. Then ϱ_2 will be equal to $5\varrho_1$ since the gas in container 2 occupies only one-fifth the volume of the gas in container 1.

If the simple expression $I = I_0e^{-\mu x}$ is used, both μ and x will be different for the two volumes. If $I = I_0e^{-(\mu/\varrho)\varrho x}$ is used, then since the product ϱx is constant, μ/ϱ is also the same for both containers, thus showing that it is determined by the types of molecule and not their number density. Of course both equations will show that beam attenuation is the same in both volumes. The units of mass attenuation coefficient are $m^2\ kg^{-1}$.

It is important to emphasize that in radiology the linear attenuation coefficient is the more relevant quantity.

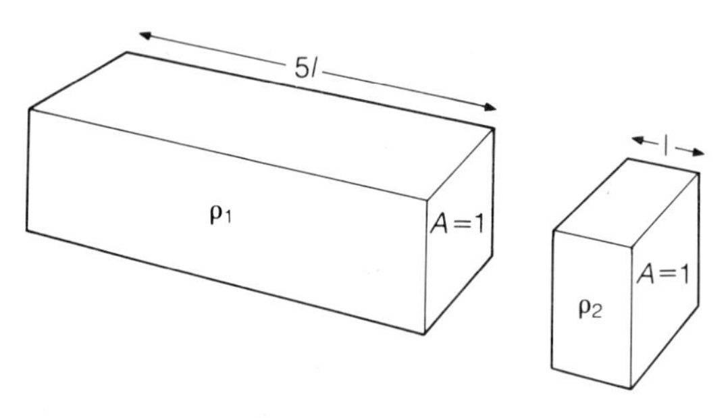

Fig. 3.2. Demonstration that the mass attenuation coefficient of a gas is independent of its density.

3.3 Introduction to the interaction processes

To understand why μ and $H_{\frac{1}{2}}$ vary with photon energy and atomic number in the manner shown in Table 3.1, it is necessary to consider in greater detail the nature of the interaction processes between X and gamma rays and matter. A large number of different processes have been postulated. However, only four have any relevance to diagnostic radiology and need be considered here. As shown in Chapter 1, these interactions are essentially collisions between electromagnetic photons and the orbital electrons surrounding the nuclei of matter through which the radiation is passing.

Before considering any interactions in detail, it is useful to discuss some general ideas.

3.3.1 Bound and free electrons

All electrons are 'bound' in the sense that they are held by positive attractive forces to their respective nuclei, but the binding energy is very variable, being much higher for the K shell electrons than for electrons in other shells. When an interaction with a passing photon occurs, the forces of interaction between the electron and the photon may be smaller than the forces holding the electron to the nucleus, in which case the electron will remain 'bound' to its nucleus and will behave accordingly. Conversely, the forces of interaction may be much greater than the binding forces, in which case the latter may be discounted and the electron behaves as if it were 'free'. Since the energy of one photon of X-rays is much higher than the binding energy of even K shell electrons for the low atomic number elements found in the body, most electrons can behave as if they are free when the interaction is strong enough. However, the interaction is frequently much weaker, a sort of glancing blow by the photon which involves only a fraction of its energy, and thus in many interactions electrons behave as if they were bound. Hence interactions that involve both bound and free electrons will occur under all circumstances and it is frequently the relative contribution of each type of interaction that is important.

This simple picture allows two general statements to be made. First, the higher the energy of the bombarding photons, the greater the probability that the interaction energy will exceed the binding energy. Thus the proportion of interactions involving free electrons can be expected to increase as the quantum energy of the radiation increases. Secondly, the higher the atomic number of the bombarded atom, the more firmly its electrons are held by electrostatic forces. Hence interactions involving bound electrons are more likely when the mean atomic number of the stopping material is high.

3.3.2 Attenuation, scatter and absorption

It is important to distinguish between these three processes and this can also be done on the basis of the simple model of interaction that has already been described (Fig. 1.8).

When a beam of collimated X-ray photons interacts with matter, some of the X-rays may be **scattered**. This simply means they no longer travel in the same direction as the collimated beam. Such photons have the same energy after the interaction as they had before and lose no energy to the medium. Other photons lose some energy when they are scattered. This energy is transferred to electrons and, as already noted, is dissipated locally. Such a process results in energy **absorption** as well as scattering in the medium. Finally, some photons may undergo interactions in which they are completely destroyed and all their energy is transferred to electrons in the medium. Under the conditions normally obtaining in diagnostic radiology, all this energy is usually dissipated locally and thus the process is one of **total absorption**. Both scatter and absorption result in beam **attenuation**, i.e. a reduction in the intensity of the collimated beam.

If a new term, the **mass absorption coefficient** μ_a/ϱ is introduced, it follows that μ_a/ϱ is always less than the mass attenuation coefficient μ/ϱ although the difference is small at low photon energies. From the viewpoint of good radiology, only absorption is desirable since scatter results in uniform irradiation of the film, which, as will be shown in Chapter 5, reduces contrast. Note also that only the absorbed energy contributes to the radiation dose to the patient. This is an undesirable but unavoidable side-effect if good radiographic images are to be obtained.

3.4 The interaction processes

Four processes will be considered. Two of these, the photoelectric effect and the Compton effect are the most important in diagnostic radiology. However, it may be more helpful to discuss the processes in a more logical order, starting with one that is only important at very low photon energies and ending with the one that dominates at high photon energies. Low photon energies are sometimes referred to as 'soft' X-rays, higher photon energies as 'hard' X-rays.

3.4.1 Elastic scattering

When X-rays pass close to an atom, they may cause electrons to take up energy of vibration. The process is one of resonance such that the electron vibrates at a frequency corresponding to that of the X-ray photon. This is an unstable state and the electron quickly reradiates this energy in all directions and at exactly the same frequency as the incoming photons. The process is one of scatter and attenuation without absorption.

The electrons that vibrate in this way must remain bound to their nuclei, thus the process is favoured when the majority of the electrons behave as bound electrons. This occurs when the binding energy of the electrons is high, i.e. the atomic number of the scattering material is high, and when the quantum energy of the bombarding photons is relatively low. The probability of elastic scattering can be expressed by identifying a mass attenuation coefficient with this particular process, say ε/ϱ. Numerically, ε/ϱ is expressed as a cross-section area. If the effective cross-section area for elastic scattering is high, the process is more likely to occur than if the effective cross-section area is low. ε/ϱ increases with increasing atomic number of the scattering material ($\varepsilon/\varrho \propto Z^2$) and decreases as the quantum energy of the radiation increases ($\varepsilon/\varrho \propto 1/hf$).

Although a certain amount of elastic scattering occurs at all X-ray energies, it never accounts for more than 10% of the total interaction processes in diagnostic radiology.

3.4.2 Photoelectric effect

At the lower end of the diagnostic range of photon energies, the photoelectric effect is the dominant process. From a radiologist's viewpoint this is the most important interaction that can occur between X-rays and bound electrons. In this process the photon is completely absorbed, dislodging an electron from its orbit around a nucleus. Part of the photon energy is used to overcome the binding energy of the electron, the remainder is given to the electron as kinetic energy and is dissipated locally (see section 3.2). The following equation describes the energy changes

hf	=	W	+	$\frac{1}{2}m_e v^2$
photon energy		binding energy of electron to nucleus		kinetic energy of electron

However, this is not a stable state because the atom of the target material now has a vacancy in one of its orbital electron shells and when this vacancy is filled by an electron of higher energy, 'characteristic radiation' is produced in exactly the same way as characteristic radiation is produced as part of the X-ray spectrum from, say, a tungsten anode (see section 2.2.4). The electron filling the K shell vacancy usually comes from the L shell and only occasionally from outer shells. The process is summarized in Fig. 3.3.

The final outcome now depends on the atomic number of the absorbing material. For the low atomic number elements in soft tissue, the binding energies of even K shell electrons are very small (about 0.3 keV for carbon), hence the emitted X-rays are so soft that they are rapidly absorbed and the process is one of attenuation and total absorption. However, when the photoelectric effect occurs in metals, quite appreciable reradiation of characteristic radiation may occur and this factor is important in the choice of suitable materials for X-ray beam filtration (see section 3.8).

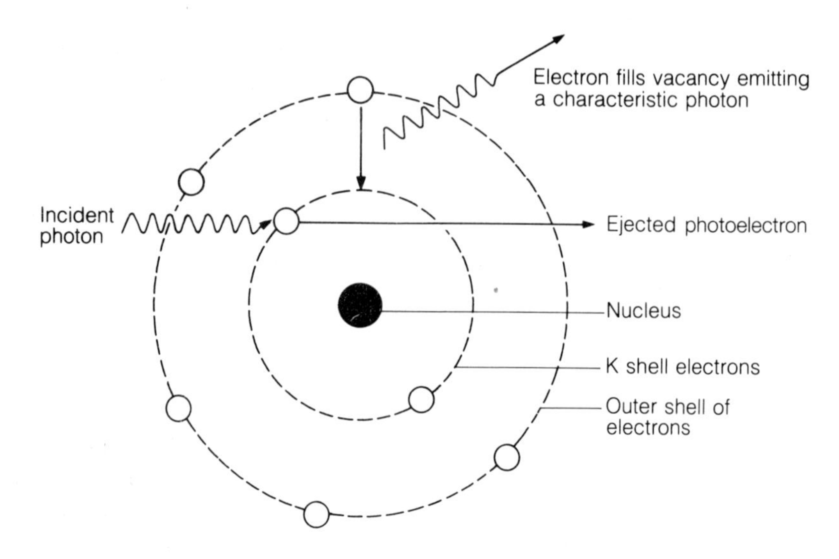

Fig. 3.3. Schematic representation of the photoelectric effect.

Since the process is again concerned with bound electrons, it is favoured in materials of high mean atomic number and the photoelectric mass attenuation coefficient τ/ϱ is proportional to Z^3. The process is also favoured by low photon energies with τ/ϱ proportional to $1/(hf)^3$. Notice that, as a result of the Z^3 factor, at the same photon energy lead ($Z = 82$) has a 300 times greater photoelectric coefficient than bone ($Z = 12.3$). This explains the big difference in μ values for these two materials at low photon energies as shown in Table 3.1.

The chance of a photoelectric interaction falls steeply with increasing photon energy although the decrease is not entirely regular because of **absorption edges** (see section 3.6).

Thus the photoelectric effect is the major interaction process at the low end of the diagnostic X-ray energy range.

3.4.3 The Compton effect

The most important effect in radiology involving unbound electrons is inelastic scattering or the Compton effect. This process may be thought of most easily in terms of classical mechanics in which the photon has energy h*f* and momentum h*f*/c and makes a billiard-ball type collision with a stationary free electron, with both energy and momentum conserved (Fig. 3.4).

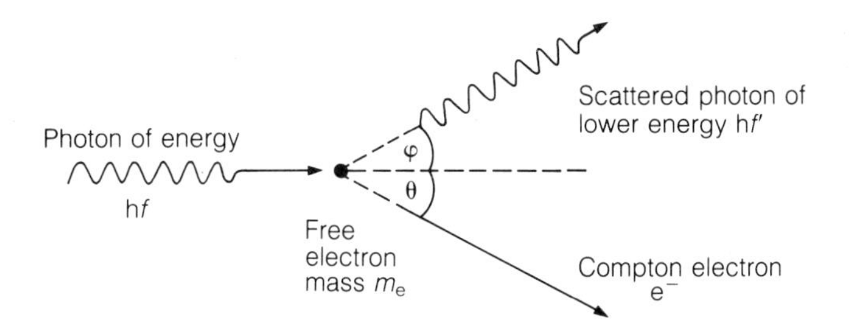

Fig. 3.4. Schematic representation of the Compton effect.

The proportions of energy and momentum transferred to the scattered photon and to the electron are determined by θ and ϕ. The kinetic energy of the electron is rapidly dissipated by ionization and excitation and eventually as heat in the medium and a scattered photon of lower energy than the incident photon emerges from the medium—assuming no further interaction occurs. Thus the process is one of scatter and partial absorption of energy.

The equation used most frequently to describe the Compton process is

$$\lambda' - \lambda = \frac{h}{m_e c}(1 - \cos\phi)$$

where λ′ is the wavelength of the scattered photon and λ is the wavelength of the incident photon. This equation shows that the change in wavelength Δλ when the photon is scattered through an angle ϕ is independent of photon energy. However, it may be shown that the change in energy of the photon Δ*E* is given by*

$$\Delta E = \frac{E^2}{m_e c^2}(1 - \cos\phi) \qquad \text{(equation 3.1)}$$

Thus the loss of energy by the scattered photon does depend on the incident photon energy. For example when the photon is scattered through

*If $\lambda = hc/E$
Then $\lambda + \Delta\lambda = hc/(E - \Delta E)$
Where Δλ is a small change in wavelength of the photon and Δ*E* is the corresponding energy change.

So $$\Delta\lambda = \frac{hc}{E - \Delta E} - \frac{hc}{E} \simeq \frac{hc}{E^2}\cdot\Delta E$$

or $$\Delta E = \frac{E^2}{hc}\cdot\Delta\lambda = \frac{E^2}{m_e c^2}(1 - \cos\phi)$$

60°, the proportion of energy taken by the electron varies from about 2% at 20 keV to 9% at 200 keV and 50% at 1 MeV.

Since the process is one of attenuation with partial absorption, the variation in the amount of energy absorbed in the medium, averaged over all scattering angles, with initial photon energy depends on:

1 the probability of an interaction (Fig. 3.5a);

2 the fraction of the energy going to the electron (Fig. 3.5b);

3 the fraction of the energy retained by the photon (Fig. 3.5b).

To find out how much energy is absorbed in the medium, the Compton cross-section must be multiplied by the percentage of energy transferred to the electron (Fig. 3.5c). This shows that there is an optimum X-ray energy for energy absorption by the Compton effect. However, it is well above the diagnostic range.

As shown by equation 3.1, the amount of energy transferred to the electron depends on the scattering angle ϕ. At diagnostic energies, the proportion of energy taken by the electron, i.e. absorbed, is always quite small. For example, even a head-on collision ($\phi = 180°$) only transfers 8% of the photon energy to the electron at 20 keV. Thus at low energies Compton interactions cause primarily scattering and this will have implications when the effects of scattered radiation on film contrast are considered.

One further consequence of equation 3.1 is that when E is small, quite large values of ϕ are required to produce appreciable changes ΔE. This is important in nuclear medicine where pulse height analysis is used to detect changes in E and hence to discriminate against scattered radiation.

Direction of scatter

After Compton interactions, photons are scattered in all directions and this effect may be displayed by using **polar diagrams** similar to those used in section 2.4 to demonstrate the directions in which X-rays are emitted from the anode (Fig. 3.6). As the photon energy increases, the scattered photons travel increasingly in the forward direction but this change is quite small in the diagnostic energy range where a significant proportion of X-rays may be back-scattered. Note that, as discussed above, the mean energy of back-scattered photons is lower than the mean energy of forward scattered photons.

For thicker objects, for example a patient, the situation is further complicated by the fact that both the primary beam and the scattered radiation will be attenuated. Thus, although in Fig. 3.7 the polar diagrams of Fig. 3.6 could be applied to each slice in turn, because of body attenua-

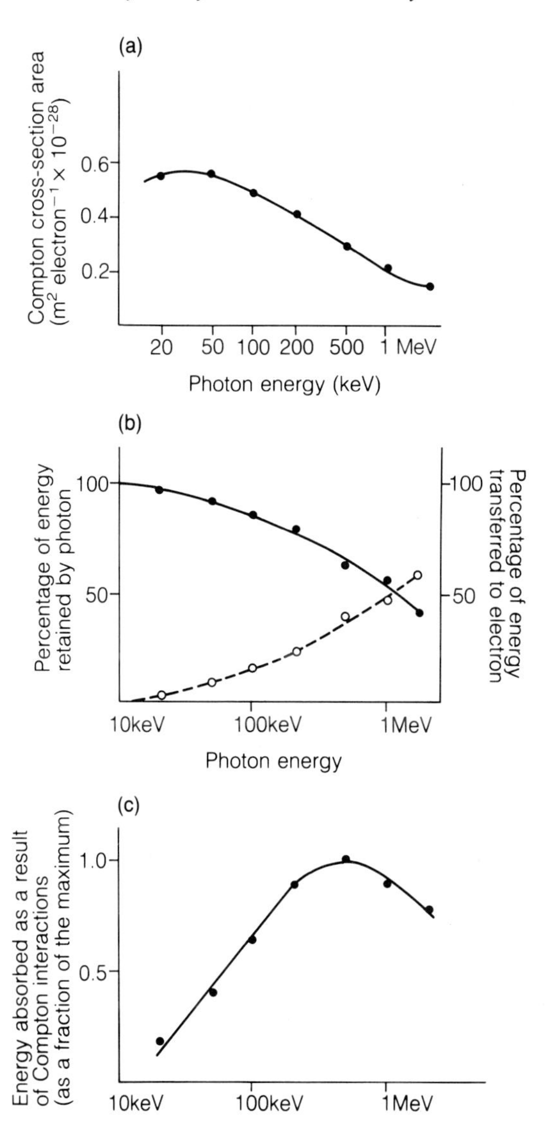

Fig. 3.5. (a) Variation of Compton cross-section with photon energy. (b) Percentage of energy transferred to the electron (*dotted line*) and percentage retained by the photon (*solid line*) per Compton interaction as a function of photon energy. (c) Product of (a) and (b) to give variation in total Compton energy absorption as a function of photon energy.

tion the X-ray intensity on slice Z may be only 1% of that on slice A. In the example shown in Fig. 3.7, which is fairly typical, the intensity of radiation scattered back at 150° is 10 times higher than that scattered forward at 30° and is comparable with the intensity in the primary transmitted beam.

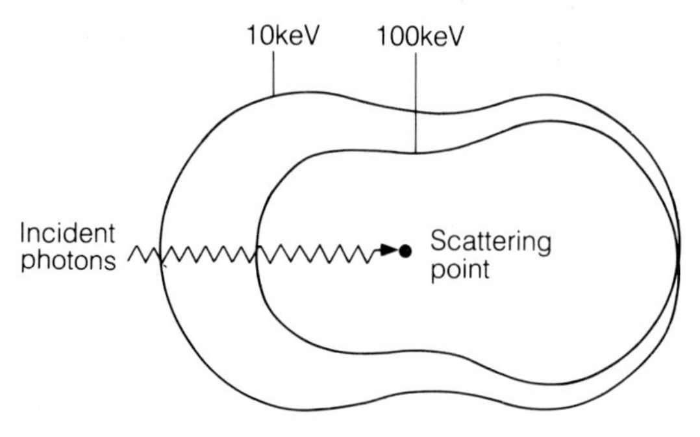

Fig. 3.6. Polar diagrams showing the spatial distribution of scattered X-rays around a free electron at two different energies.

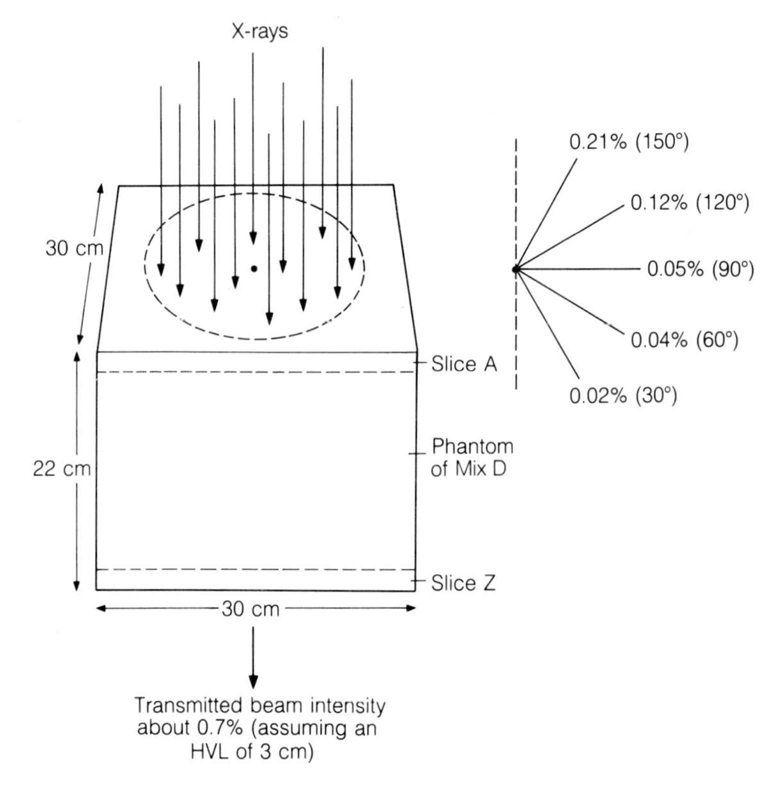

Fig. 3.7. Distribution of scattered radiation around a patient-sized phantom of tissue equivalent material. The phantom measured 30 × 30 cm by 22 cm deep and a 400 cm^2 field was exposed to 100 kVp X-rays. Intensities of scattered radiation at 1 m are expressed as a percentage of the incident surface dose (after Bomford and Burlin 1963; ICRP 1982).

Hence a high proportion of the scattered radiation emerging from the patient travels in a backwards direction and this has implications for radiation protection, for example when using an over-couch tube for fluoroscopy.

The effect of scattered radiation on image quality is considered in Chapter 5.

Variation of the Compton coefficient with photon energy and atomic number

For free electrons, the probability of a Compton interaction (σ/ϱ) decreases steadily as the quantum energy of the photon increases. However, when the electrons are subject to the forces of other atoms, low energy interactions frequently do not give the electron sufficient energy to break away from these other forces. Thus in practice the Compton mass attenuation coefficient is approximately constant in the diagnostic range and only begins to decrease ($\sigma/\varrho \propto 1/hf$) for photon energies above about 100 keV.

σ/ϱ is almost independent of atomic number. To understand why this should be so, recall the information from Chapter 1 on atomic structure. The Compton effect is proportional to the number of electrons in the stopping material. Thus if the Compton coefficient is normalized by dividing by density, σ/ϱ should depend on electron density. Now for any material the number of electrons is proportional to the atomic number Z and the density is proportional to the atomic mass A. Hence

$$\frac{\sigma}{\varrho} \propto \frac{Z}{A}$$

Examination of Table 3.2 shows that Z/A is almost constant for a wide range of elements of biological importance, decreasing slowly for higher atomic number elements.

Hydrogen is an important exception to the 'rule' for biological elements. Materials that are rich in hydrogen exhibit elevated Compton

Table 3.2. Values of charge/mass ratio for the atoms of various elements in the periodic table

	H	C	N	O	P	Ca	Cu	I	Pb
Z	1	6	7	8	15	20	29	53	82
A	1	12	14	16	31	40	63	127	208
Z/A	1	0.5	0.5	0.5	0.48	0.5	0.46	0.42	0.39

interaction cross-sections and this explains, for example, the small but measurable difference in mass attenuation coefficients in the Compton range of energies between water and air, even though their mean atomic numbers at 7.42 and 7.64 respectively are almost identical.

The Compton effect is a major interaction process at diagnostic X-ray energies, particularly at the upper end of the energy range.

3.4.4 Pair production

When a photon with energy in excess of 1.02 MeV passes close to a heavy nucleus, it may be converted into an electron and a positron. This is one of the most convincing demonstrations of the equivalence of mass and energy. The well-defined threshhold of 1.02 MeV is simply the energy equivalence of the electron and positron masses m_{e^-} and m_{e^+} respectively according to the equation:

$$E = m_{e^-}c^2 + m_{e^+}c^2$$

Any additional energy possessed by the photon is distributed equally to the two particles as kinetic energy.

The electron dissipates its energy locally and has a range given by Table 1.2. The positron dissipates its kinetic energy but when it comes to rest it undergoes the reverse of the formation reaction, annihilating with an electron to produce two 0.51 MeV gamma rays which fly away simultaneously in opposite directions.

$$e^+ + e^- \rightarrow 2\gamma \text{ (0.51 MeV)}$$

These gamma rays (or **annihilation radiation**) are penetrating radiations which escape from the absorbing material. Thus 1.02 MeV of energy is re-irradiated and only the energy in excess of this is absorbed. Hence the process is one of attenuation with partial absorption.

Pair production is the only one of the four processes considered that shows an increase in the chance of an interaction with increasing photon energy [$\pi/\varrho \propto (hf - 1.02)$MeV]. Since a large, heavy nucleus is required to remove some of the photon momentum, the process is also favoured by high atomic number materials ($\pi/\varrho \propto Z$).

Although pair production has no direct relevance in diagnostic radiology, the subsequent annihilation process is important when positron emitters are used for *in vivo* imaging in nuclear medicine. The characteristic 0.51 MeV gamma rays are detected and since two are emitted simultaneously, coincidence circuits may be used to discriminate against stray background radiation.

3.5 Combining interaction effects and their relative importance

Table 3.3 summarizes the processes that have been considered.

Table 3.3. A summary of the four main processes by which X-rays and gamma rays interact with matter

Process	Process	Type of interaction	Variation with photon energy (h*f*)	Variation with atomic number (*Z*)
Elastic	ε/ϱ	Bound electrons	$\propto 1/hf$	$\propto Z^2$
Photoelectric	τ/ϱ	Bound electrons	$\propto 1/(hf)^3$	$\propto Z^3$
Compton	σ/ϱ	Free electrons	Almost constant 10–100 keV; $\propto 1/hf$ above 100 keV	Almost independent of *Z*
Pair production	π/ϱ	Promoted by heavy nuclei	$\propto (hf-1.02)$ MeV	$\propto Z$

Each of the processes occurs independently of the others. Thus for the photoelectric effect:

$$I = I_0 e^{-(\tau/\varrho)\varrho x}$$

and for the Compton effect:

$$I = I_0 e^{-(\sigma/\varrho)\varrho x}$$

etc.

Hence effects can be combined simply by multiplying the exponentials to give

$$I = I_0 e^{-\tau/\varrho \cdot \varrho x}\, e^{-\sigma/\varrho \cdot \varrho x} \ldots$$

or

$$I = I_0 e^{-(\tau/\varrho + \sigma/\varrho + \ldots)\cdot \varrho x}$$

leading to the simple relationship

$$\mu/\varrho = \tau/\varrho + \sigma/\varrho + \ldots$$

where additional interaction coefficients can be added if they contribute significantly to the value of μ/ϱ. Hence the total mass attenuation coefficient is equal to the sum of all the component mass attenuation coefficients obtained by considering each process independently.

In the range of energies of importance in diagnostic radiology, the photoelectric effect and the Compton effect are the only two interactions that need be considered. Since the latter process generates unwanted scattered photons of lower energy but the former does not, and the former process is very dependent on atomic number but the latter is not, it is clearly important to know the relative contributions of each in a given situation. Figure 3.8a,b shows photoelectric and Compton cross-sections for nitrogen, which has approximately the same atomic number as soft tissue, and for aluminium, with approximately the same atomic number as bone, respectively. Because the photoelectric coefficient is decreasing rapidly with photon energy (note the logarithmic scale) there is a sharp transitional point. By 30 keV the Compton effect is already the more

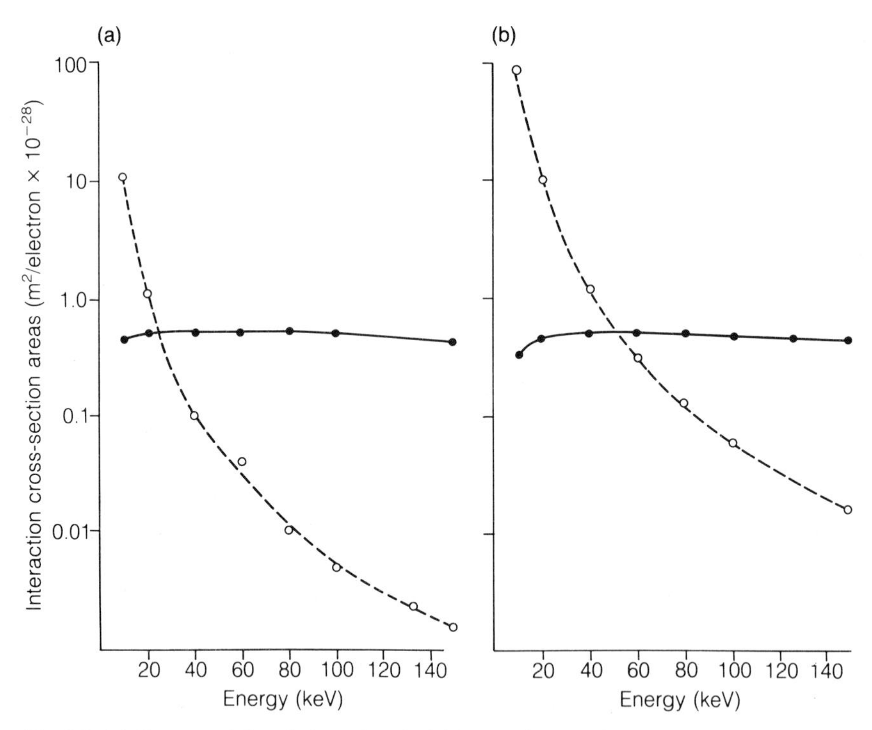

Fig. 3.8. (a) Compton (●) and photoelectric (○) interaction coefficients for nitrogen ($Z = 7$). (b) Compton (●) and photoelectric (○) interaction coefficients for aluminium ($Z = 13$).

important process in soft tissue. Thus if soft tissue contrast depends primarily on the photoelectric effect (i.e. differences in atomic number rather than density) as it does for example in mammography, very small changes in kV can make a big difference to the degree of contrast in the image.

For the higher atomic number material the photoelectric curve is shifted to the right but the Compton curve remains almost unchanged and the cross-over point is above 50 keV. This trend continues throughout the periodic table. For iodine which is the major interaction site in a sodium iodide scintillation detector, the cross-over point is about 300 keV and for lead it is about 500 keV. Note that even for lead, pair production does not become comparable with the Compton effect until 2 MeV and for soft tissues not until about 20 MeV.

It is now possible to interpret more fully the data in Table 3.1. For any material, the linear attenuation coefficient will decrease with increasing energy, initially because the photoelectric coefficient is decreasing and subsequently because the Compton coefficient is decreasing. At low energy there is a very big difference between the μ values for water and lead, partly because of the density effect but also because the photoelectric effect, which depends on Z^3, dominates.

3.6 Absorption edges

Whenever the photon energy is just slightly greater than the energy required to remove an electron from a particular shell around the nucleus, there is a sharp increase in the photoelectric absorption coefficient. This is known as an absorption edge, and absorption edges associated with K shell electrons have a number of important applications in radiology—the edges associated with the L shell and outer shells are at energies that are too low to be of any practical significance.

As shown in Fig. 3.9 there will be a substantial difference in the attenuating properties of a material on either side of the absorption edge. There are two reasons for the sudden increase in absorption with photon energy. First the number of electrons available for release from the atom increases. However, in the case of lead the number available only increases from 80 to 82 since the K shell only contains two electrons, and the increase in absorption is proportionately much bigger than this. Thus a more important reason is that a resonance phenomenon occurs whenever the photon energy just exceeds the binding energy of a given shell. Since at

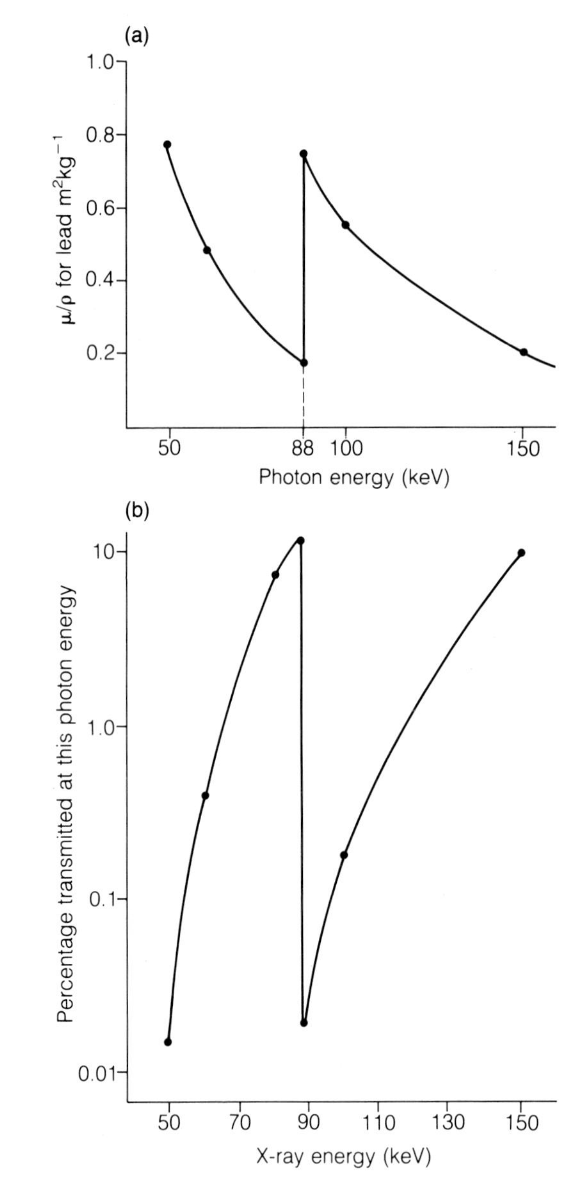

Fig. 3.9. (a) Variation in the mass attenuation coefficient for lead across the K-edge boundary. (b) Corresponding variation in transmission through 1 mm of lead. Note the use of a logarithmic scale; at the absorption edge the transmitted intensity falls by a factor of about 500.

88–90 keV the photon energy is almost exactly equal to that required to remove K shell electrons from lead, a disproportionately large number of K shell interactions will occur and absorption by this process will be high.

Because of absorption edges, there will be limited ranges of photon energies for which a material of low atomic number actually has a higher absorption coefficient than a material of higher atomic number and this has a number of practical applications. For example the presence of K absorption edges has an important influence on the selection of suitable materials for intensifying screens, where a high absorption efficiency by the photoelectric process is required. Although tungsten in a calcium tungstate screen has a higher atomic number than the rare earth elements and therefore has an inherently higher mass absorption coefficient, careful comparison of the appropriate absorption curves (Fig. 3.10) shows that in the important energy range from about 40 to 70 keV where for many investigations there will be a high proportion of photons, absorption by rare earth elements is actually higher than for tungsten.

Other examples of the application of absorption edges are:

1 in the use of iodine ($Z = 53$, K-edge = 33 keV) and barium ($Z = 56$, K-edge = 37 keV) as contrast agents. Typical diagnostic X-ray beams contain a high proportion of photons at or just above these energies, thus ensuring high absorption coefficients.

2 the use of a selenium plate ($Z = 34$, K-edge = 13 keV) for xeroradiography. This K-edge value makes selenium a good absorber for the

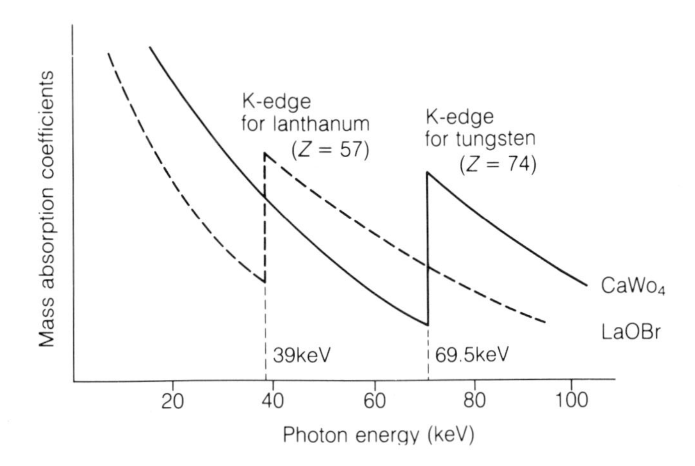

Fig. 3.10. Curves showing the relative absorptions of lanthanum oxybromide (LaOBr) and calcium tungstate ($CaWO_4$) as a function of X-ray energy in the vicinity of their absorption edges (not to scale).

low energy radiation (~ 20 keV) used for mammography.
3 the presence of absorption edges also has a significant effect on the variation of sensitivity of photographic film with photon energy (see section 7.17).

An important consequence of the absorption edge effect is that a material is relatively transparent to radiation that has a slightly lower energy than the absorption edge, including the material's own characteristic radiation. This factor is important when choosing materials for X-ray beam filtration (see section 3.8). It should also be considered when examining the properties of materials for shielding. For example, although lead is normally used for shielding, if a particular X-ray beam contains a high proportion of photons of energy approaching the K-edge for lead ($Z = 82$, K-edge = 88 keV), some other material, e.g. tin ($Z = 50$, K-edge = 29 keV), may be a more effective attenuator on a weight-for-weight basis in the given situation.

3.7 Broad beam and narrow beam attenuation

As already explained, the total attenuation coefficient is simply the sum of all the interaction processes and this should lead to unique values for μ, the linear attenuation coefficient, and $H_{\frac{1}{2}}$, the half value thickness.

However, if a group of students were each asked to measure $H_{\frac{1}{2}}$ for a beam of radiation in a given attenuating material, they would probably obtain rather different results. This is because the answer would be very dependent on the exact geometrical arrangement used and whether or not any scattered radiation reached the detector.

Two extremes, narrow beam and broad beam conditions, are illustrated in Fig. 3.11a,b. For narrow beam geometry it is assumed that the primary beam has been collimated so that the scattered radiation misses the detector. For broad beam geometry, radiation scattered out of the primary beam that would otherwise have reached the detector, labelled A and A′, is not recorded. However, radiation such as B and B′ which would normally have missed the detector is scattered into it and multiple scattering may cause a further increase in detector reading. Hence a broad beam does not appear to be attenuated as much as a narrow beam and, as shown in Fig. 3.12, the value of $H_{\frac{1}{2}}$ will be different. Absorbed radiation is of course stopped equally by the two geometries. $H_{\frac{1}{2}}$ values are always quoted for narrow beam conditions unless otherwise stated.

An important application of this phenomenon in radiology is the practice of reducing the field size to the smallest value consistent with the

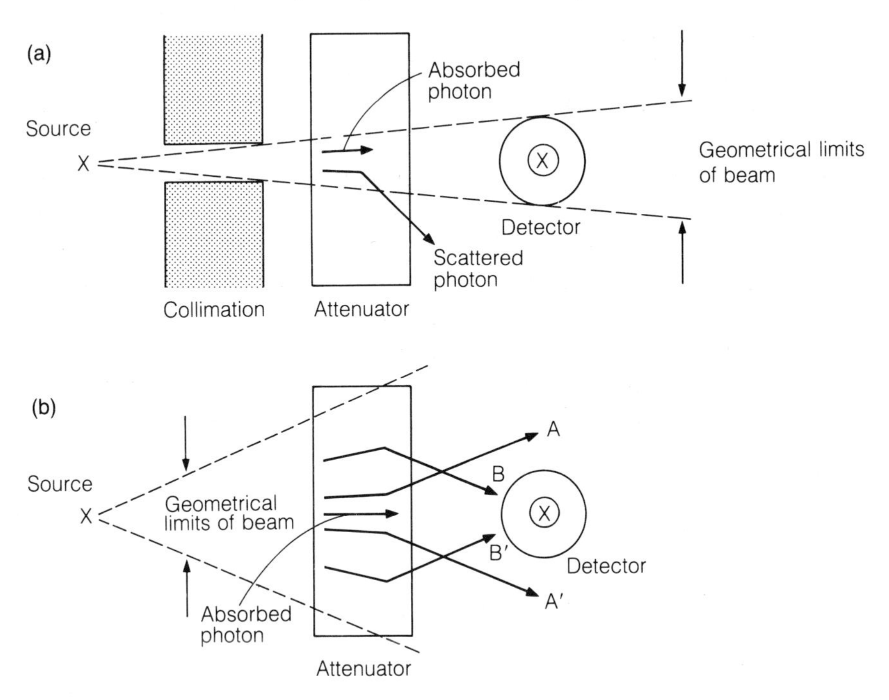

Fig. 3.11. Geometrical arrangements for the study of (a) narrow beam and (b) broad beam attenuation.

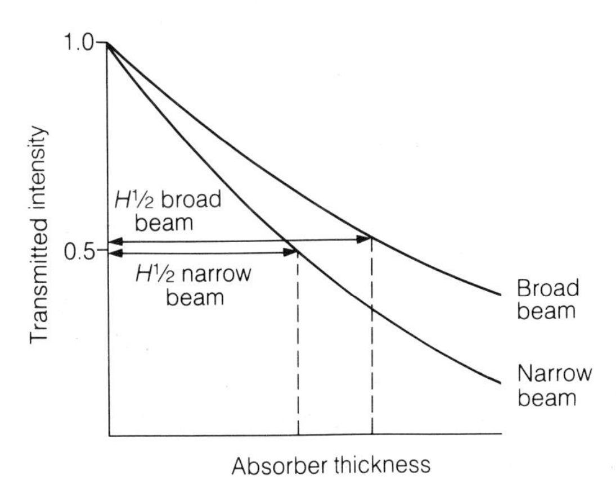

Fig. 3.12. Attenuation curves and values of $H\frac{1}{2}$ corresponding to the geometries shown in Fig. 3.11.

required image. The detector is now an X-ray film and if a broader beam than necessary is used extra scattered radiation reaches the film thereby reducing contrast (see section 5.5). Reducing the field size of course also reduces the total radiation energy absorbed in the patient.

3.8 Filtration and beam hardening

Thus far in this chapter, care has been taken to use the term gamma rays and to talk about attenuation of a beam of single energy. Radiation from an X-ray set has a range of energies and these radiations will not all be attenuated equally. Since, in the diagnostic range, the lower energy radiations are the least penetrating, they are removed from the beam more quickly. In other words the attenuating material acts like a filter (Fig. 3.13).

The choice of a suitable filter is important to the performance of an X-ray set because this provides a mechanism for reducing the intensity of low

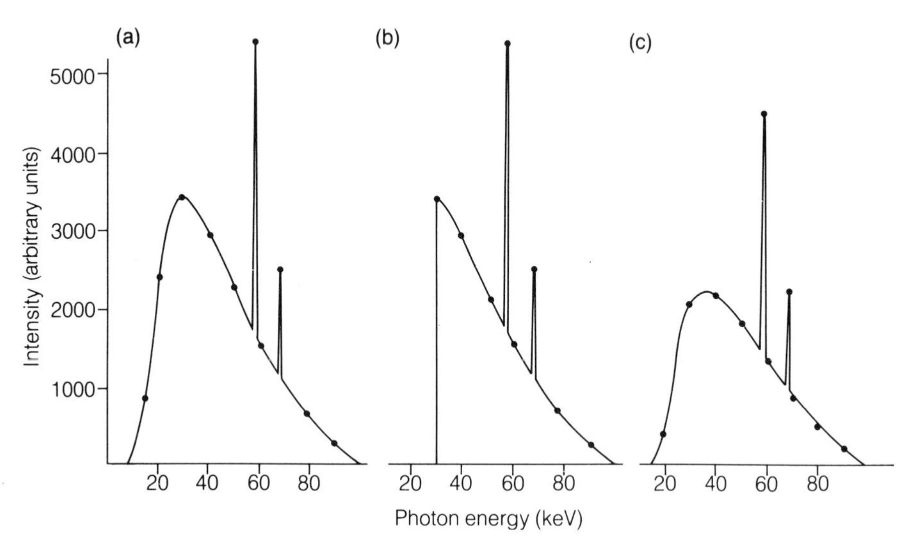

Fig. 3.13. Curves showing the effect of filters on the quality (spectral distribution) of an X-ray beam. (a) Spectrum of the emergent beam generated at 100 kVp with inherent filtration equivalent to 0.5 mm Al. (b) Effect of an ideal filter on this spectrum. (c) Effect of total filtration of 2.5 mm aluminium on the spectrum.

energy X-ray photons. These photons would be absorbed in the patient, thereby contributing nothing to the image but increasing the dose to the patient. The effect of an ideal filter is shown in Fig. 3.13a,b, but, in practice, no filter completely removes the low energy radiation or leaves the high energy component unaffected.

The position of the K absorption edge must also be considered when choosing a filter material. For example tin, with a K-edge of 29 keV, will transmit 25–29 keV photons rather efficiently and these would be undesirable in, for example, a radiographic exposure of the abdomen. Aluminium

($Z = 13$, K-edge 1.6 keV) is the material normally chosen for filters in the diagnostic range. Aluminium is easy to handle, and 'sensible' thicknesses of a few mm are required. Photons of energy less than 1.6 keV, including the characteristic radiation from aluminium, will either be absorbed in the X-ray tube window or in the air gap between filter and patient. The effect of 2.5 mm of aluminium on a 120 kV beam from a tungsten target is shown in Fig. 3.13c.

Finally, the thickness of added filtration will depend on the operating kVp and the inherent filtration. This is the filtration caused by the glass envelope, insulating oil and bakelite window of the X-ray tube itself and is usually equivalent to about 0.5–1 mm of aluminium. The total filtration should be at least 2.0 mm Al up to 100 kVp and at least 2.5 mm Al above 100 kVp.

When a beam passes through a filter, it becomes more penetrating or 'harder' and its $H\frac{1}{2}$ increases. If log (intensity) is plotted against the thickness of absorber, the curve will not be a straight line as predicted by $I = I_0 e^{-\mu x}$. It will fall rapidly at first as the 'soft' radiation is removed and then more slowly when only the harder, high energy component remains (Fig. 3.14a). Note that the beam never becomes truly monochromatic but after about four half value layers, i.e. when the intensity has been reduced to about one-sixteenth of its original value, the range of photon energies in the beam is quite small. The value of $H\frac{1}{2}$ then becomes constant within the accuracy of measurement and *for practical purposes* the beam is monochromatic (Fig. 3.14b).

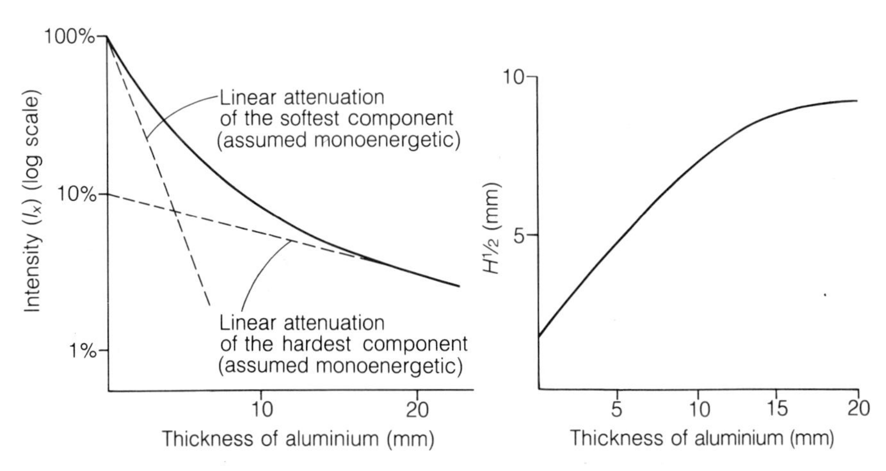

Fig. 3.14. (a) Variation of intensity (I_x) with absorber thickness (x) as an heterogeneous beam of X-rays becomes progressively harder on passing through attenuating material (plotted on a log scale). (b) Corresponding change in the value of $H\frac{1}{2}$.

The effect of filtration on skin dose is shown in Table 3.4. There is clearly a substantial benefit to be gained in terms of patient dose from the use of filters. Note, however, the final column of the table which shows that there is a price to be paid. Although heavy filtration (3 mm Al) reduces the skin dose even more than 1 mm filtration, the X-ray tube output begins to be affected and this is reflected in the increased exposure time. A prolonged exposure may not be acceptable but if an attempt is made to restore the exposure time to its unfiltered value by increasing the tube current, there may be problems with the tube rating. At 130 kVp, increasing the filtration from 1 mm to 3 mm aluminium would have virtually no effect on exposure time.

Table 3.4. Exposure dose to the skin and exposure time for comparable density radiographs of a pelvic phantom (18 cm thick) using a 60 kVp beam with different filtrations (adapted from Trout *et al.* 1952)

Aluminium filtration (mm)	Skin dose in air (mGy)	Exposure time at 100 mA (s)
None	20.7	1.4(1)
0.5	16.1	1.6(1)
1.0	11.0	1.6(4)
3.0	4.1	2.1(4)

Because materials are relatively transparent to their own characteristic radiation, the effect of filtration can be rather dramatic when the filter is of the same material as the target anode producing the X-rays. This effect is exploited in mammography and Fig. 3.15 shows how a 0.05 mm molybdenum filter changes the spectrum from a tube operating at 35 kV constant potential using a molybdenum target. Note that the output is not only near-monochromatic, it also consists almost entirely of characteristic radiation. Thus the photon energy of the intensity peak cannot drift, for example with generator performance, and soft tissue contrast will remain constant.

3.9 Conclusions

In this chapter both experimental and theoretical aspects of the interaction of X-rays and gamma rays with matter have been discussed. In attenuating materials the intensity of such beams decreases exponentially, provided

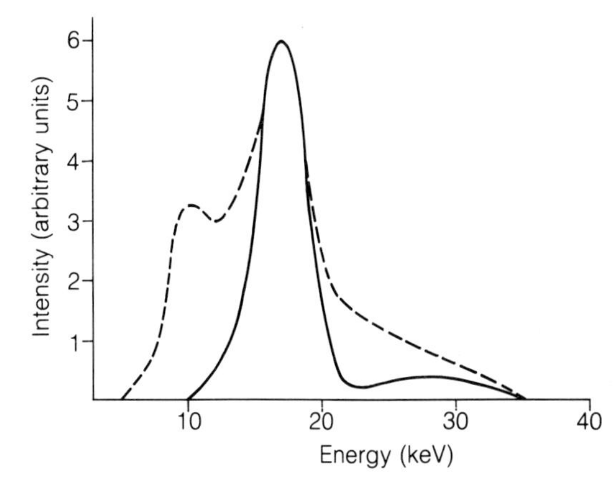

Fig. 3.15. The effect of a 0.05 mm molybdenum filter on the spectrum from a tube operating at 35 kV constant potential using a molybdenum target. (*Dotted line* = no filter; *solid line* = with filter.)

they are monochromatic or near-monochromatic, at a rate determined by the density and mean atomic number of the attenuator and the photon energy. In the diagnostic range the amount of attenuation always decreases with increasing photon energy.

The two most important interaction processes are the photoelectric effect and the Compton effect. The former is primarily responsible for differences in attenuation (contrast) at low photon energies and its effect is very dependent on atomic number. However, the photoelectric effect decreases rapidly with increasing photon energy and when the Compton effect dominates, only differences in density cause an appreciable difference in attenuation.

The Compton effect produces scattered photons of lower mean energy than the primary beam. This scattered radiation is undesirable both because it reduces contrast in the radiograph and also because it constitutes a radiation hazard to staff.

Although the attenuation coefficient generally decreases with increasing photon energies, there are sharp discontinuities, known as absorption edges. The location of an edge along the photon energy axis depends on the atomic number of the absorber, and absorption edges have a number of important consequences, notably in the choice of material for intensifying screens and in determining the variation in sensitivity of photographic film with X-ray energy.

Because the attenuation coefficient decreases with increasing photon energy, low energy photons are preferentially removed from an inhomogeneous or heterochromatic beam—a phenomenon known as beam hardening. This feature is used in filtration to remove soft X-rays that would increase the patient dose.

The effect of the basic interaction processes considered here on contrast and image quality will be discussed in detail in Chapter 5.

References and further reading

Bomford C. K. and Burlin T. E. (1963) The angular distribution of radiation scattered from a phantom exposed to 100–300 kVp X-rays. *Br. J. Radiol.* **36**, 436–439.

Hospital Physicists' Association (1979) *Catalogue of Spectral Data for Diagnostic X-rays.* (Scientific Report Series 30) The Hospital Physicists' Association, London.

International Commission on Radiological Protection (1982) *Protection against Ionizing Radiation from External Sources Used in Medicine* (ICRP publication 33). *Ann. ICRP,* **9**(1).

Johns H. E. and Cunningham J. R. (1983) *The Physics of Radiology*, 4th edn. Thomas, Springfield.

Meredith W. J. and Massey J. B. (1977) *Fundamental Physics of Radiology*, 3rd edn. Wright, Bristol.

Trout E. D., Kelley J. P. and Cathey G. A. (1952) The use of filters to control radiation exposure to the patient in diagnostic radiology. *Am. J. Roentgenol.* **67**, 946–963.

Exercises

1 Explain the terms
  (a) inverse square law,
  (b) linear attenuation coefficient,
  (c) half value thickness,
  (d) mass absorption coefficient.
Indicate the relationships between them, if any.

2 Describe the process of Compton scattering, explaining carefully how both attenuation and absorption of X-rays occur.

3 Describe the variation of the Compton attenuation coefficient and Compton absorption coefficient with scattering angle in the energy range 10–200 keV.

4 How does the process of Compton scattering of X-rays depend on the nature of the scattering material and upon X-ray energy? What is the significance of the process in radiographic imaging?

5 An X-ray beam loses energy by the processes of absorption and/or scattering. Discuss the principles involved at diagnostic X-ray wavelengths and explain how the processes are modified by different types of tissue.

6 Explain why radiographic exposures are usually made with an X-ray tube voltage in the range 50–110 kVp.

7 If the mass attenuation coefficient of aluminium at 60 keV is

$0.028\ m^2\ kg^{-1}$ and its density is $2.7 \times 10^3\ kg\ m^{-3}$, estimate the fraction of a monoenergetic incident beam transmitted by 2 cm of aluminium.

8 A parallel beam of monoenergetic X-rays impinges on a sheet of lead. What is the origin of any lower energy X-rays which emerge from the other side of the sheet travelling in the same direction as the incident beam?

9 What is meant by characteristic radiation? Describe briefly three situations in which characteristic radiation is produced.

10 How would a narrow beam of 100 kV X-rays be changed as it passed through a thin layer of material? What differences would there be if the layer were

(a) 1 mm lead ($Z = 80$, $\varrho = 1.1 \times 10^4\ kg\ m^{-3}$),

(b) 1 mm aluminium ($Z = 13$, $\varrho = 2.7 \times 10^3\ kg\ m^{-3}$)?

11 Before the X-ray beam generated by electrons striking a tungsten target is used for radiodiagnosis it has to be modified. How is this done and why?

12 What factors determine whether a particular material is suitable as a filter for diagnostic radiology?

13 A narrow beam of X-rays from a diagnostic set is found experimentally to have a half value thickness of 2 mm of aluminium. What would happen to

(a) the half value thickness of the beam,

(b) the exposure rate,

if an additional filter of 1 mm aluminium were placed close to the X-ray source?

14 Discuss the advantages and disadvantages of using aluminium as the filter material in X-ray sets at different generating potentials.

15 Compare the output spectra produced by a tungsten target and a copper target operating at 60 kVp. What would be the effect on these spectra of using

(a) an aluminium filter,

(b) a lead filter?

	Atomic number	K shell	L shell (both in keV)
Al	13	1.6	—
Cu	29	9.0	1.1
W	74	69.5	11.0
Pb	82	88.0	15.0

16 The dose rate in air at a point in a narrow beam of X-rays is $0.3\ Gy\ min^{-1}$. Estimate, to the nearest whole number, how many half value thicknesses of lead are required to reduce the dose rate to

10^{-6} Gy min^{-1}. If $H_{\frac{1}{2}}$ at this energy is 0.2 mm, what is the required thickness of lead?

17 What is the 'lead equivalence' of a material?

18 Explain what you understand by the homogeneity of an X-ray beam and describe briefly how you would measure it.

19 What is meant by an inhomogeneous beam of X-rays and why does it not obey the law of exponential attenuation with increasing filtration?

4

The Image Receptor

4.1 Introduction—band structure in solids

Roentgen discovered X-rays when he noticed that a thin layer of barium platinocyanide on a cardboard screen would fluoresce even when the discharge tube (a primitive X-ray tube) was covered by black paper. Simultaneously he had discovered the first X-ray receptor! Although barium platinocyanide is no longer used, the first stage in the detection process for over 95% of all X-ray imaging devices is now a fluorescent screen or scintillation crystal. For some image receptors, such as film, it is not essential although generally used. In some systems, for example in computerized axial tomography and in nuclear medicine, an essential feature of the scintillation crystal is that the first stage of the X-ray detection system produces a flash of light of intensity proportional to the energy of the X-ray photon interacting with the crystal.

To understand adequately the scintillation mechanism, it is necessary to consider briefly the electron levels in a typical solid. The simple model of discrete energy levels introduced in Chapter 1 and used to explain features of X-ray spectra in Chapter 2 is strictly only true for an isolated atom. In a solid, where the atoms are close together, it applies reasonably well for the innermost K and L shells but for the outer shells the close proximity of other electrons permits each electron to occupy a range or 'band' of energies (see Fig. 4.1).

The highest energy level is the conduction band. Here the electrons have sufficient energy to move freely through the crystalline lattice and in particular to conduct electricity. Next is the forbidden band. In pure materials there are no permitted energy levels here and electrons cannot exist in this region. The next highest energy band is the valence band which

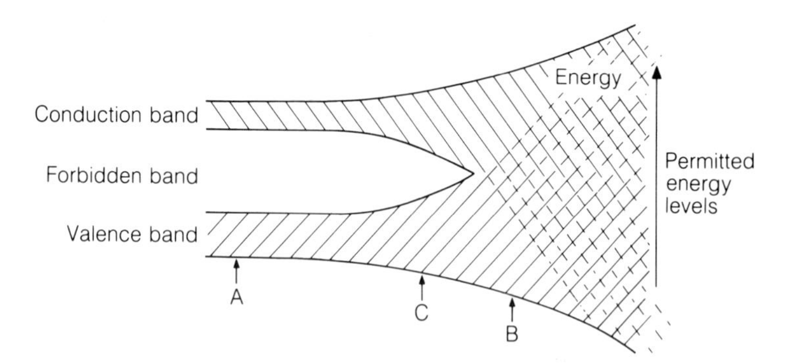

Fig. 4.1. The band structure of energy levels found in solids.

contains the valence electrons. Depending on the element, there are further bands of even lower energy. These bands carry their full complement of electrons, with forbidden zones and filled bands alternating like layers in a sandwich, but these are of no interest here.

With this simple model, three types of material may be identified. If the forbidden band between the valence and conduction bands is wide there are no electrons in the conduction band and the material is an insulator (point A in Fig. 4.1). If the bands overlap the material is a conductor because electrons can move freely from one atom to the next as the conduction band is continuous between atoms (point B). When the forbidden band is narrow (point C) materials can be made to change from non-conducting to conducting under specific voltage conditions and are termed semi conductors.

Although this is the idealized pattern of a pure crystal, in practice real crystals always contain imperfections and impurities which manifest themselves, amongst other ways, as additional energy levels or electron traps in the forbidden energy band. These traps play an important role in all scintillation processes and the manufacture of materials with these properties depends on the production of very pure crystals to which impurities are added under carefully controlled conditions to produce the right number and type of traps. The imperfections are called luminescent centres and can be thought of in terms of one discrete energy level close to the conduction band, called an electron trap, and one close to the valence band called a hole trap (Fig. 4.2a).

With this simple picture of the band structure of a solid, the details of three slightly different luminescent mechanisms are summarized in the next section.

4.2 Fluorescence, phosphorescence and thermoluminescence

The essential features of **fluorescence** are as follows:

1 Electron traps are normally occupied.

2 An X-ray photon interacts by the photoelectric or Compton process to produce a photoelectron which dissipates energy by exciting other electrons to move from the valence band to the conduction band.

3 Holes are thus created in the filled valence band.

4 The hole, which has a positive charge numerically equal to that of the electron, moves to a hole trap at a luminescent centre in the forbidden band.

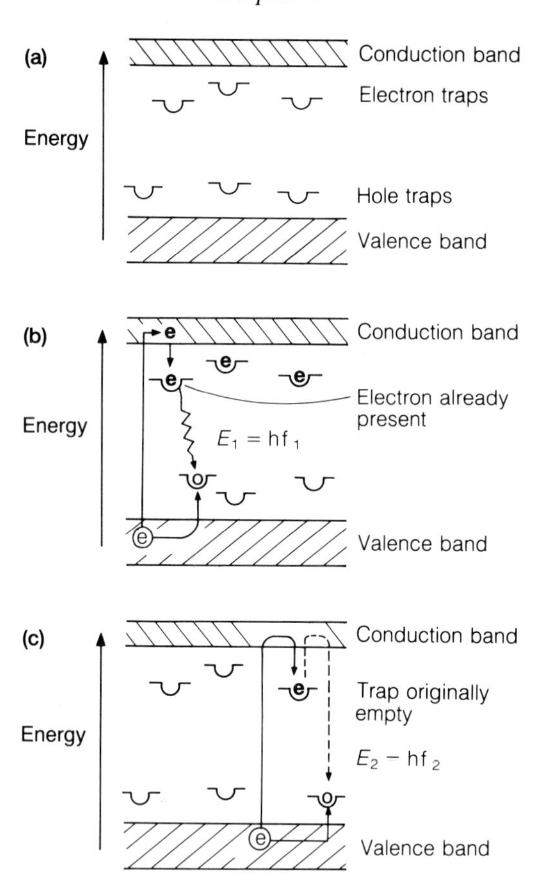

Fig. 4.2. (a) Schematic representation of electron and hole traps in the 'forbidden' energy band in a solid. (b) The change in energy level resulting in light emission in fluorescence. (c) The change in energy level resulting in light emission in phosphorescence.

5 When both an electron and a hole are trapped at a luminescent centre, the electron may fall into the hole emitting visible light of characteristic frequency f where $E_1 = hf_1$ (see Fig. 4.2b).
6 The electron trap is refilled by an electron that has been excited up into the conduction band.

In fluorescence, the migration of electrons and holes to the fluorescent centres and the emission of a photon of light happens so quickly that it is essentially instantaneous. Not all transitions of electrons at luminescent centres produce light. The efficiency of the transfer can vary enormously between different materials.

The phenomenon of **phosphorescence** also depends on the presence of traps in the forbidden band but differs in the following respects from fluorescence:

1 The electron traps are now normally empty.

2 X-ray interactions stimulate transitions from the valance band to the conduction band, and, for reasons too complex to discuss here, these electrons fall into the traps rather than back into the valence band.

3 Furthermore, these electrons must return to the conduction band before they can descend to a lower level.

4 Visible light is, therefore, only emitted when the trapped electron acquires sufficient energy to escape from the trap up into the conduction band.

5 Visible light of a different characteristic frequency f_2 (see Fig. 4.2c) is now emitted. (The electron is falling through a larger energy gap.)

If the electron trap is only a little way below the conduction band, the electron eventually acquires this energy by statistical fluctuations in its own kinetic energy. Thus light is emitted after a time delay (phosphorescence). The time delay that distinguishes phosphorescence from fluorescence is somewhat arbitrary and might range from 10^{-10} s to 10^{-3} s. The two processes can be separately identified by heating the material. Light emission by phosphorescence is facilitated by heating, because the electrons more readily acquire the energy required to escape. Fluorescence is temperature independent. In radiology phosphorescence is sometimes called 'afterglow' and, unless the light is emitted within a very short time when it may contribute to the quantum yield, its presence in a fluorescent screen is detrimental.

If the electron trap is a long way below the conduction band, the chance of the electron acquiring sufficient energy by thermal vibrations at room temperature is negligible and the state is metastable. If the electron is given extra kinetic energy by heating, however, it may be released and then light is emitted. This is essentially the process of **thermoluminescence**.

4.3 Phosphors and fluorescent screens

4.3.1 Properties of phosphors

The spectral output required from a phosphor depends on the spectral response of the next stage in the imaging system. It is also desirable for a phosphor to have a high average Z value so that it absorbs a high percentage of the energy from an X-ray beam. However, this is not a sufficient condition, since the overall 'luminescent radiant efficiency' of the phosphor, or light output per unit X-ray beam intensity, will also depend on its efficiency in converting this energy to light output. The chemical properties of a phosphor can also limit how the phosphor is used, e.g. a hygroscopic phosphor must always be used either encapsulated or inside a vacuum. Finally phosphors must also be commercially available in known crystal sizes of uniform sensitivity.

Aside from CT scanning, dealt with in detail in Chapter 10, fluorescent screens may be used as the input for three basic imaging systems:

1 coupled to film in radiography,
2 viewed directly by eye during fluoroscopy,
3 in an image intensifier coupled to a photocathode.

Each of these light receptors has a different spectral response (Fig. 4.3) and, as already stated, the spectral output of the phosphor must be matched as closely as possible to the spectral response of the light receptor to which it is coupled.

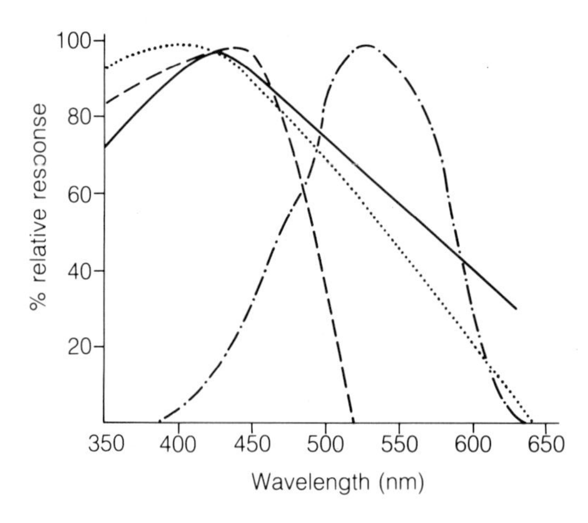

Fig. 4.3. The spectral response of different light receptors. *Solid line:* 520 photocathode. *Dashed line:* X-ray film. *Dot-dash line:* the eye. *Dotted line:* S11 photocathode.

Until the mid 1960s, the most widely used phosphors were calcium tungstate in radiography and zinc cadmium sulphide in fluoroscopy and image intensifiers. The spectral outputs of these phosphors are shown in Fig. 4.4. As a direct result of the American space programme, new phosphors were developed which have since been adopted for medical use. They are mainly crystals of salts of the rare earth elements or crystals of barium salts activated by rare earth elements. A selection of spectral outputs is shown in Fig. 4.4. Rare earth phosphors containing metallic

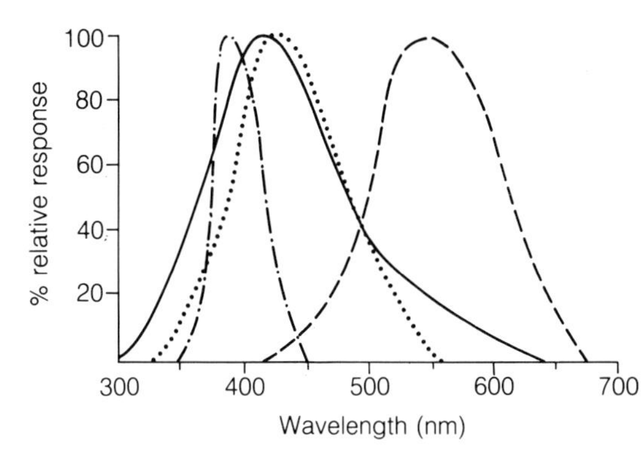

Fig. 4.4 The spectral output of different phosphors. *Solid line:* CsI : Na. *Dashed line:* (ZnCd)S : Ag. *Dot-dash line:* BaFCl : Eu^{2+}. *Dotted line:* $CaWO_4$.

elements from the lanthanide series emit light at several discrete wavelengths between 380 nm and 620 nm with the strongest emissions at about 550 nm. Note that the major exception to the above groups is caesium iodide activated by sodium.

The reason why the rare earth phosphors have replaced calcium tungstate for many purposes can be seen in Table 4.1. Although the Z value of the new rare earth phosphors is slightly lower than that of the heavy elements in calcium tungstate phosphor, and thus slightly less energy may be absorbed from an X-ray beam, the luminescent radiant efficiency for the

Table 4.1. Atomic numbers and luminescent radiant efficiencies for some important phosphors

Phosphor	Z of heavy elements	Luminescent radiant efficiency (%)
BaFC1 : Eu^{2+}	56	13
$BaSO_4$: Eu^{2+}	56	6
$CaWO_4$	74	3.5
CsBr : Tl	35/55	8
CsI : Na	53/65	10
CsI : Tl	53/55	11
Gd_2O_2S : Tb	64	15
La_2O_2S : Tb	57	12
Y_2O_2S : Tb	39	18
(ZnCd)S : Ag	30/48	18
NaI : Tl	53	10

The element behind the : sign is the activator to the phosphor salt before the : sign.

rare earth phosphors is at least three times higher. The same light output from the phosphor can thus be achieved with a much lower X-ray dose. In practice, because of absorption edges (section 3.6), at certain photon energies X-ray absorption would also be higher.

Figures are also given for comparison for thallium doped sodium iodide NaI : Tl, the primary detector used almost universally in nuclear medicine.

4.3.2 Production of fluorescent screens

Most fluorescent screens are produced by laying down a phosphor crystal/ binder suspension onto a paper or metallic substrate and then drying it on the substrate. The size of the crystals affects screen performance with larger crystals producing more light for a given screen thickness. If small crystals are used, the effective crystal area is small because there is a high proportion of interstitial space. However, larger crystals produce a screen with a much lower resolution. For a given thickness of screen, with an acceptable resolution, the best packing density (ratio of phosphor to interstitial space) of phosphor crystals that can be achieved is approximately 50%. For high resolution screens the ratio is lower.

A very limited number of phosphors based on halide salts, e.g. CsI : Na, can be vapour deposited in sufficiently thick layers to enable a useful detector to be made. The packing density is almost 100% thus giving a gain of approximately two over crystal deposited screens with a consequent higher X-ray absorption per unit thickness. It is also possible to get better resolution and less noise from these phosphors, and this is important in the construction and performance of, for example, an image intensifier (see section 4.10).

4.3.3 Film–phosphor combinations in radiography

For medical applications, radiographic film is almost always used in combination with a fluorescent screen, with only a very few specialized exceptions. However, the properties of the combination are essentially governed by the properties of the film.

Film responds to the light emitted from a fluorescent screen in the same way as it does to X-rays except in one or two minor respects. Thus the many advantages and few disadvantages of using a screen can best be understood by first examining the response of film to X-rays and then considering how the introduction of a screen changes this response.

4.4 X-ray film

4.4.1 Film construction

The basic film construction is shown in Fig. 4.5. To maintain rigidity and carry the emulsion, which is the sensitive part of the detector, all radiographic films use a transparent base material approximately 0.15 mm thick. The material used is generally 'polyester' and is normally tinted blue although other colours are used for some films. It is completely stable during development. Normally the polyester base has an emulsion bound to both sides (as in Fig. 4.5), although some special films have emulsion on one side only. For example, as described in section 6.6, during mammography a single screen is used on the side of the film remote from the breast. A single emulsion is used next to the screen to eliminate parallax effects. If transparency film is used in nuclear medicine it is single-sided because by that stage in the imaging process only visible light is being recorded. Both sides of the film are coated with a protective layer to prevent mechanical damage.

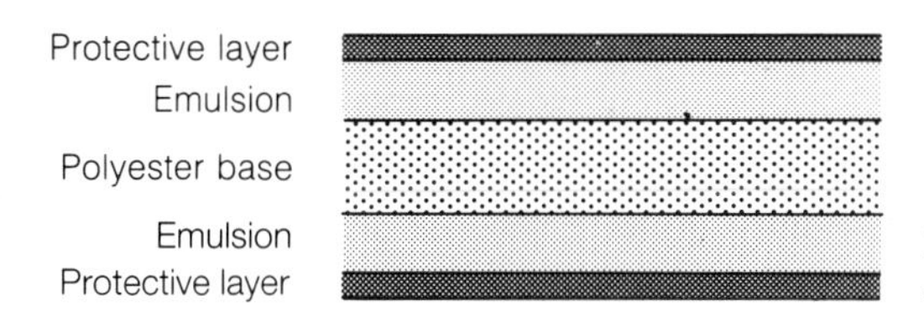

Fig. 4.5. The basic construction of an X-ray film.

As well as being sensitive to X-rays, the emulsion is sensitive to light and must be kept in a light-tight container. It must only be loaded into a cassette using a special daylight loading system or in a dark-room illuminated by a safe light. Short exposures to the wavelength of light in the safe light do not affect the emulsion.

The emulsion is composed of crystals of silver (Ag^+) and bromine (Br^-) ions which are in a cubic lattice and would, in the pure crystal, be electrically neutral. However, the presence of impurities distorts the lattice and produces on the surface of the crystal, a spot, called the sensitivity speck, which will attract any free electrons produced within the crystal. When exposed to X-radiation, free electrons are produced by the Compton or photoelectric effect. These electrons, or in the case of visible light from, say, an intensifying screen, light photons, are able to displace further electrons from bromine ions

$$\underset{}{Br^-} + \underset{\text{light}}{hf} \rightarrow \underset{\text{atoms}}{Br} + e^-$$

Removal of the electron from the bromine ion produces a free bromine atom which is absorbed by the gelatin used to make the emulsion stick to the base.

The electrons move through the crystal and are trapped by the sensitivity speck. An electron at a sensitivity speck then attracts a positively charged silver ion to the speck and neutralizes it to form a silver atom. This occurs many times and the result is an area of the crystal with a number of neutral silver atoms on the surface. This crystal is then said to constitute a latent image. For the crystal to be developable, between 10 and 80 atoms of silver must be produced. During development a reducing alkaline agent is used. Crystals with a latent image in them allow the rest of the silver ions present to be reduced and thus form a dark silver grain speck on the film. Where many X-rays have hit the film, several crystals are affected and thus many silver grains are produced. If the crystals are large it is relatively easy to produce large black specks. If the crystals are small many more X-rays have to hit the film to produce the same amount of blackening. The film is fixed and hardened at the same time using a weakly acidic solution. The crystals which did not contain a latent image are washed off at the fixation stage leaving a light area on the film.

If some time elapses between the production of a latent image and developing the radiograph, some of the crystals revert back to their original state and are no longer developed during processing. The latent image is said to have 'faded'. In practical terms, latent image fading is of no significance in radiology as the radiographs are developed immediately following exposure. It can, however, have a small effect when film is used in a personnel dosimetry service. For further discussion on this point see section 4.7.

If the developing agent is too strong it will develop crystals in which no latent image is present. Even in an unexposed film some crystals will be developed to produce a low level of blackening called fog. The 'fog' level can be increased by using inappropriate developing conditions, e.g. too strong a developer or too high a developing temperature.

4.4.2 Characteristic curve and optical density

The amount of blackening produced on a film by any form of radiation—visible light or X-rays—is measured by its density. Note the use of density here relates to optical density and must not be confused with mass density (ϱ) used elsewhere in the book. The preface 'optical' will not be used elsewhere unless there is some possibility of confusion.

The density of a piece of blackened film is measured by passing visible light through it (Fig. 4.6a). Density is defined by the equation

$$D = \log_{10} \frac{I_0}{I}$$

where I_0 is the incident intensity of visible light and I is the transmitted intensity.

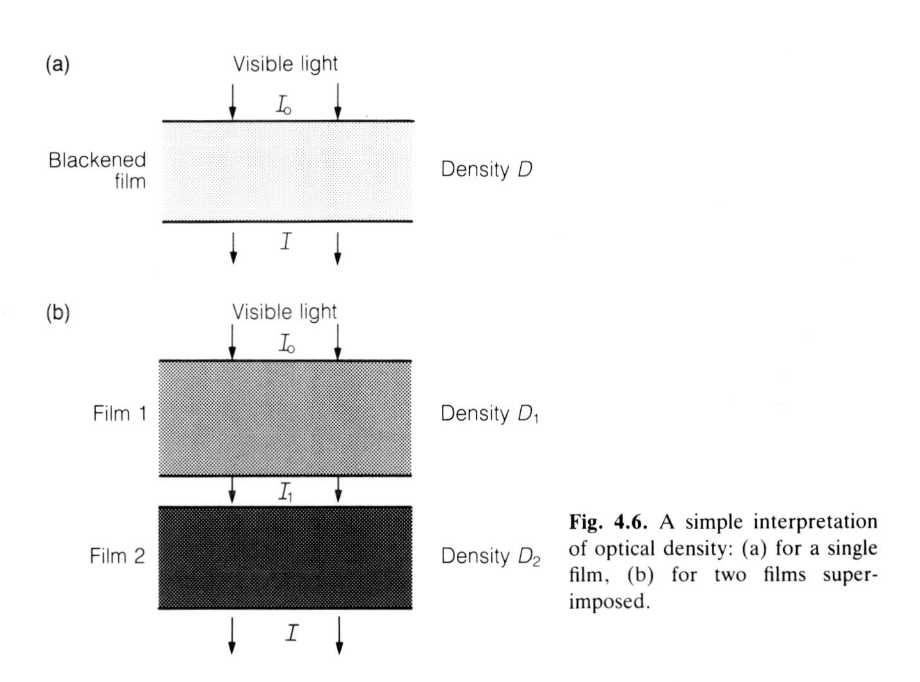

Fig. 4.6. A simple interpretation of optical density: (a) for a single film, (b) for two films superimposed.

Basing the definition on the log of the ratio of incident and transmitted intensities has three important advantages:

1 It represents accurately what the eye sees, since the physiological response of the eye is also logarithmic to visible light.

2 A very wide range of ratios can be accommodated and the resulting number for the density is small and manageable (see Table 4.2).

Table 4.2. Relationship between optical density and transmitted intensity

Transmitted intensity as a percentage of I_0	$OD = \log_{10} I_0/I$
10%	$\log_{10} 10 = 1.0$
1%	$\log_{10} 100 = 2.0$
0.01%	$\log_{10} 10000 = 4.0$

3 The total density of two films superimposed is simply the sum of their individual densities. From Fig. 4.6b,

$$\text{Total } D = \log \frac{I_0}{I} = \log \frac{I_0}{I_1} \cdot \frac{I_1}{I}$$

$$= \log \frac{I_0}{I_1} + \log \frac{I_1}{I}$$

$$= D_1 + D_2$$

When different amounts of light are transmitted through different parts of the film (Fig. 4.7), the difference in density between the two parts of the film is called the contrast. Hence.

$$C = D_2 - D_1 = \log_{10} \frac{I_0}{I_1} - \log_{10} \frac{I_0}{I_2} = \log_{10} \frac{I_2}{I_1} \qquad \text{(equation 4.1)}$$

The eye can easily discern differences in density ranging from approximately 0.25 to 2.5, the minimum discernible difference being about 0.02.

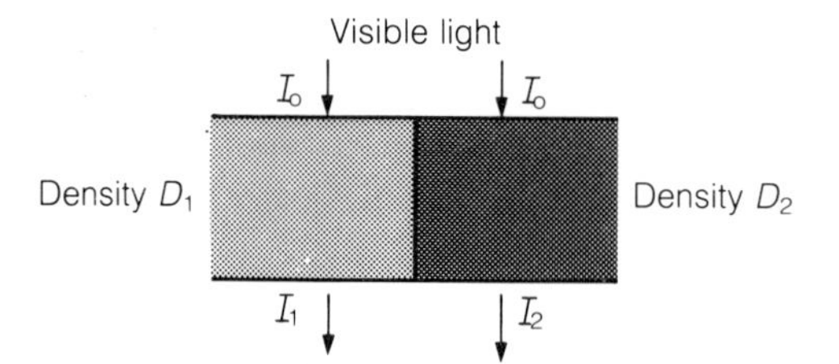

Fig. 4.7. Representation of contrast between two parts of a blackened film as a difference in transmitted light intensities.

If the density produced on a film is plotted against the log of the radiation exposure producing it, the characteristic curve of the film is generated (Fig. 4.8). The log scale again allows a wide range of exposure to be accommodated. Each type of film has its own characteristic curve although all have the same basic shape.

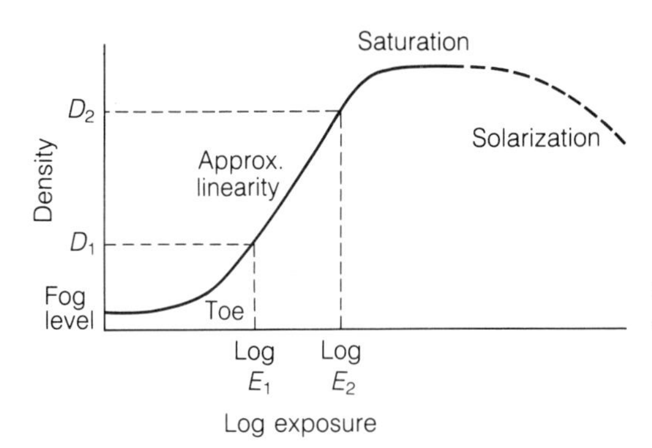

Fig. 4.8. A typical characteristic curve for an X-ray film.

The finite density at zero exposure is due to 'fogging' of the film, i.e. the latent images produced during manufacture, by temperature, humidity and other non-radiation means. This can be kept to a minimum but never completely removed. Note that the apparently horizontal initial portion of the curve arises primarily because one logarithmic quantity (D) has been plotted against another (log E). This has the effect of compressing the lower end of the curve. If the data were replotted on linear axes, there would be a steady increase in film blackening with exposure but when using the characteristic curve, a finite dose is required to be given to the film before densities above the fog level are recorded.

The initial curved part of the graph is referred to as the 'toe' of the characteristic curve and this leads into the approximately linear portion of the graph covering the range of densities and doses over which the film is most useful. Eventually, after passing over the shoulder of the curve, the graph is seen to saturate and further exposure produces no further blackening. This decrease in additional blackening is due to the black spots from developed crystals overlapping until eventually the production of more black silver spots has no further effect on the overall density.

At very high exposures (note that the scale is logarithmic), film blackening begins to decrease again, a process known as solarization. The mechanism by which this occurs is not fully understood although it may be a result of the release of excess bromine at these high exposures. After the exposure is completed, this bromine may recombine with free silver to reconstitute silver bromide which will protect the latent images against development.

Solarization provides a method for obtaining a negative from a negative since the amount of blackening decreases as the intensity of the light transmitted increases. Hence it is used for film copying. The film is specially treated by the manufacturer ('solarized') so that this effect is produced when it is exposed to ultraviolet light. The film to be copied is simply placed over the solarized film and, by adjusting the exposure, it is possible to produce a copy that is lighter or darker than the original if required.

4.4.3 Film gamma and film speed

The gamma of the film is the maximum slope of the approximately linear portion of the characteristic curve and from Fig. 4.8 is defined as

$$\gamma = \frac{D_2 - D_1}{\log E_2 - \log E_1}$$

If no part of the curve is approximately linear, the average gradient may be calculated between defined points on the steepest part of the curve.

The gamma of a film depends on the type of emulsion present, principally the distribution and size of the silver bromide crystals, and secondly on how the film is developed. If the crystals are all the same size a very 'contrasty' film is produced with a large gamma. A wide range of crystal sizes will produce a much lower gamma (Fig. 4.9). A 'fast' film with large crystals generally also has a wide range of crystal sizes. Finally, increased grain size reduces resolution although unsharpness in the film itself is rarely a limiting factor.

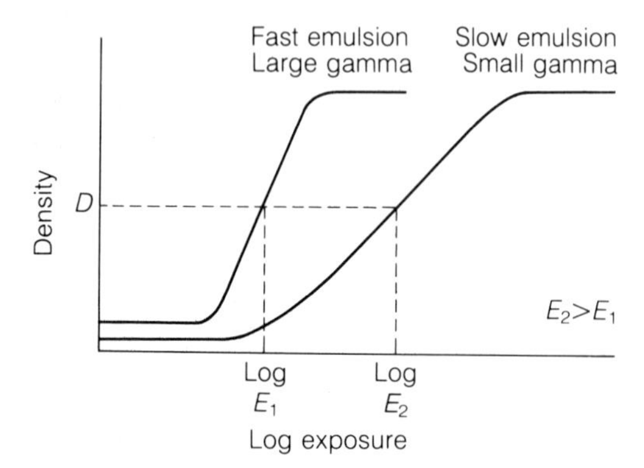

Fig. 4.9. Characteristic curves for films of different gammas and different speeds.

The correct characteristic curve for a film can only be obtained by using the developing procedure recommended by the film manufacturer, including the concentration of developer, the temperature of the developer, the period of development and even the amount of agitation to be applied to the film. An increase in any of these factors will result in over-development of the film, a decrease will under-develop the film.

Within realistic limits, over-development increases the fog level, the film gamma and the saturation density. Under-development has the opposite effect (Fig. 4.10).

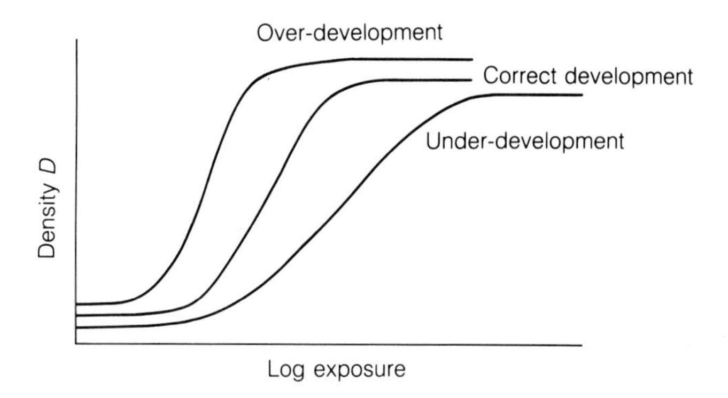

Fig. 4.10. Variation of the characteristic curve for a film, and in particular film gamma and fog level for different development conditions.

The amount of radiation required to produce a given density is an indication of film speed. The speed is usually taken to be the reciprocal of the exposure that causes unit density above fog so a fast film requires less radiation than a slow film. The speed of the film depends on the size of the crystals making up the emulsion and on the energy of the X-rays striking the film. If the crystals are large then fewer X-ray interactions are required to blacken a film and, because of this, fast films are often called 'grainy' films as the crystals when developed give a 'grainy' picture. This is because the energy deposited by a single X-ray photon is sufficient to produce a latent image in a large crystal as well as a small crystal. Fewer large crystals need be developed to obtain a given density. The speed of the film varies with the energy of the X-ray photon, a property that will be considered in detail in section 7.17. In practice, due to the wide range of photon energies present in a diagnostic X-ray beam, this variation in sensitivity can be neglected during subjective assessment of X-ray films.

4.4.4 Latitude

Two distinct, but related aspects of latitude are important, film latitude and exposure latitude. Consider first film latitude. The optimum range of densities for viewing, using a standard light box, is between 0.3 and 2.25. Between these two limits the eye can see small changes in contrast quite easily. The latitude of the film refers to the range of exposures that can be given to the film such that the density produced is within these limits. The higher the gamma of the film, the smaller the range of exposures it can tolerate and thus the lower the latitude. For general radiography a film with a reasonably high latitude is used. There is an upper limit however because if the gamma of the film is made too small the contrast produced is too small for reasonable evaluation.

Tc understand exposure latitude, refer to Fig. 4.8. If a radiograph is produced in which all film densities are on the linear portion of the curve, the exposure may be altered, shifting these densities up or down the linear portion, without loss of contrast. In othe words there is 'latitude' or some freedom of choice over exposure. If, on the other hand, the range of densities on a film covers the whole of the linear range, exposure cannot be altered without either pushing the dark regions into saturation or the light regions into the fog level. There is no 'latitude' on choice of exposure.

Exposure latitude can be restored by choosing a film with a lower gamma (greater film latitude), but only at the expense of loss of contrast.

4.5 Film used with a fluorescent screen

Most properties of film described to this point have assumed an exposure to X-radiation. Film also responds to light photons but as the quantum energy of one light photon is only 2–3 eV (see Chapter 1) several tens of light photons have to be absorbed to produce one latent image. In contrast, the energy from just one X-ray photon is more than sufficient to produce a latent image.

The advantages of using film in conjunction with fluorescent screens are two-fold. First a much greater number of X-ray photons are absorbed by the screen compared to the number absorbed by film alone. The ratio varies between 20 and 40 depending on the screen composition.

Secondly, by first converting the X-ray photon energy into light photons, the full blackening potential is realized. If an X-ray photon is absorbed directly in the film, it will sensitize only one or two silver grains. However, each X-ray photon absorbed in an intensifying screen will release at least 400 photons of light—some screens will release several thousand photons. Thus, although tens of light photons are required to produce a latent image, the final result is that the density on the film for a given exposure is between 30 and 300 times blacker (depending on the type of screen) when a screen is used than when a film alone is exposed.

The increase in blackening when using a fluorescent screen is quantified by the **intensification factor**, defined as:

$$\frac{\text{The exposure required when a screen is not used}}{\text{The exposure required when a screen is used}}$$

for the same film blackening.

A value for the intensification factor is only strictly valid for one density and one kVp. This is because, as shown in Fig. 4.11, when used in conjunction with a screen, the characteristic curve of the film is altered. Not only is it moved to the left, as one would expect, but the gamma is also increased.

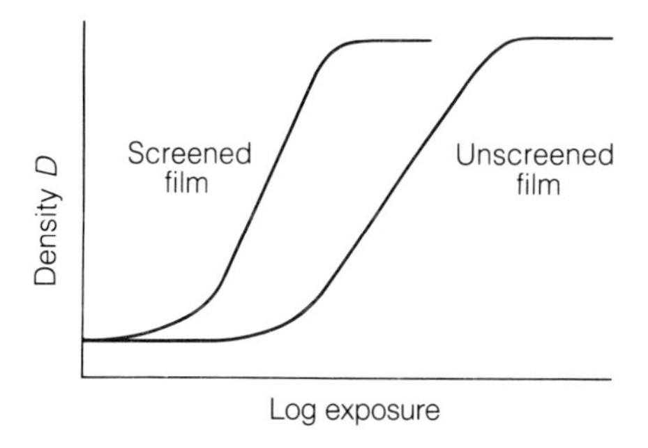

Fig. 4.11. Change in the characteristic curve of a film when using an intensifying screen.

When using screens, double-sided film has radiological advantages as well as for processing since two emulsion layers enable double the contrast to be obtained for a given exposure. This is because the super-position of two densities (one on each side of the film base) produces a density equal to their sum. Although, theoretically, the same contrast could be achieved by doubling the thickness of a single emulsion, this is not possible in practice due to the limited range in emulsion of the light photons from the screens. Note that, because the range of X-ray photons is not limited in this way, the contrast for unscreened films would be almost the same for a one-sided film of double emulsion thickness. The mechanical advantages of double-sided film of course remain.

4.6 Cassettes

Radiographic film must be used and contained in a light-tight cassette, otherwise it is fogged by ambient light. The front of the cassette is made of a low atomic number material, generally either aluminium or plastic. The back of the cassette is either made of, or lined with, a high atomic number material. This high atomic number material is more likely to absorb totally the X-ray photons passing through both screens and film by the photoelectric effect, than to undergo a Compton scatter reaction, which could backscatter photons into the screen.

The two screens are kept in close contact with the film by the felt pad exerting a constant pressure as shown in Fig. 4.12. If close contact is not maintained then resolution is lost.

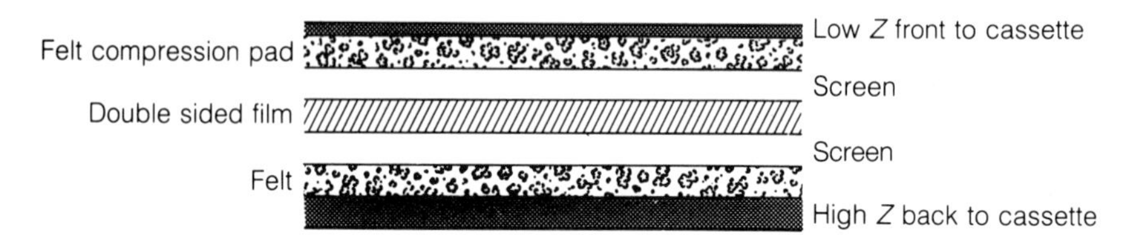

Fig. 4.12. Construction of a 'sandwich' of intensifying screens and double-sided film.

The light emitted from a screen does not increase indefinitely if the screen thickness is increased to absorb more X-ray photons. A point is reached where increasing the thickness produces no more light because internal absorption of light photons in the screen takes place. The absorption of X-ray photons produces an intensity gradient through the screen and significantly fewer leave the screen than enter. The light produced at a given point in the screen is directly proportional to the X-ray intensity at that point.

In the case of the back screen, the reduction in intensity through it is of no significance. This screen must be thick enough to ensure that the maximum amount of light for a given X-ray intensity is produced, but once this is achieved it need be no thicker. The front screen thickness is a compromise between achieving the maximum light output through X-ray photon absorption and not reducing the X-ray photon intensity by too much in the area of effective maximum light production, i.e. in the layers of the screen closest to the film. The compromise thickness for the front screen, giving maximum light production, is in fact somewhat less than the optimum for the back screen. Since unsharpness is less from the back screen, attempts to optimize light production can improve image sharpness.

If the screens are of unequal thickness they must never be reversed but for some modern screens there is no difference between front and back thicknesses.

4.7 Reciprocity

The exposure received by a cassette can be considered to depend on two basic parameters, the intensity of the radiation beam striking the cassette and the time of exposure. The intensity at a fixed kVp is proportional to the milliamps (mA) of the exposure, and the exposure is thus proportional to the milliampseconds. For unscreened film the same milliampseconds will always give the same blackening of the film regardless of the period of exposure. When using screened film, however, it is found that for very short periods of exposure and very long periods of exposure, although the same milliampseconds are given, the blackening of the film is less.

For long exposures this effect is known as latent image fading and the amount of fading depends to a large extent on whether the image has been produced by X-ray interactions or by visible light photons. In general, a single X-ray photon will form a developable grain because it has so much energy. Hence image fading does not occur. Conversely, many visible light photons are required to produce a latent image and if their rate of arrival is slow the silver halide lattice may revert to its normal state before sensitization of the speck is completed. This effect is called failure of the reciprocity law.

4.8 Film-screen unsharpness

When screens are used there is some loss of resolution. This is because light produced in the screen travels in all directions (see Fig. 4.13). Because the screen is of finite thickness, those photons travelling in the

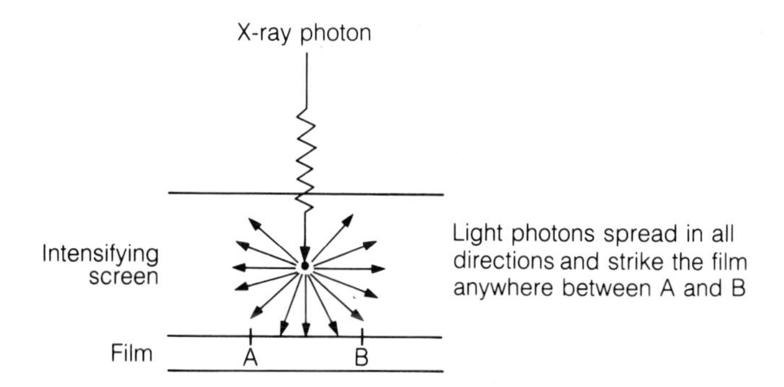

Fig. 4.13. Schematic representation of image unsharpness created by the use of an intensifying screen.

direction of the film spread out a little before reaching it. Note that very good contact must be maintained between the screen and film. If they are even slightly separated, the light will be allowed to spread out even further.

A major cause of unsharpness in double-sided film is 'cross over' where light from the upper intensifier screen sensitizes the lower film layer and vice versa. The development of emulsions containing grains that are flat or tubular in shape, with the flat surface facing the X-ray beam, rather than pebble-shaped, has helped to reduce this source of image unsharpness with no loss in sensitivity.

4.9 Eye-phosphor combination in fluoroscopy

Direct viewing of a fluorescent screen during screening is now undertaken in few radiology departments and the facility will not be available on new machines. A fluoroscopic screen is shown in Fig. 4.14. The resolution of the screen varies with crystal size in exactly the same way and for the same reasons as the resolution of a radiographic screen. The wavelength of emitted light is of course independent of the X-ray photon energy. If this technique, which is not recommended, is used, it is essential that certain basic preparatory steps are taken, otherwise all the information available may not be seen and excessive patient doses may result. For example, the room must be darkened and the eye fully dark-adapted. However, this causes further problems. The spectral output of a cadmium activated zinc sulphide screen closely matches the spectral response of the eye using cone vision at high light intensities (Fig. 4.15) but the eye is poorly equipped for detecting changes in light levels or resolution at the low intensities of light given out by a fluoroscopic screen. Furthermore, these levels cannot be

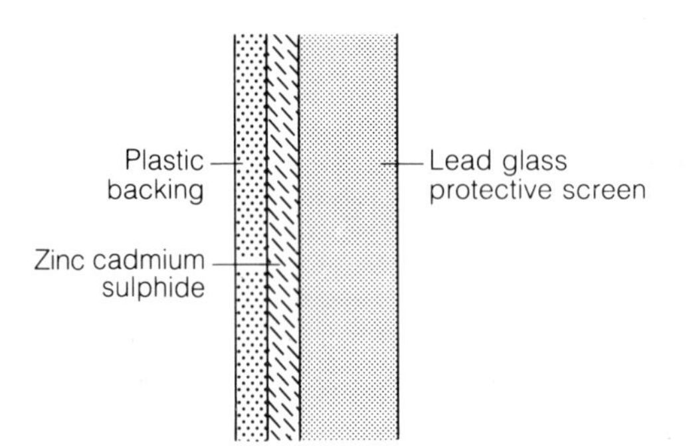

Fig. 4.14. A cross-sectional view through a fluoroscopy viewing screen.

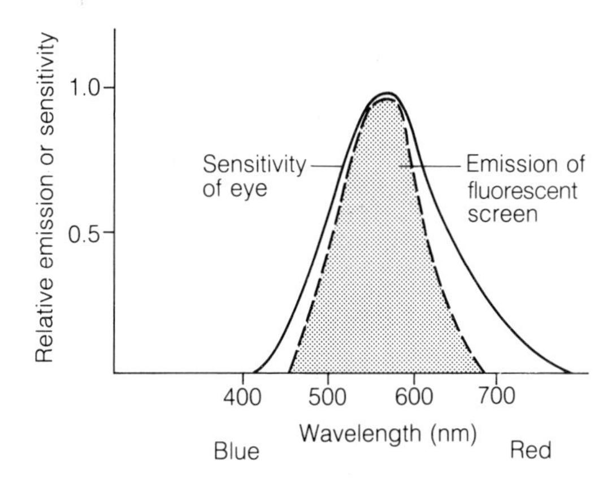

Fig. 4.15. Comparison of the emission spectrum of visible light from a fluorescent screen with the sensitivity spectrum of the eye.

increased by increasing the X-ray intensity without delivering an unacceptable dose to the patient. At the levels of light involved, 10^{-3}–10^{-4} millilamberts, the resolution of the eye is no better than 1 mm and only changes in light levels of approximately 20% can be detected. This is because vision at these levels is by rod vision alone, cone vision having a threshold of brightness perception of 0.1 millilamberts. Also the spectral response of rods peaks at a wavelength of 500 nm, somewhat higher than the peak wavelength for screen output. For further discussion see section 9.4.2.

All of the figures quoted above are for a well dark-adapted eye, so the eye must be at low light levels for 20–40 min before the fluoroscopic screen is viewed. This allows the visual purple produced by the eye to build up and sensitize the rods. Alternatively, because the rods are insensitive to red light, red goggles can be worn in ambient light, producing the same effect. Exposure to ambient light for even a fraction of a second completely destroys the build-up of visual purple.

4.10 Phosphors used with image intensifiers

4.10.1 Construction and mode of operation

The light intensity emitted from a fluorescent screen can only be increased by increasing the exposure rate from the X-ray tube or by increasing the screen thickness. Neither of these methods is acceptable, as the first increases the dose to the patient and the second reduces the resolution. Any increase in signal strength must, therefore, be introduced after the light has been produced. This can now be achieved by using an image intensifier which increases the light level to such a degree that cone vision rather than rod vision may be used.

The basic image intensifier construction is shown in Fig. 4.16. The intensifier is partially evacuated and the fluorescent screen is protected by a thin metal housing. This housing also excludes all fluorescent ambient light. The fluorescent screen is laid down on a very thin metal substrate and is now generally CsI. This has two advantages over most other fluorescent materials, as discussed in section 4.3. It can be laid down effectively as a solid layer thus allowing a much greater X-ray absorption per unit thickness to be achieved. It is also laid down in needle-like crystals (see Fig. 4.17). These crystals act as light pipes and any light produced in them is internally reflected along them and does not spread out to cover a large area. CsI can thus be used in thicker layers without a significant loss in resolution.

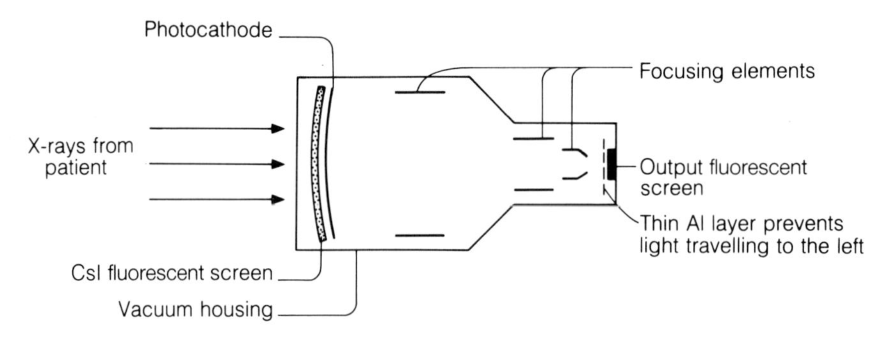

Fig. 4.16. Construction of a simple image intensifier.

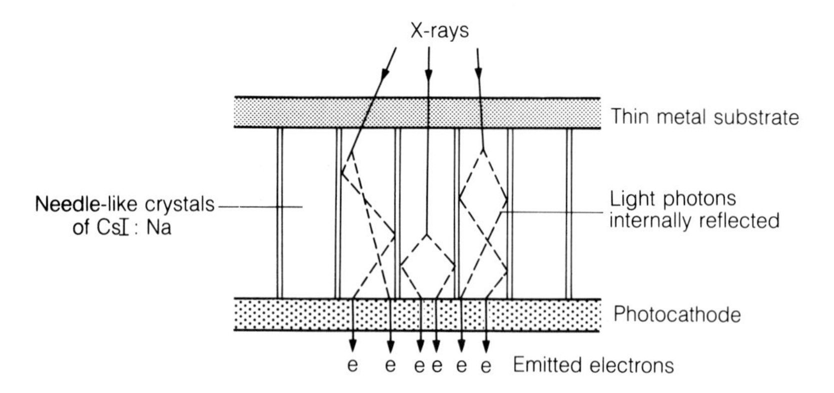

Fig. 4.17. Greatly enlarged view of the needle-like crystalline structure of a CsI : Na screen.

Intimately attached to the CsI is the photocathode. Just as in radiography and fluorography, the output of the fluorescent screen must be closely matched to the photo response of the photocathode and CsI is very closely matched to two commonly used photo-electric materials, S11 and S20 as can be seen in Fig. 4.18. The output of the intensifier is a fluorescent screen shielded from the internal part of the intensifier by a very thin piece of aluminium. This stops light from the screen entering the image intensifier. This screen is much smaller than the input phosphor. A voltage of approximately 25 kV is maintained between the input and output phosphors.

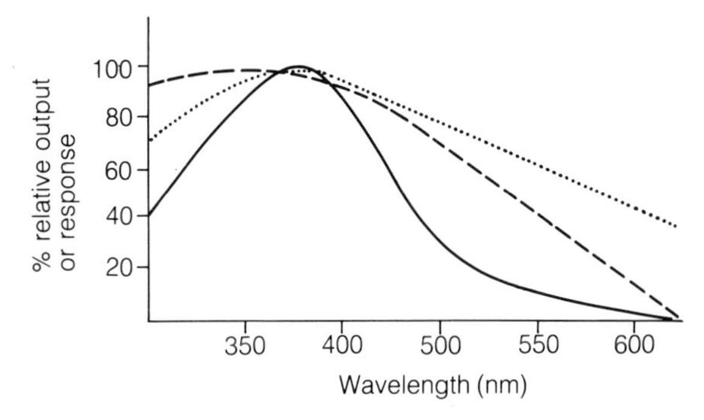

Fig. 4.18. Comparison of the spectral output of a CsI : Na screen (*solid line*) with the sensitivity response of S11 (*dashed line*) and S20 (*dotted line*) photocathodes.

The mode of action of an image intensifier will now be described. When light falls on the photocathode it is converted into electrons, and for the present discussion electrons have two important advantages over photons:

1 they can be accelerated, and

2 they can be focused.

Thus under the influence of the potential difference of 25 kV, the electrons are accelerated and acquire kinetic energy as they travel towards the viewing phosphor. This increase in energy is a form of amplification. Secondly, by careful focusing, so as not to introduce distortion, the resulting image can be minified, thereby further increasing its brightness. Even without the increase in energy of the electrons, the brightness of the output screen would be greater by a factor equal to the ratio of the input and output screen areas.

The amplification resulting from electron acceleration is usually about 50. In other words, for every light photon generated at the input fluorescent screen, approximately 50 light photons are produced at the output screen. The increased brightness resulting from reduction in image size depends of course on the relative areas of the input and output screens. If the input screen is approximately 25 cm (10 inches) in diameter and the output screen is 2.5 cm (1 inch), the increase in brightness is $(25/2.5)^2 = 100$ and the overall gain of the image intensifier about 5000.

The use of image intensifiers has not only produced an important improvement in light intensity, it has also opened up the way for the introduction of high technology into fluoroscopy. For example, direct viewing of the output of the image intensifier is not very convenient and somewhat restrictive, and although it is possible to increase the size of the image by viewing it through carefully designed optical systems, even with this arrangement there may be a considerable loss of light photons and hence information. In the worst cases this loss in light gathering by the eye can result in a system not much better than conventional fluoroscopy. Thus current practice is generally to view the output directly by a television camera as shown in Fig. 4.19a. Alternatively, a television camera is used after dividing the output using a partially reflecting mirror as in Fig. 4.19b. This system allows other recording media to be used such as a cine camera or a 100 mm still camera.

The use of a television viewing system has several advantages. The first is convenience. The amplification available through the intensifier/tele-

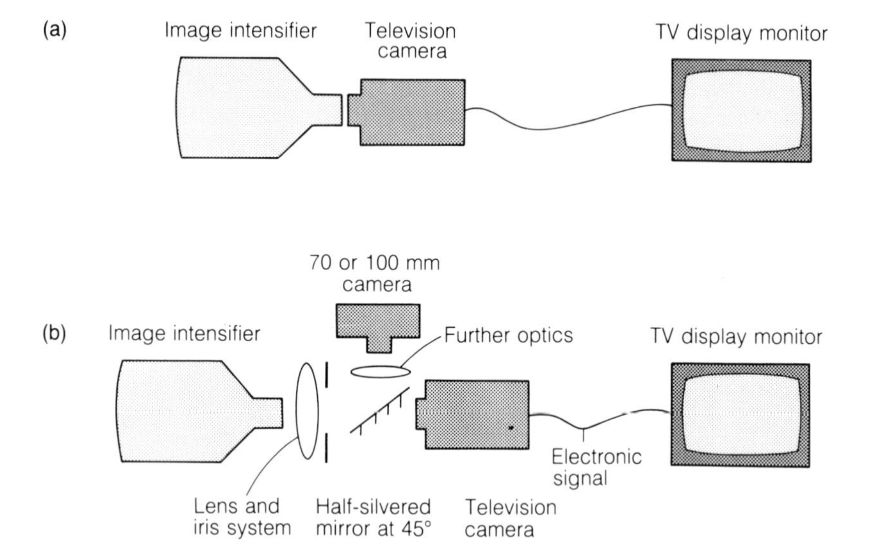

Fig. 4.19. Methods of coupling an image intensifier with a television camera either (a) directly or (b) using a half silvered mirror to deflect part of the image to a cine camera.

vision system allows a large image to be viewed under ambient lighting conditions thus eliminating the necessity for dark adaptation. The light output from the television monitor is well above the threshold for cone vision and no limit on resolution is imposed because of the limitation of the human eye. It is also a very efficient system allowing good optical coupling which results in little loss of information after the input stage of the image intensifier. The video signal can be recorded allowing a permanent record of the investigation to be kept. These records are available for immediate play-back but there is, unfortunately, some loss of information and thus a reduction in image quality during the recording/play back sequence.

The use of 70 mm or 100 mm film or cine film as a substitute for a full sized radiograph has some advantages but also some drawbacks. The quality of the images is high and it is possible to take rapid sequences of films using a motor driven camera. The radiation dose to the patient per film is lower than for a conventional radiograph but this can be negated if large numbers of films are taken. It is also much cheaper to use 70 mm or 100 mm film than standard radiographs. The major drawback is that the film size is small and requires a magnifying viewing system. Separate developing procedures are also required.

If the still camera is replaced by a cine camera, it is possible, because of the half-silvered mirror, to take a cine film and view the image on the television screen simultaneously. This allows recording to be restricted to the information that is really required. Although the final image on cine film is better than the image on a video recorder, it does take some time for it to be developed and thus it cannot be viewed immediately following the recording.

4.10.2 Quantum mottle

An amplification of several thousand is available through a modern image intensifier/TV system. However, this does not mean that the dose of radiation delivered to the patient can be reduced indefinitely. Although the brightness of the image can be restored by electronic means, image quality will be lost.

To understand why this is so, it is important to appreciate that image formation by the interaction of X-ray photons with a receptor is a random process. A useful analogy is rain drops coming through a hole in the roof. When a lot of rain has fallen, the shape of the hole in the roof is clearly outlined by the wet patch on the floor. If only a few drops of rain have fallen it is impossible to decide the shape of the hole in the roof.

Similarly, if sufficient X-ray quanta strike a photoreceptor, they produce enough light photons to provide a detailed image but when fewer X-ray quanta are used the random nature of the process produces a mottled effect which reduces image quality. For example a very fast rare earth screen may give an acceptable overall density before the mottled effect is completely eliminated. Thus the information in the image is related to the number of quanta forming the image.

When there are several stages to the image formation process, as in the image intensifier/TV system, overall image quality will be determined at the point where the number of quanta is least, the so-called 'quantum sink'. This will usually be at the point of primary interaction of the X-ray photons with the fluorescent screen. If insufficient X-ray photons are used to form this image, further amplification is analogous to empty magnification in high power microscopy, being quite unable to restore to the image detail that has already been lost.

This subject will be considered again under the heading 'quantum noise' in section 9.6.

4.11 The vidicon camera

This is responsible for converting the visual information in the image into electronic form. The two main components (Fig. 4.20a) are a focused electron gun and a specially constructed light sensitive surface. As shown in Fig. 4.20b, the light sensitive surface is actually a double layer, the lower of which is the more important. It consists of a photoconductive material, usually antimony trisulphide, but constructed in such a way that very small regions of the photoconductor are insulated one from another by a matrix of mica.

When the camera is directed at visible light, photoelectrons are released from the antimony trisulphide matrix and positive charges are trapped there. The amount of charge trapped at any one point is proportional to the light intensity that has fallen on it. Thus the image information has now been encoded in the relative sizes of the positive charges stored at different points in the image plate matrix.

The pencil beam of electrons now scans across the image plate. As it strikes each pocket of charge, a current pulse proportional to the size of the charge flows through the conducting signal plate. Thus the image data are rapidly converted into a sequence of electrical pulses, each of which corresponds to an exact point on the image plate.

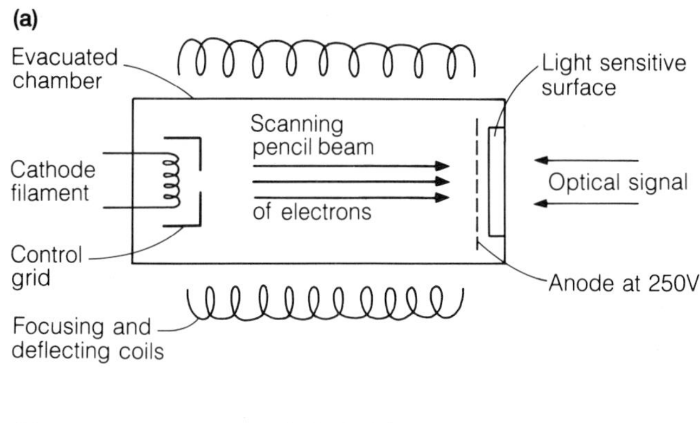

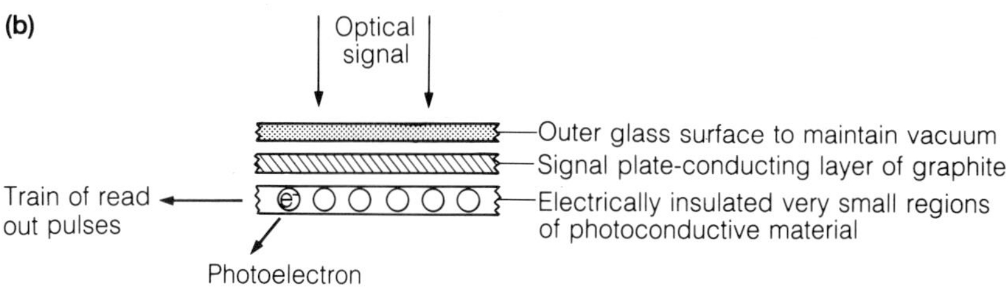

Fig. 4.20. (a) Basic features of the construction of a vidicon camera. (b) The light-sensitive surface shown in more detail.

The reverse process of image reconstruction is very similar. The photoconductive material is now replaced by a fluorescent screen and, as the electron beam scans across the screen, its intensity is modulated by the information in the electrical pulses representing the image.

The following additional points should be noted about a television system:

1 Irrespective of the resolving capability of the image intensifier system, the television system will impose its own resolution limit. This is because, whatever the size of the image, the electron beam only executes a fixed number scan lines (625 in the United Kingdom, equivalent to about 313 line pairs). The effective number of line pairs, for the purpose of determining vertical resolution, is somewhat less than this—probably about 200. Now for a small image, say 5 cm in diameter, this represents four line pairs per mm which is better than the resolving capability of an image intensifier (see section 4.14). However, any attempt to view a full 230 mm diameter (9 inch) screen would provide only about 0.8 line pairs per mm and would severely limit the resolving capability of the complete system. To display large fields of view at high resolution it is necessary to use 35 mm cine film or 100 mm spot film.

2 Contrast is modified by the use of a TV system. It is reduced by the vidicon camera but increased by the television display monitor, the net result being an overall improvement in contrast (for a discussion of this point see section 5.3).

3 Rapid changes in brightness seriously affect image quality when using a television system. Thus a photo cell is incorporated between the image intensifier and the television camera with a feedback loop to the X-ray generator. If there is a sudden change in brightness as result of moving to image a different part of the patient, the X-ray output is quickly adjusted to compensate.

For further information on the Vidicon camera and TV systems the reader should consult *Christensen's* (Curry *et al.* 1984).

4.12 Cinefluorography

The main image forming components of a cine fluorographic system were shown in Fig. 4.19b. Not shown in this diagram are the X-ray tube and generator and the electro mechanical connection which exists between the cine camera and the X-ray generator.

Cinefluorography may place stringent demands on the X-ray tube. For example, although the cine camera may only need to operate at 10 frames per second to see dynamic movements in the stomach, up to 60 frames per second may be required for coronary angiography. If the X-ray tube were emitting X-rays continuously during the period the camera was operating (probably several seconds), the patient would receive an excessively large dose of radiation and there would be rating problems especially if a fine focus spot were being used. Both these problems can be reduced by arranging for the X-rays to be pulsed with a fixed relationship between the pulse and the frame movement in the cine camera as shown in Fig. 4.21. The X-ray pulses are normally between 2 and 5 ms long but the light output from the image intensifier does not fall to zero immediately the X-ray pulse is terminated due to an after-glow in the tube. The after-glow has, however, decreased to zero before the next frame is taken.

The dose in air at the front face of the image intensifier required to produce an acceptable image on the film is usually of the order of 0.2 μGy per frame. Modern units operating at 0.1 μGy per frame have been

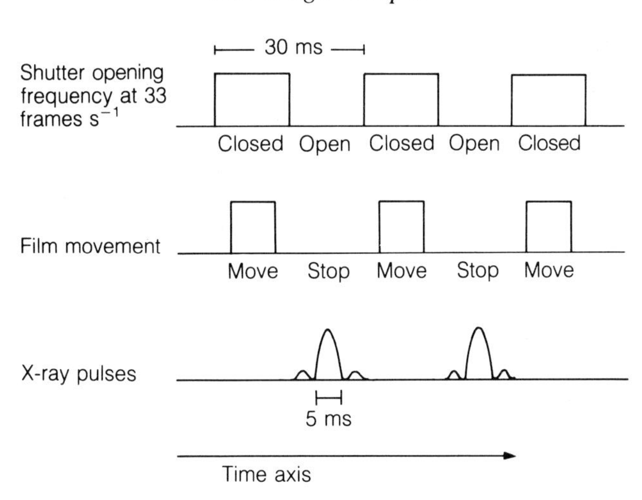

Fig. 4.21. Synchronization of X-ray pulse with film frame exposure during cinefluorography.

reported to give acceptable results but below 0.15 μGy per frame quantum mottle effects generally start to degrade the image.

The images produced during cinefluorography may suffer from all the artefacts of conventional images (see section 5.10). Depending on the investigations being performed, it is sometimes possible to use a small image intensifier with a correspondingly small field of view. This allows an X-ray tube with an anode angle of only 6° to be used, thus giving a greater output from a smaller effective focal spot. This reduces some of the geometric artefacts as well as reducing the heel effect.

In addition to the electronic imaging parts of the system (image intensifier, cine camera), optical imaging processes are also important. Two lens systems are included, the first between the image intensifier output and the half-silvered mirror and the second in front of the cine camera. For a detailed discussion of lens systems the reader is referred elsewhere (e.g. *Christensen's*, Curry *et al.* 1984).

In general terms the function of the optical system is to transfer the image, at appropriate magnification, from the image intensifier screen via the mirror on to the film in the camera. Since the image intensifier screen is circular and a frame of film is rectangular, some mis-match is inevitable. Figure 4.22 shows two extreme examples. Figure 4.22a represents what is known as 'exact framing' with the intensifier screen totally inside the film frame. Figure 4.22b shows the situation known as 'total overframing' where the whole of the film frame is within the screen. Clearly total

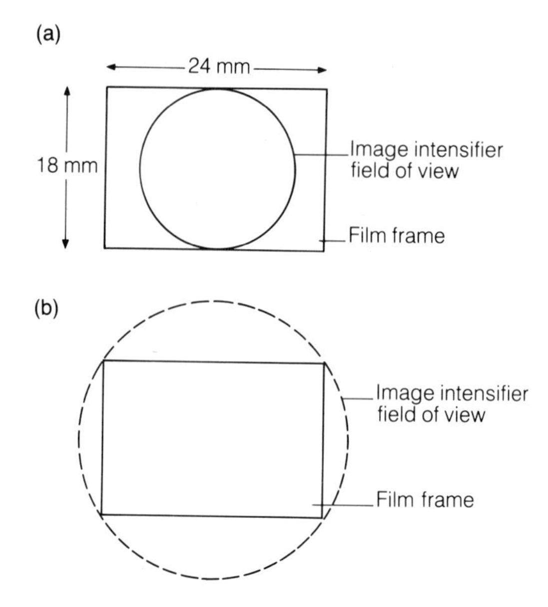

Fig. 4.22. Different framing arrangements with cinefluorography: (a) exact framing, (b) total overframing.

overframing provides a larger image than exact framing but it is restricted to a smaller field of view. Total overframing has the further advantage that the area of the patient exposed to radiation can be reduced. The most probable arrangement is intermediate between these two extremes.

The image intensifier uses a caesium iodide input screen (as discussed in section 4.10) and an output phosphor with maximum output in the wavelength range (550 ± 25) nm. The film used must be sensitive in this wavelength range.

Since cine film has rather a small dynamic range, some form of brightness control is required to compensate for variations in patient thickness. The sensing device, which is analogous to a phototimer, can either measure the current flowing across the image intensifier or the brightness of the output phosphor. This detector then provides feedback control of either the tube kilovoltage, the tube current or the effective exposure time per frame (pulse width). Each control mechanism has disadvantages. If the kilovoltage is driven too high contrast is lost, whereas adjustment of the tube current or the pulse width has only a limited brightness range. Variation of pulse width has a much faster response time than either of the other two methods but in practice a combination of controls is normally chosen.

4.13 Spot films

If spot films are required, the cine camera is replaced by a single frame film camera using either 70 mm or 100 mm film (see section 4.10.1). As this film is larger than cine film, the lens on the front of this camera has a longer focal length than that on the cine camera. The advantages and disadvantages of using spot films for, say, serial angiography can be summarized as follows:

1 There is a reduction in procedure time but because the medical personnel stay close to the patient throughout the investigation, their radiation dose can be greater.

2 Production and processing of spot films is easier and the images can be constantly monitored during the study. More experience is required, however, to learn the panning technique.

3 Reduced patient exposure occurs because shorter exposure times are used. A consequential secondary effect is that equipment wear is reduced.

4 The film is much cheaper and there is less of a film storage problem. The smaller film is however not always easy to read.

5 The technique cannot be used in some investigations such as peripheral angiography or where the inferior resolution of the image intensifier precludes it.

An approximate comparison of patient exposures from the different filming methods is summarized in Table 4.3.

Table 4.3. Approximate exposures (see Chapter 7) for various image-recording systems

		Dose rate in air	
Imaging system	Film size	per exposure or per frame at imager	at the skin surface*
Cine	16–35 mm	0.2–0.5†	20–50
Spot film	70–100 mm	1–3	100–300
Serial radiography	35×35 cm	3–10	300–1000

* An attenuation factor of 100 has been assumed between skin dose and film exposure (this is of the right order for the head). Other factors may be deduced from Table 7.1.

† In each instance the smaller number is an approximate lower limit if perceptible image degradation due to quantum mottle (see sections 4.10.2 and 9.6) is to be avoided.

4.14 Quality control of recording media and image intensification systems

One of the major causes of inferior quality X-ray films is poor processing and strict attention to developing parameters must be paid at all stages of processing. The development temperature must be controlled to better than 0.2°C and the film must be properly agitated to ensure uniform development. All chemicals must be replenished at regular intervals, generally after a given area of film has been processed, and care must be taken to ensure that particulate matter is removed by filtration. Thorough washing and careful drying are essential if discolourations, streaks and film distortion are to be avoided. The whole process can be controlled by the preparation, at regular intervals, of test film strips obtained using a suitable sensitometer. This sensitometer consists of a graded set of filters of known optical density (Fig. 4.23) which is placed over the film and exposed to a known amount of visible light, in a darkened room of course. After processing, the film can be densitometered and a characteristic curve can be constructed (see Fig. 4.8). This may be compared with previous curves and the manufacturer's recommendations. Since the construction of complete characteristic curves is a rather time-consuming process, it is normal only to make three or four spot checks routinely.

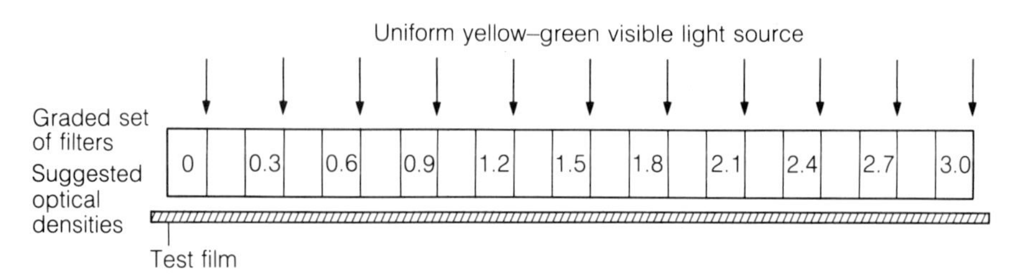

Fig. 4.23. Use of a sensitometer, consisting of a graded set of filters of known optical density, for quality control of film processing.

If sensitometry is used for quality control of the intensifying screen and film in combination, there are a number of additional problems. For example, it is not possible to use a wide range of timed exposures to X-rays to produce different film densities because of failure of the reciprocity law. Perhaps the best method is to use the inverse square law to vary the X-ray intensity reaching the screen, exposing at different distances and plotting film density versus exposure. Note the need for very reproducible exposure times and a very steady tube output. A slightly less accurate method is to use a step wedge of aluminium or copper filters to provide a range of X-ray intensities. If this method is chosen, fairly heavy filtration must be placed in front of the step wedge to minimize change in beam quality with additional filtration, and lead masks must be used to reduce scatter from the step wedge. As a third possibility, a simulated light sensitometer and a range of neutral density filters may be used, provided that the light spectrum is the same as that from the intensifier screen.

Assessment of the performance of image intensification systems introduces a number of problems in addition to those considered in section 2.7 for standard X-ray sets. Among the more important measurements that should be made are:

Field size

A check that the field of view seen on the television monitor is as big as that specified by the manufacturer. The area of the patient exposed to radiation must not exceed that required for effective screening.

Image distortion

The heavy dependence on electronic focusing in the image intensifier and TV system makes distortion much more likely than in a simple radiograph. Measurements on the image of a rectangular grid permit distortion to be checked.

Conversion factor

This is the light output per unit exposure rate, measured in candela per square metre for each $\mu Gy\ s^{-1}$. In general this will be markedly lower for older intensifiers, partly due to ageing of the phosphor and partly because better phosphors are being used in the more modern intensifiers. There is little evidence that the conversion factor affects image quality directly, but of course to achieve the same light output a bigger radiation dose must be given to the patient.

Contrast capability and resolution

A number of test objects, for example the Leeds Test Objects, have been devised to facilitate such measurements. One of them consists of a set of discs, each approximately 1 cm in diameter with a range of contrasts from 12% to 1.8% (for a definition of contrast in these terms see section 9.5). They should be imaged using a 70 kVp beam with 1 mm copper filtration to simulate the patient. At a dose rate of about 0.3 $\mu Gy\ s^{-1}$, a contrast difference of 2–4% (certainly better than 5%) should be detectable. It is important to specify the input dose rate because, as shown in Fig. 4.24, the minimum perceptible contrast difference is higher for both sub-optimum and supra-optimum dose rates. When the dose rate is too low this occurs because of quantum mottle effects. When it is too high there is loss of contrast because the video output voltage begins to saturate.

The limiting resolution is measured by using a test pattern consisting of line pairs, in groups of three or four, separated by different distances, and searching the TV monitor for the minimum resolvable separation. Under high contrast conditions (50 kVp), 1.2 line pairs per mm should be resolvable. Note that the contrast conditions must be controlled carefully since, in common with other imaging systems, the finest resolvable detail varies with contrast level for the image intensifier (see section 9.4.3 and Fig. 9.3).

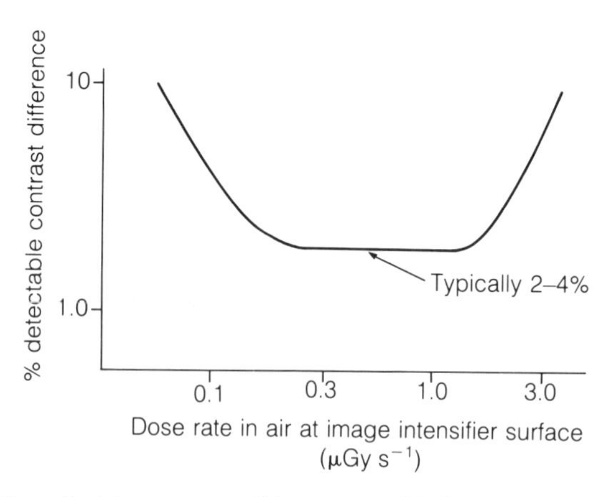

Fig. 4.24. Variation of minimum perceptible contrast with dose rate for an image intensifier screening system.

Automatic brightness control

It is important to ensure that screening procedures do not result in unacceptably high dose rates to the patient. One way to achieve this is by monitoring the light output from the image intensifier, changing the X-ray tube output (kV and/or mA) by means of a feed-back loop, whenever attenuation in the patient changes.

The following additional points should be noted:

1 The light output from an appreciable area of the image intensifier screen, as predetermined by the manufacturer, must be measured.

2 The brightness of the final image can also be controlled by a technique known as automatic gain control, which uses the video signal to adjust amplification factors in the TV monitor without altering exposure factors or dose rate to the patient.

3 Automatic control may mask deterioration in performance somewhere in the system. For example loss of light output from the image intensifier screen could be compensated by increasing mA. X-ray output should therefore be checked directly.

4 Automatic brightness control should be distinguished from automatic exposure control (see section 2.3.6). The latter term is generally used to refer to 'hard copy' control. For example, when using 100 mm film the control might be a photo cell looking at the image intensifier output.

Most systems should be capable of operating in the range 0.5–1 $\mu Gy\ s^{-1}$ but dose rates as high as 5 $\mu Gy\ s^{-1}$ have been reported for incorrectly adjusted equipment! Note that the skin dose to the patient will be, typically, 100 times greater than this—about 500 $\mu Gy\ s^{-1}$ or 30 $mGy\ min^{-1}$.

Viewing screen performance

A calibrated grey-scale step wedge may be used to check that the contrast and brightness settings on the television monitor are correctly adjusted.

Further information on quality control is given in references at the end of the chapter. A good general rule is that the more sophisticated the unit, the greater the care that must be taken over quality control of the images. In many situations a slow, but steady deterioration in picture quality may take place over a period of weeks and this can be difficult to detect unless quantitative measurements are made on a regular basis.

4.15 Summary

A primary X-ray photon image cannot be viewed directly by the human eye and in clinical practice nearly all imaging systems convert this X-ray photon image to a light photon image using one of the many phosphors now available. This light photon image is rarely viewed directly by the human eye and, indeed, on modern X-ray units this undesirable practice is no longer possible.

In general, the light photon image is either recorded on film or is viewed through an image intensifier/TV viewing system and both of these media have been considered in detail in this chapter. Other forms of recording medium are also used, especially in diagnostic imaging techniques other than plane field radiology, and Table 4.4 summarizes the

Table 4.4. Summary of information on different recording media

Recording medium	Use	Special features
Duplitized X-ray film	General purpose radiography	Conventional double-sided X-ray film used with a pair of intensifying screens
Single emulsion film (screen type)	Mammography (sect. 6.6) High definition extremity film	Less sensitive but improved resolution Anti-halo backing prevents light reflection at film base–air interface
Non-screen film	Dental Kidney surgery Radiation monitoring	Direct exposure film for ultra-high resolution Greatly reduced sensitivity
Single emulsion film	Photo fluorography	Spot film/rapid sequence camera film—cut and roll Film sensitive to green light—maximum output of image intensifier screen is 500–600 nm
Film for video imaging	Cathode ray tube imaging Radionuclide imaging (Chapt. 8) Digital subtraction imaging (Chapt. 9) CT (Chapt. 10) Ultrasound (Chapt. 12) MRI (Chapt. 13)	Single emulsion negative film generally best for good resolution and a wide density range. Positive film and Polaroid are alternatives
Subtraction-duplication film	**1** Subtraction masks (section 6.5) **2** Print film **3** Duplication	**1** Gamma must be 1.0 **2** Gamma generally greater than 1.0 to increase contrast **3** Gamma of -1.0
Miscellaneous	Miniaturization Microfilm Video tape	Information reproduction in a smaller format may require special fine-grain development to retain resolution Usually panchromatic for maximum grey-scale range Film speed is not a problem—no patient exposure involved
Special techniques	Xeroradiography (section 6.7) Ionography (section 6.8)	

options available. Some of the other methods for recording images are considered in detail elsewhere in the book.

In all viewing systems the spectral output of the phosphor must be closely matched to the spectral response of the next stage of the viewing system. This allows maximum transfer of image information with minimum radiation dose to the patient. Quantum mottle effects do however impose a lower limit below which further dose reduction produces an image of unacceptable quality. This will occur even if the overall light or density levels are nominally satisfactory.

Each imaging process has its own limits on resolution and contrast and these will be discussed in greater detail in Chapters 5 and 9. However, it is important to emphasize here that the inter-relationship between object size, contrast and patient dose is a matter of everyday experience and not a peculiarity of quality control measurements or sophisticated digital techniques. Furthermore, the choice of imaging process cannot always be governed solely by resolution and contrast considerations. The speed at which the image is produced and is made available for display must sometimes be taken into account.

Imaging equipment using optical and electronic imaging systems must be carefully maintained and all imaging systems should be subject to quality control. With many phosphors it is possible for a slow degradation of the final image to occur. This degradation may only be perceptible if images of a test object taken at regular intervals are carefully compared, preferably using quantitative methods.

References and further reading

Chesney D. N. and Chesney M. O. (1981) *Radiographic Imaging*, 4th edn. Blackwell Scientific, Oxford.

Curry T. S., Dowdey J. E. and Murry R. C. Jr. (1984) *Christensen's Introduction to the Physics of Diagnostic Radiology*, 3rd edn. Lea and Febiger, Philadelphia.

Dressler G., Eriskat H., Schibilla H., Haybittle J. L. and Secretan L. F. (eds) (1985) Criteria and methods of quality assurance in medical X-ray imaging. *Br. J. Radiol.* Suppl. 18.

Hay G. A., Clark O. F., Coleman N. J., Cowen A. R. and Craven D. M. (1979) *Test Equipment for Quality Control in Diagnostic Radiology*. Department of Medical Physics, University of Leeds.

Hospital Physicists' Association (1979) *Quality Assurance Measurements in Diagnostic Radiology*. (Conference Report Series 29) The Hospital Physicists' Association, London.

Hospital Physicists' Association (1981) *Measurement of the Performance Characteristics of Diagnostic X-ray Systems used in Medicine*. Part II 'X-ray image intensifier television systems'. (Topic Group Report 32) The Hospital Physicists' Association, London.

Hospital Physicists' Association (1983) *Measurement of the Performance Characteristics of Diagnostic X-ray Systems used in Medicine*. Part VI 'X-ray image intensifier fluorography systems'. (Topic Group Report 32) The Hospital Physicists' Association, London.

Kodak (1985) *Fundamentals of Radiographic Photography*, Vols 1–6. Kodak Ltd, London.

Exercises

1 Discuss the use of intensifying screens in the cassettes used for radiography.

2 Explain briefly the effect of increasing the kVp from 50 to 100 on the intensification factor of calcium tungstate screens.

3 Explain how the intensification factors of a set of radiographic screens might be compared. Summarize and give reasons for the main precautions that must be taken in the use of such screens.

4 Draw a labelled cross-section of an X-ray photographic film. What features make for high sensitivity?

5 Draw on the same axes the characteristic curves for

(a) a fast film held between a pair of calcium tungstate plates,

(b) the same film with no screens,

and explain the difference between them.

6 Why is it desirable for the gamma of a radiographic film to be much higher than that of a film used in conventional photography and how is this achieved?

7 Explain what is meant by the speed of an X-ray film and discuss the factors on which the speed depends.

8 A radiograph is found to lack contrast. Under what circumstances would increasing the current on the repeat radiograph increase contrast, and why?

9 Make a labelled diagram of the intensifying screen film system used in radiology. Discuss the physical processes that occur from the emergence of X-rays at the anode to the production of the final radiograph.

10 Given that the gamma of an idealized radiographic film–screen combination is 3.5, and the range of acceptabe film densities is 2.8, what is the maximum ratio of exposures for which the combination can be used?

11 Discuss the factors which affect the sensitivity and resolution of a screen–film combination used in radiography and their dependence upon each other.

12 How does the difference in diameter of the input and output screens of an X-ray intensifier contribute to the performance of the system?

13 What advantages are associated with a caesium iodide input phosphor in an image intensifier?

14 Discuss the uses made of the brightness amplification available from a modern image intensifier, paying particular attention to any limitations.

15 Compare and contrast the use of fluorescent screens in radiography and fluoroscopy.

16 Discuss the limitations 'quantum mottle' imposes on both image intensifying systems and sensitive film–screen systems. Why are these limits not always reached?
17 Explain how an image intensifier may be used in conjunction with a photoconductive camera to produce an image on a TV screen.
18 Explain what is meant by cinefluorography and give a brief descripton of the functions of the various parts of such a system.

5

The Radiological Image

5.1 Introduction—the meaning of image quality

As shown in Chapter 3, the fraction of the incident X-ray intensity transmitted by different parts of a patient will vary due to variations in thickness, density and mean atomic number of the body. This pattern of transmitted intensities therefore contains the information required about the body and can be thought of as the primary image.

However, this primary image cannot be seen by the eye and must first be converted to a visual image by interaction with a secondary imaging device. This change can be undertaken in several ways, each of which has its own particular features. The definition of quality for the resultant image in practical terms depends on the information required from it. In some instances it is resolution that is primarily required, in others the ability to see small increments in contrast. More generally the image is a compromise combination of the two, with the dominant one often determined by the personal preference of the radiologist. (This preference can change; the 'contrasty' crisp chest radiographs of several years ago are now rejected in favour of lower contrast radiographs which appear much flatter but are claimed to allow more to be seen.)

The quality of the image can depend as much on the display system as on the way it was produced. A good quality image viewed under poor conditions such as inadequate non-uniform lighting may be useless. The quality actually required in an image may also depend on information provided by other diagnostic techniques or previous radiographs.

This chapter extends the concept of contrast introduced in Chapter 4 to the radiological image and then discusses the factors that may influence or degrade the quality of the primary image. Methods available for improving the quality of the information available at this stage are also considered. Other factors affecting image quality in the broader context are discussed in Chapter 9.

5.2 The primary image

The primary image produced when X-ray photons pass through a body depends on the linear attenuation coefficient (μ) and the thickness of the tissue they traverse. At diagnostic photon energies μ is dependent on the photoelectric and Compton effects. For soft tissue, fat and muscle the effective atomic number varies from approximately 6 to 7.5. In these materials μ is thus primarily dependent on the Compton effect, which falls relatively slowly with increasing photon energy (see Chapter 3). The photoelectric effect is not completely absent however and at low photon energies forms a significant part of the attenuation process. In mammography low energy X-ray photons are used to detect malignant soft tissue which has a very similar *Z* value to breast tissue. The difference in attenuation between the two is due to the higher photoelectric attenuation of the higher *Z* material. For bone with a *Z* of approximately 14, most of the attenuation is by the photoelectric effect. This falls rapidly with increasing photon energy.

In absolute terms the Compton effect also decreases with increasing energy and the resultant fall in attenuation coefficient with photon energy for tissue and for bone is shown in Fig. 5.1.

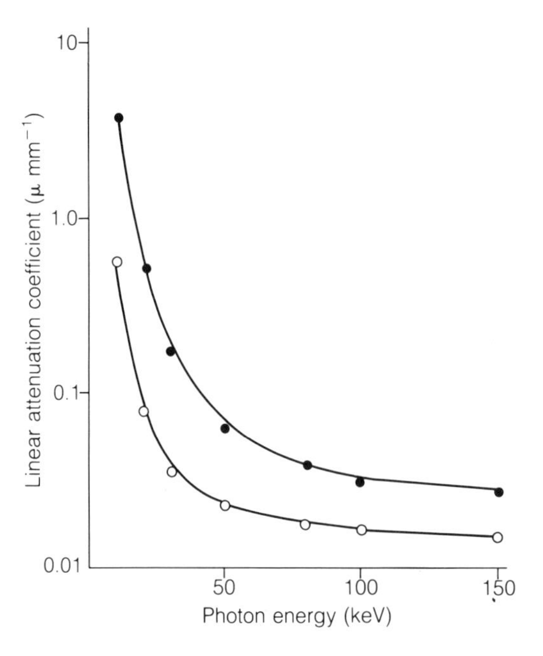

Fig. 5.1. Variation of linear attenuation coefficient with photon energy for (○) muscle and (●) bone in the diagnostic range.

5.3 Contrast

The definition of contrast differs somewhat depending on the way the concept is being applied. For conventional radiology and fluoroscopy, the normal definition is an extension of the definition introduced in Chapter 4. In Chapter 9 an alternative approach will be considered.

Consider the situation shown in Fig. 5.2. This is clearly very similar to that in Fig. 4.6. Contrast in the primary image will be due to any difference between X_1 and X_2 and by analogy with equation 4.1 may be defined as

$$C = \log_{10} \frac{X_2}{X_1}$$

Converting to Naperian logs:

$$C = 0.43 \ln \frac{X_2}{X_1} = 0.43 (\ln X_2 - \ln X_1)$$

Since, from Chapter 3

$$X_1 = X_0 \exp (-\mu_1 x_1)$$

and

$$X_2 = X_0 \exp (-\mu_2 x_2)$$

thus

$$C = 0.43 (\mu_1 x_1 - \mu_2 x_2)$$

If μ_1 and μ_2 were the same, the difference in contrast would be due to differences in thickness. If $x_1 = x_2$ the contrast is due to differences in linear attenuation coefficient. It is conceivable that the product $\mu_2 x_2$ might be exactly equal to $\mu_1 x_1$ but this is unlikely. Note from Fig. 5.1 that the difference in μ values decreases on moving to the right, thus contrast between two structures always decreases with increasing kVp.

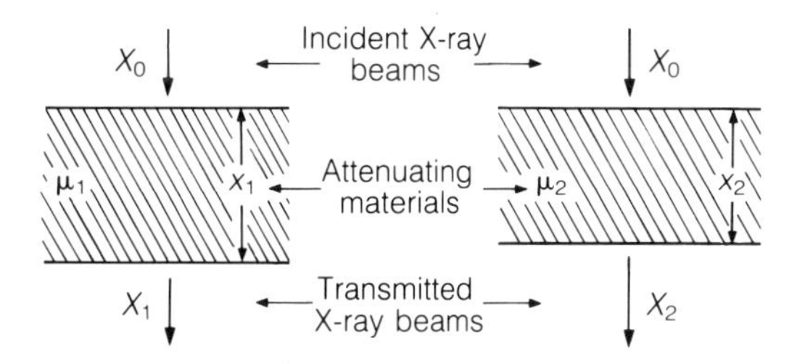

Fig. 5.2. X-ray transmission through materials that differ in both thickness and linear attenuation coefficient.

5.3.1 Contrast on a fluorescent screen

If this primary image is allowed to fall on a fluorescent screen, the light emitted from those parts of the screen exposed to X_1 and X_2 say L_1 and L_2 will be directly proportional to X_1 and X_2. Hence

$$L_1 = kX_1 \text{ and } L_2 = kX_2$$

The contrast

$$C\,(\text{screen}) = \log \frac{L_1}{L_2} = \log \frac{kX_1}{kX_2} = \log \frac{X_1}{X_2}$$

Hence the contrast on the screen, $C = 0.43\,(\mu_1 x_1 - \mu_2 x_2)$, is exactly the same as in the primary image. As this is how the eye perceives the image, this is known as the radiation contrast and will be denoted by C_R.

Note that simple amplification, in a fluorescent or intensifying screen, does not alter contrast.

5.3.2 Contrast on a radiograph

If the transmitted intensities X_1 and X_2 are converted into an image on radiographic film, the contrast on the film will be different from that in the primary image because of the imaging characteristics of the film.

As shown in section 4.4.2 the imaging characteristics of film are described by its characteristic curve. By definition

$$\gamma = \frac{D_2 - D_1}{\log E_2 - \log E_1}$$

So for X radiation

$$\gamma = \frac{D_2 - D_1}{\log X_2 - \log X_1} \qquad \text{(equation 5.1)}$$

Now from section 5.3

$$\log X_2 - \log X_1 = 0.43\,(\mu_1 x_1 - \mu_2 x_2)$$

and from section 4.4.2

$$D_2 - D_1 = C$$

Hence substituting in equation 5.1

$$C = \gamma\, 0.43\,(\mu_1 x_1 - \mu_2 x_2)$$

Thus the contrast on film C_F differs from the contrast in the primary image by the factor γ which is usually in the range 3–4. Gamma is often termed the film contrast, thus

$$\begin{array}{ccccc} \text{Radiographic contrast} & = & \text{Radiation contrast} & \times & \text{Film contrast} \\ C_F & = & C_R & \times & \gamma \end{array}$$

Note that contrast is now modified because the characteristic curve relates two logarithmic quantities. Film can be said to be a 'logarithmic amplifier'. A TV camera can also act as a logarithmic amplifier.

5.3.3 Origins of contrast for real and artificial media

As discussed in Chapter 3, attenuation and hence contrast will be determined by differences in atomic number and density. Typical values for normal tissues are shown in Table 5.1.

Table 5.1. Mean atomic number and density for the major body constituents

Material	Mean atomic number Z	Density kg $m^{-3} \times 10^3$
Bone	13.8	1.8
Soft tissue } Muscle }	7.4	1.0
Fat	6.0	0.9
Lung	7.4	0.25
Air	7.6	Almost 0

Contrast may be changed artificially by introducing materials with either a different atomic number or with a different density. Thus barium compounds ($Z = 56$) are used for studies of the stomach and colon, iodine compounds ($Z = 53$) for the kidney or ureter, and the density may be modified by the introduction of gas, e.g. air or CO_2 for pneumography of ventricles in the brain.

Note that modification of atomic number is a very kVp dependent process whereas modification of density is not.

The number of 'contrast' materials available is limited by the requirements for such materials. They must have a suitable viscosity and persistence and must be miscible or immiscible as the examination requires. Most importantly they must be non-toxic. Iodine-based contrast materials do carry a risk for certain individuals. One contrast material, used for many years, contained thorium which is a naturally radioactive substance. It has been shown in epidemiological studies that patients investigated using this contrast material had an increased chance of contracting cancer or leukemia.

5.4 Effects of overlying and underlying tissue

Under scatter free conditions, it may be demonstrated that a layer of uniformly attenuating material either above or below the region of differential attenuation has no effect on contrast. Consider the situation shown in Fig. 5.3 where the two regions are shown separated for clarity.

Radiation contrast

$$C_R = \log_{10} \frac{X_2'}{X_1'} = 0.43 \ln \frac{X_2'}{X_1'}$$

Now

$$X_1' = X_0' \, e^{-\mu_1 x_1}$$

and

$$X_2' = X_0' \, e^{-\mu_2 x_2}$$

Hence

$$C_R = 0.43 \ln \frac{X_0' e^{-\mu_2 x_2}}{X_0' e^{-\mu_1 x_1}}$$

Thus

$$C_R = 0.43 \, (\mu_1 x_1 - \mu_2 x_2) \text{ as before.}$$

The fact that some attenuation has occurred in overlying tissue and that

$$X_0' = X_0 e^{-\mu_0 x_0}$$

is irrelevant because X_0' cancels.

A similar argument may be applied to uniformly attenuating material below the region of interest.

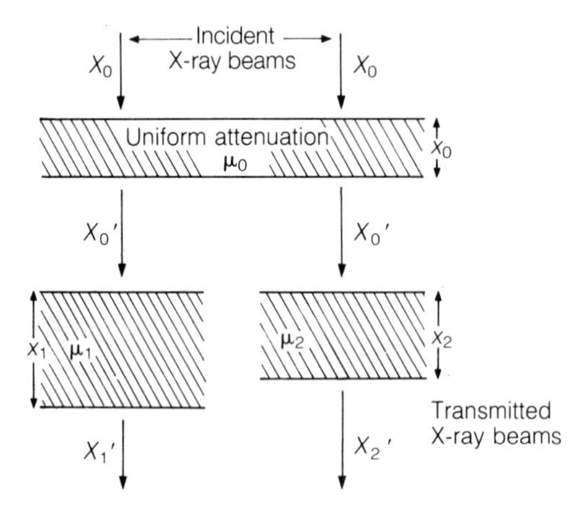

Fig. 5.3. Diagram showing the effect of an overlying layer of uniformly attenuating material on the X-ray transmitted beams of Fig. 5.2.

5.5 Reduction of contrast by scatter

In practice, contrast is reduced by the presence of overlying and underlying material because of scatter. The scattered photons arise from Compton interactions. They are of reduced energy and travel at various angles to the primary beam.

The effect of this scatter, which is almost isotropic, is to produce a uniform increase in blackening across the film. It may be shown, quite simply, that the presence of scattered radiation of uniform intensity invariably reduces the radiation contrast.

If $C_R = \log_{10}(X_2/X_1)$ and a constant X_0 is added to the top and bottom of the equation to represent scatter,

$$C_R' = \log_{10} \frac{X_2 + X_0}{X_1 + X_0}$$

The value of C_R' will be less than the value of C_R for any positive value of X_0.

The presence of scatter will almost invariably reduce contrast in the final image for the reason given above. The only condition under which scatter might increase contrast would be for photographic film if X_1 and X_2 were so small that they were close to the fog level of the characteristic curve. This is a rather artificial situation.

The amount of scattered radiation can be very large relative to the unscattered transmitted beam. This is especially true when there is a large thickness of tissue between the organ or object being imaged and the film. The ratio of scatter to primary beam in the latter situation can be as high as 8:1 but in more normal examinations is probably in the range 2–4.

5.6 Variation in scatter with photon energy

If it is necessary to increase the kVp to compensate for loss of intensity, the amount of scatter reaching the film increases. This is the result of a complex interaction of factors, some of which increase the scatter, others decrease it.

1 The amount of scatter actually produced in the patient is reduced because:

(a) the amount of scatter produced decreases as the kVp is increased, although the Compton interaction coefficient only decreases slowly in the diagnostic range;

(b) a smaller amount of primary radiation is required to produce a given density on the film as film density is proportional to $(kVp)^4$.

2 However, the forward scatter leaving the patient will be increased because:

(a) the fraction of the total scatter produced going in a forward direction increases as the kVp rises;

(b) the mean energy of the scattered radiation increases and thus less of it is absorbed by the patient.

In practice, as the kVp rises from 50 to 100 kVp, the fall in linear attenuation coefficient of the low energy scattered radiation in tissue is much more rapid than the fall in the scatter-producing Compton cross-section. Thus factor **2**(b) is more important than factor **1**(a) and this is the prime reason for the increase in scatter reaching the film.

The increase in scatter is steep between 50 and 100 kVp but there is no increase at higher kVp and above 140 kVp the amount of scatter reaching the film does start to fall slowly.

5.7 Reduction of scatter

There are several ways in which scatter can be reduced.

Careful choice of beam parameters

A reduction in the size of the beam to the minimum required to cover the area of interest reduces the volume of tissue available to scatter X-ray photons.

A reduction of kVp will not only increase contrast but will also reduce the scatter reaching the film. This reduction is however limited by the patient penetration required.

Orientation of the patient

The effect of scatter will be particularly bad when there is a large thickness of tissue between the region of interest and the film. When the object is close to the film it prevents both the primary beam and scatter reaching the film. The object stops scatter very effectively since the energy of these photons is lower than those in the primary beam. Thus the region of interest should be as close to the film as possible—see also section 5.10 on geometric effects

Compression of the patient
This is a well-known technique that requires some explanation. It is important to appreciate that the process a physicist would call compression, for example a piston compressing a volume of gas, will not reduce scatter. Reference to Fig. 5.4 will show that the X-rays encounter exactly the same number of molecules in passing through the gas on the right as on the left. Hence there will be the same amount of attenuation and the same amount of scatter.

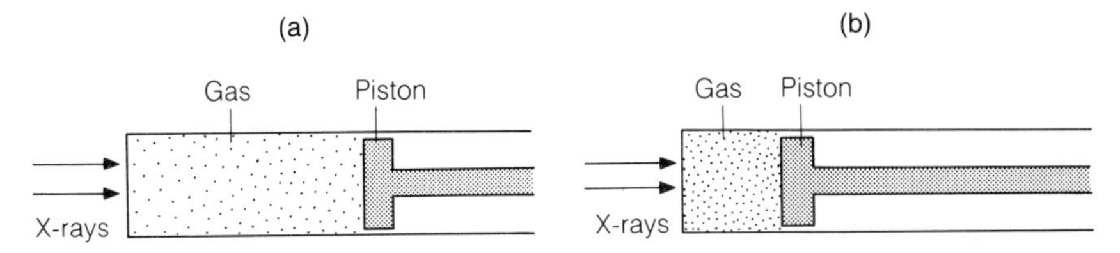

Fig. 5.4. Demonstration that compression in the physical sense will not alter the attenuating properties of a fixed mass of gas. (a) Gas occupies a large volume at low density. (b) Gas occupies a much smaller volume at high density.

When a patient is 'compressed', soft tissue is actually forced out of the primary beam, hence there is less scattering material present and contrast is improved.

Use of grids
This is the most effective method for preventing the scatter leaving the patient from reaching the film and is discussed in the next section.

Air gap technique
If the patient is separated from the film, some obliquely scattered rays may miss the film. This technique is discussed more fully in section 6.3.

Design of intensifying screen and film holder
Since some radiation will pass right through the film, it is important to ensure that no X-ray photons are back-scattered from the film holder. A high atomic number metal backing to the film cassette will ensure that all transmitted photons are totally absorbed by the photoelectric effect at this point.

The slightly greater sensitivity of intensifying screens to the higher energy primary photons is a marginal benefit in reducing the effect of scatter on the film.

5.8 Grids

5.8.1 Construction

A grid is an array of long parallel lead strips held an equal distance apart by a material with a very low *Z* value. Most of the scattered photons, travelling at an angle to the primary beam, will not be able to pass through the grid but will be intercepted and absorbed as shown in Fig. 5.5. Some of the rays travelling at right angles, or nearly right angles, to the grid are also stopped due to the finite thickness of the grid strips, again shown in Fig. 5.5. Both the primary beam and the scatter are stopped in this way but the majority of the primary beam passes through the grid along with some scattered radiation. Scatter travelling at an angle of $\theta/2$ or less to the primary beam is able to pass through the grid to point P (Fig. 5.6).

As grids can remove up to 90% of the scatter, there is a large increase in the contrast in radiographs when a grid is used. This increase is expressed in the 'contrast improvement factor' *K* where

$$K = \frac{\text{X-ray contrast with grid}}{\text{X-ray contrast without grid}}$$

K normally varies between 2 and 3 but can be as high as 4. The higher values of *K* are normally achieved by increasing the number of grid strips per centimetre. As these are increased, more of the primary beam is removed due to its being stopped by the grid. The proportion stopped is

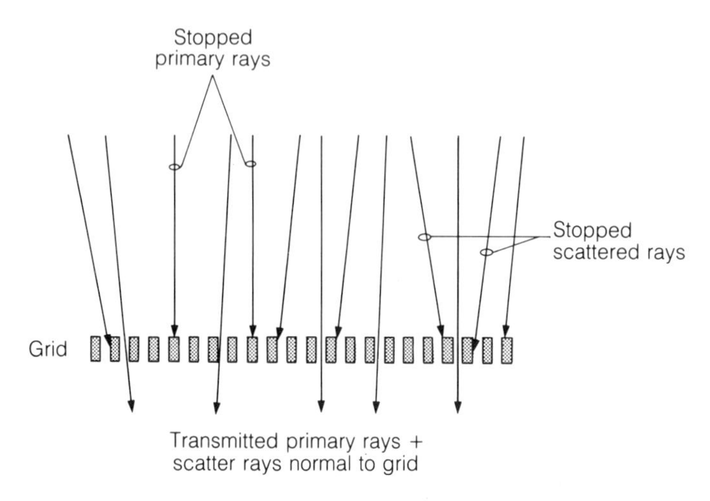

Fig. 5.5. Use of a grid to intercept scattered radiation.

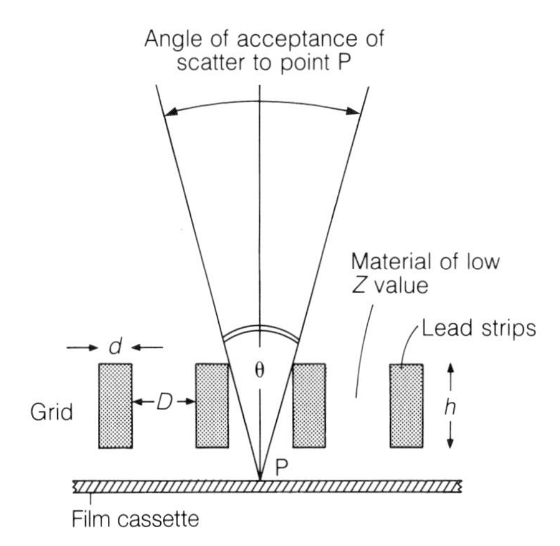

Fig. 5.6. Grid geometry. Number of strips per mm, $N = 1/(D + d)$; typically N is about 4 for a good grid. Grid ratio $r = h/D$; typically r is about 10. Fraction of primary radiation removed from the beam $= d/(D + d)$. Since d might be 0.075 mm and $(D + d)$ 0.25 mm, $d/(D + d)$ will be about 0.3. Tan $\theta/2 = D/h$.

given by

$$\frac{d}{d + D}$$

where d is the thickness of a lead strip and D is the distance between them (Fig. 5.6). Reduction of the primary beam intensity means that the exposure must be increased to compensate. The use of a grid therefore increases the radiation dose to the patient and thus there is a limit on the number of strips per centimetre that can be used.

5.8.2 Use of grids

As grids are designed to stop photons travelling at angles other than approximately normal to them, it is essential that they are always correctly positioned with respect to the central ray of the primary beam. Otherwise, as shown in Fig. 5.7, the primary beam will be stopped. The fact that the primary photon beam is not parallel but originates from a point source limits the size of film that can be exposed due to interception of the primary beam by the grid (Fig. 5.8). The limit is at ray C where

$$\tan \psi = \frac{C}{FFD}$$

Tan ψ can be calculated from the grid characteristics. By similar triangles,

$$\tan \psi = \frac{D}{h}$$

where D is the distance between strips and h is the height of a strip.

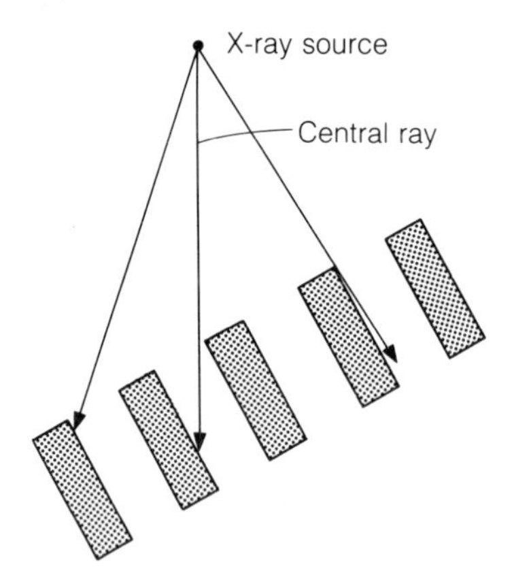

Fig. 5.7. Diagram showing that a grid which is not orthogonal to the central X-ray axis may obstruct the primary beam.

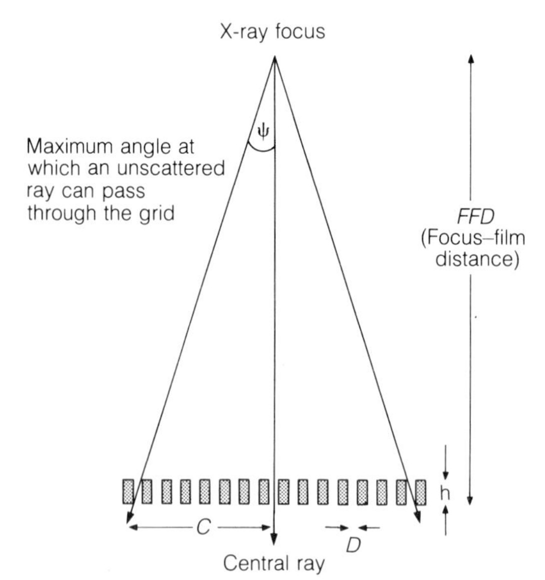

Fig. 5.8. Demonstration that the field of view is limited when using a grid.

The grids used most widely are like those shown in previous figures and are termed linear grids, but other grids are available. One is the focused grid where the lead strips are progressively angled on moving away from the central axis (Fig. 5.9). This eliminates the problem of cut off at the periphery of the grid but imposes a very restrictive condition on the *FFD*s than can be used. Crossed grids are also sometimes used with two sets of strips at right angles to each other. This combination is very effective for removing scatter but absorbs a lot more of the primary beam and requires a

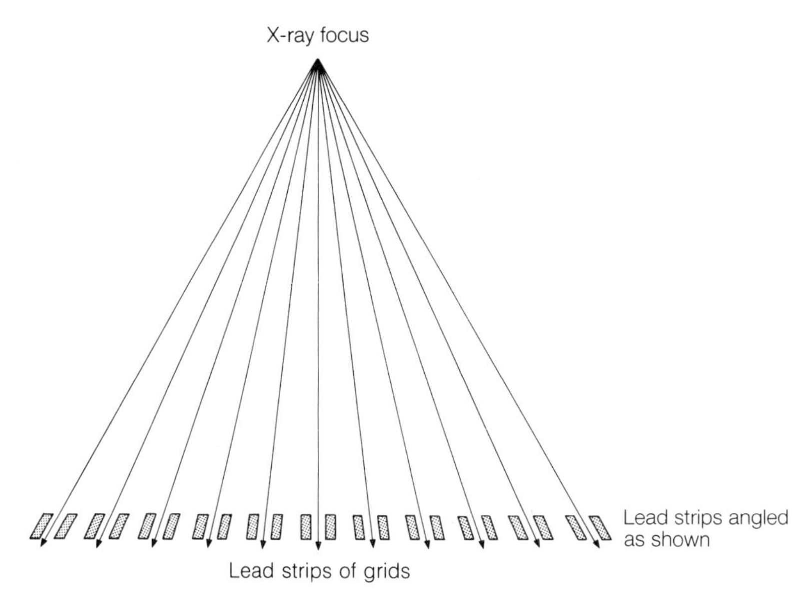

Fig. 5.9. Construction of a focused grid.

much larger increase in the exposure with consequent increase in patient dose.

5.8.3 Movement of grids

If a stationary grid is used it imposes on the patient a radiograph of the grid as a series of lines (due to the absorption of the primary beam). With modern very fine grids this effect is reduced but not removed.

The effect can be overcome by moving the grid during exposure so that the image of the grid is blurred out. Movements on modern units are generally oscillatory, often with the speed of movement in the forward direction different to that on the return. Whatever the detailed design, the movement should be such that it starts before the exposure and continues beyond the end of the exposure. Care must also be taken to ensure that, in single phase machines, the grid movement is not synchronous with the pulses of X-rays from the tube. If this occurs, although the grid has moved between X-ray pulses, the movement may be equal to an exact number of lead strips. The lead strips in the grid are thus effectively in the same position as far as the radiograph is concerned. This is an excellent example of the stroboscopic effect.

5.9 Resolution and unsharpness

The resolution of a radiological image depends on three factors, each associated with a different part of the imaging system. These are geometric unsharpness, patient unsharpness and the resolution of the final imager.

An ultimate limit on resolution is provided by the inherent resolution of the image recording system. As discussed in Chapter 9 the resolution of film is much better than even the best film–screen combination. The resolution of a fluorescent screen used in fluorography is about 0.25 mm (four line pairs per mm). That of the most up-to-date image intensifier/television systems is about 1 mm. In CAT scanning, resolution is limited by the size of the pixels used on the screen.

Except on rare occasions in the case of fluoroscopic screens, the imaging devices used in radiography are not the major cause of loss of resolution. Of greater significance are geometric unsharpness and inherent patient unsharpness.

5.9.1 Geometric unsharpness

Geometric unsharpness is produced because the focal spot of an X-ray tube has finite size (Fig. 5.10). Although the focal spot has a dimension, *b*, on the anode, the apparent size of the focal spot for the central X-ray beam, *a*, is much reduced due to the slope on the anode. The dimension normal to the plane of the paper is not altered. If, as shown in Fig. 5.10, a sharp X-ray opaque edge is placed directly under the centre of the focal spot, the image of the edge is not produced directly underneath at T but extends from S to U where S is to the left of T and U to the right of T. By analogy with optics, the shadow of the object to the left of S is termed the **umbra**, the region SU is the **penumbra**. This penumbra, in which on moving from S to U the number of X-ray photons rises to that in the unobstructed beam, is termed the **geometric unsharpness**. The magnitude of SU is given by

$$SU = b \sin \alpha \frac{d}{(FFD - d)} \qquad \text{(equation 5.2)}$$

For target angle = 13°, b = 1.2 mm, FFD = 1 m and d = 10 cm then SU = 0.06 mm. Since the focal spot size is strictly limited by rating considerations (Chapter 2), a certain amount of geometric unsharpness is unavoidable.

The actual size of the image on the radiograph is only altered significantly when the size of the object to be radiographed approaches or is less than the size of the focal spot.

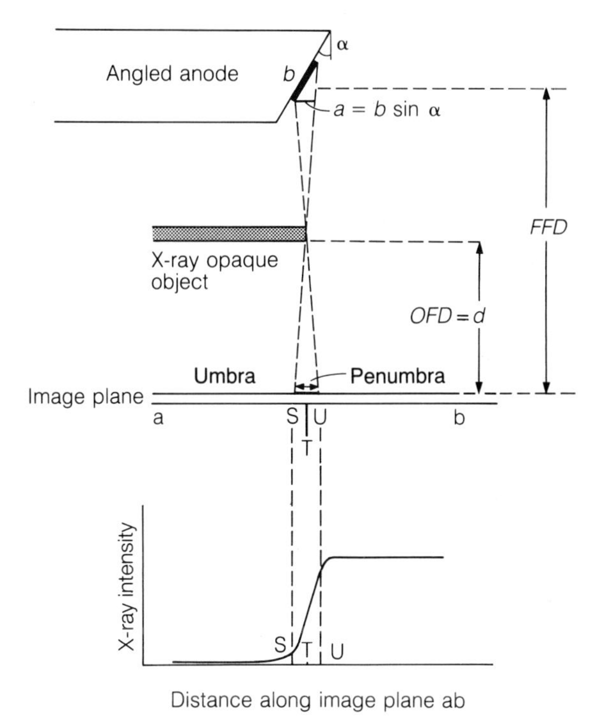

Fig. 5.10. The effect of a finite X-ray focal spot size in forming a penumbral region. *FFD* = focus–film distance. *OFD* = object–film distance.

5.9.2 Patient unsharpness

Other sources of unsharpness can arise when the object being radiographed is not idealized, i.e. infinitely thin yet X-ray opaque. When the object is a patient, or part of a patient, it has a finite thickness generally with decreasing X-ray attenuation towards the edges. These features can be considered as part of the geometric unsharpness of the resulting image and are in fact often much larger than the geometric unsharpness described in section 5.9.1. The effect on the number of photons transmitted is shown in Fig. 5.11. As can be seen there is a gradual change from transmission to absorption, producing an indistinct edge.

Another source of unsharpness arises from the fact that during a radiograph many organs within the body can move either through involuntary or voluntary motions. This is shown simply in Fig. 5.12, where the edge of the organ being radiographed moves from position A to position B during the course of the exposure. Again the result is a gradual transition

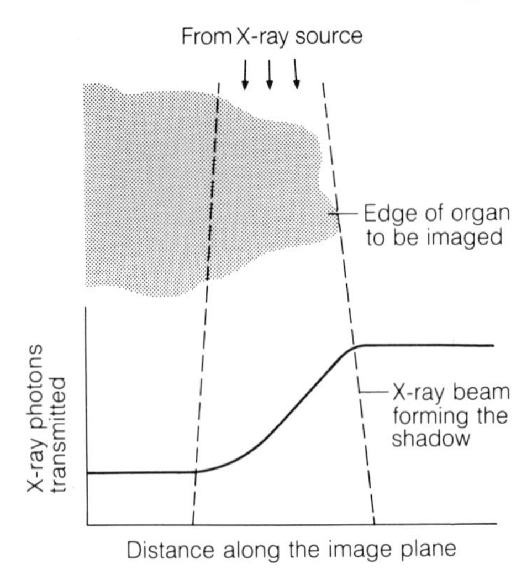

Fig. 5.11. Contribution to image blurring that results from an irregular edge to the organ of interest.

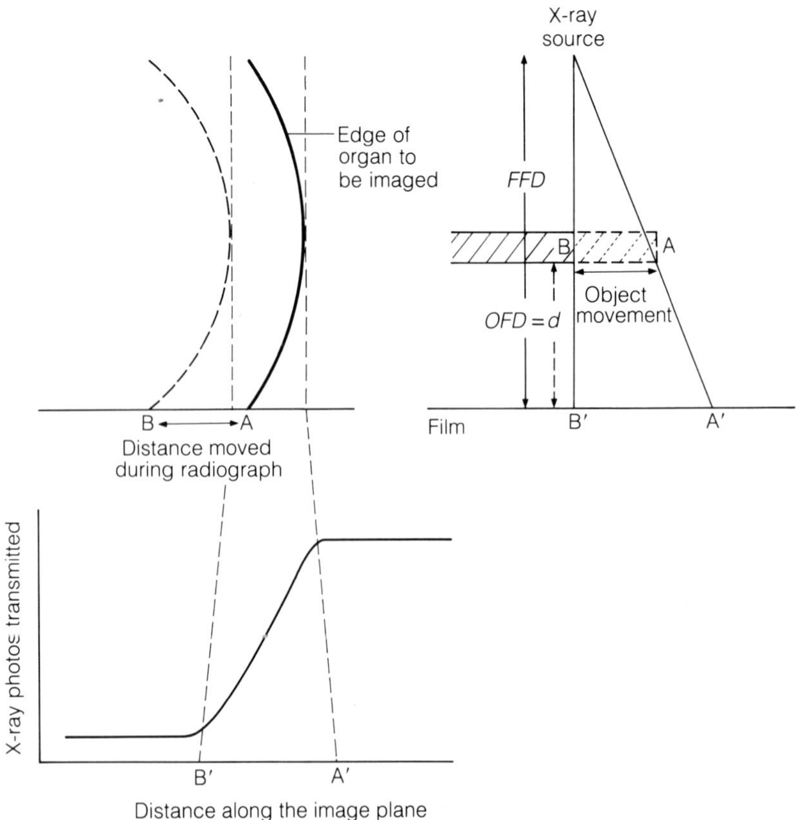

Fig. 5.12. Effect of movement on radiographic blurring. If the object moves with velocity v during the exposure time t, AB = vt and A′B′ = AB·$FFD/(FFD - d)$.

of film density resulting in an unsharp image of the edge of the organ. The main factors that determine the degree of movement unsharpness are the speed of movement of the region of interest and the time of exposure. Increasing the patient–film distance increases the effect of movement unsharpness.

5.9.3 Combining unsharpnesses

It will be apparent from the preceding discussion that in any radiological image there will be several sources of unsharpness and the overall unsharpness will be the combination of all of them.

Unsharpnesses are combined according to a power law with the power index varying between 2 and 3. The power index 2 is most commonly used and should be applied where the unsharpnesses are all of the same order. The power index 3 should be used if one of the unsharpnesses is very much greater than the rest.

Note that, because of the power law relationship, if one contribution is very large it will dominate the expression. If this unsharpness can be reduced at the expense of the others the minimum overall unsharpness will be when all contributions are approximately equal.

For example if the geometric unsharpness is U_G, the movement unsharpness is U_M, the film/screen unsharpness U_F then the combined unsharpness of the three is

$$U = \sqrt{U_G^2 + U_M^2 + U_F^2}$$

If $U_G = 0.5$ mm, $U_M = 1.0$ mm, $U_F = 0.8$ mm,

then $U = 1.37$ mm.

If $U_G = 0.7$ mm, $U_M = 0.7$ mm, $U_F = 0.8$ mm,

then $U = 1.27$ mm.

5.10 Geometric relationship of film, patient and X-ray source

The interpretation of radiographs eventually becomes second nature to the radiologist who learns to ignore the geometrical effects which can, and do, produce very distorted images with regard to the size and position of organs in the body. Nevertheless, it is important to appreciate that such distortions do occur.

Most of the effects may be easily understood by assuming that the focus is a point source and that X-rays travel in straight lines away from it.

5.10.1 Magnification without distortion

In the situation shown in Fig. 5.13 the images of three objects of equal size lying parallel to the film are shown. The images are not the same size as the object. Assume magnifications M_1, M_2 and M_3 for objects 1, 2 and 3 respectively given by

$$M_1 = \frac{\mathrm{AB}}{\mathrm{ab}} \qquad M_2 = \frac{\mathrm{XY}}{\mathrm{xy}} \qquad M_3 = \frac{\mathrm{GH}}{\mathrm{gh}}$$

Consider triangles Fxy and FXY. Angles xFy and XFY are common; xy is parallel to XY. Triangles Fxy and FXY are thus similar. Therefore

$$\frac{\mathrm{XY}}{\mathrm{xy}} = \frac{FFD}{FFD - d} \qquad \text{(equation 5.3)}$$

By the same considerations triangles aFb and AFB are similar and triangles gFh and GFH are similar

$$\frac{\mathrm{AB}}{\mathrm{ab}} = \frac{FFD}{FFD - d}$$

and

$$\frac{\mathrm{GH}}{\mathrm{gh}} = \frac{FFD}{FFD - d}$$

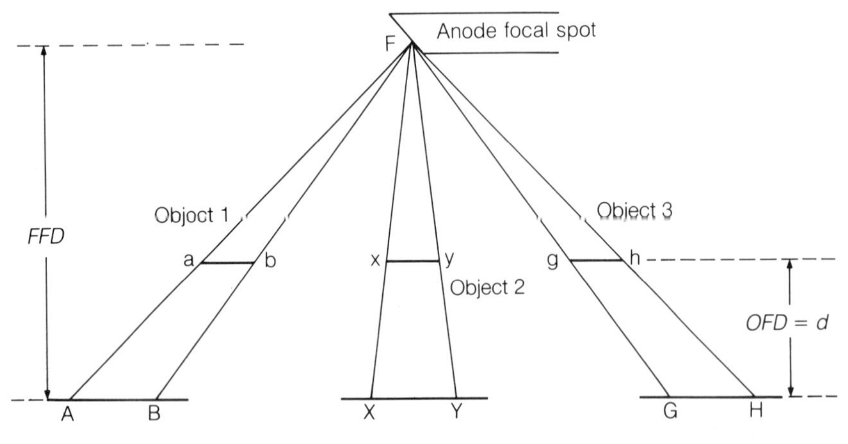

Fig. 5.13. Demonstration of a situation in which magnification without distortion will occur.

Therefore

$$M_1 = M_2 = M_3$$

i.e. For objects in the same plane parallel to the film, magnification is constant. This is magnification without distortion.

The magnification increases as:

1 the *FFD* is decreased,

2 d is increased.

The deliberate use of magnification techniques is discussed in section 6.4.

5.10.2 Distortion of shape and/or position

In general the rather artificial conditions assumed in section 5.10.1 do not apply when real objects are to be radiographed. For example, they are not infinitessimally thin, they are not necessarily orientated normal to the principal axis of the X-ray beam and they are not all at the same distance, measured along the principal axis, from the X-ray source. All of these factors introduce distortions into the resulting image. Figure 5.14a shows the distortion that results from twisting a thin object out of the horizontal plane, and Fig. 5.14b shows distortion for objects of finite thickness. Although all the spheres are of the same diameter, the cross section area projected parallel to the film plane is now greater if the sphere is off-axis and the image is enlarged more (compare A and B). Note that when the sphere is in a different plane (e.g. C) the distortion may be considerable. Figure 5.14b also shows that when objects are in different planes, distortion of position will occur. Although C is nearer to the central axis than D, its image actually falls further away from the central axis.

A certain amount of distortion of shape and position is unavoidable and the experienced radiologist learns to take such factors into consideration. The effects will be more marked in magnification radiography.

5.11 Review of factors affecting the radiological image

It is clear from this and the preceding chapter that a large number of factors can affect a radiological image and these will now be summarized.

Choice of tube kilovoltage

A high kV gives a low contrast and a large film latitude and vice versa. X-ray output from the tube is approximately proportional to $(\text{kVp})^2$ and film

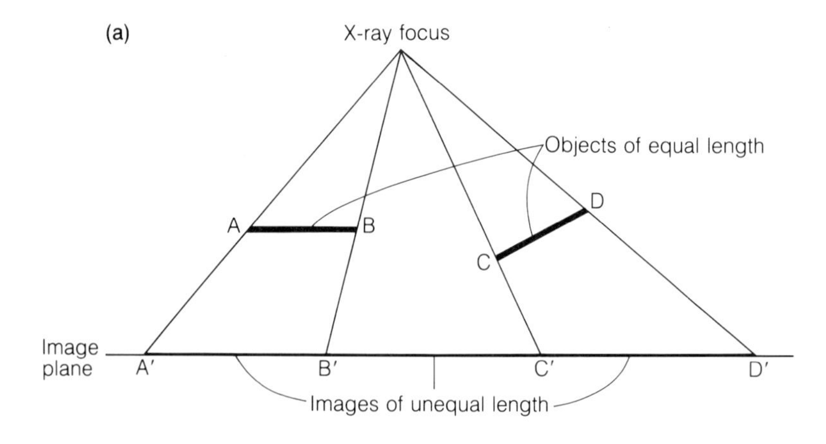

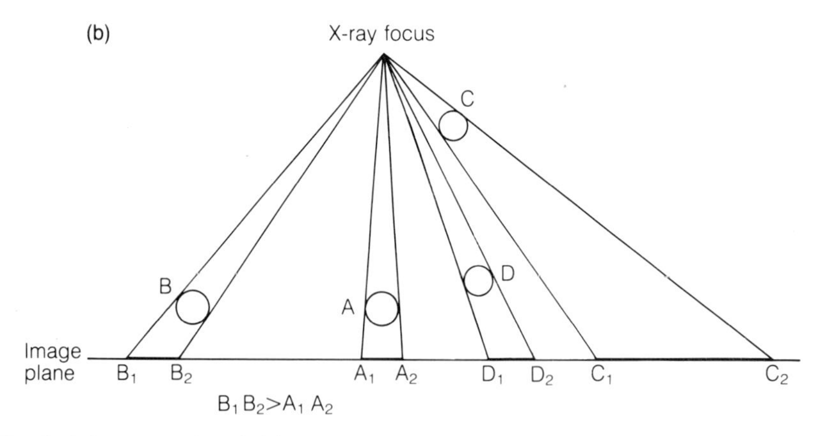

Fig. 5.14. Demonstration of: (a) distortion of shape of an object; (b) distortion of shape when objects are of finite thickness, (A, B, C) and of relative position when they are at different depths (C, D).

blackening proportional to $(kVp)^4$. The higher the kV the lower the patient dose. A higher kV however increases the amount of scatter.

Increasing the kV allows the tube current or the exposure time to be reduced. At 70 kVp an increase of 10 kVp allows tube current to be approximately halved (or the exposure time to be halved).

Exposure time

In theory, the exposure time should be as short as possible to eliminate movement unsharpness. If movement will not be a problem, exposure time may be increased so that other variables can be optimized.

Focal spot size
As the spot size increases so does geometric unsharpness. The minimum size (from a choice of two) should be chosen consistent with the choice of other factors which affect tube rating (kVp, mA, s).

Quality of anode surface
As discussed in Chapter 2, damage to the anode surface will result in a non-uniform X-ray intensity distribution and an increased effective spot size. Because of the heel effect there is always some variation in X-ray beam intensity in the direction parallel to the anode–cathode. If a careful comparison of the blackening produced by two structures is required they should be orientated at right angles to the electron flux from the cathode to anode.

Tube current
In an ideal situation all other variables would be chosen first and then an mA would be selected to give optimum film blackening. If this is not possible, for example because of rating limits, the system becomes highly interactive and the final combination of variables is a compromise to give the best end result.

Beam size
This should be as small as possible, commensurate with the required field of view, to minimize patient dose and scatter. Note that, strictly speaking, collimation reduces the integral dose, i.e. the dose × the volume irradiated, rather than the dose itself.

Grids
These must be used if scatter is significantly reducing contrast, e.g. when irradiating large volumes. Use of grids requires an increased mA s thus increasing patient dose.

Focus–film and object–film distance
A large focus–film distance reduces geometric blurring, magnification and distortion, but the intensity at the patient is reduced because of the inverse square law. The working distance is thus governed eventually by the tube rating. The object–film distance is not really controllable but is kept as small as possible by equipment manufacturers. Movement and geometric blurring and magnification can be influenced by patient orientation either anterior/posterior or posterior/anterior.

Contrast enhancement
Modification of either the atomic number of an organ, e.g. by using barium or iodine-containing contrast agents, or of its density, e.g. by introducing a gas, alters its contrast relative to the surrounding tissue.

Films and screens
There is generally only a limited choice. A fast film–screen combination will minimize patient dose, geometric and movement unsharpness. Associated screen unsharpness and quantum mottle may be high. If extremely fine detail is required, a non-screen film may be used but this requires a much higher mA s and thus gives a higher patient dose.

Film processing
Although this is now standardized, it should not be overlooked. Quality control of development is extremely important as it can have a profound effect on the radiograph. Bad technique at this stage can completely negate all the careful thought given to selecting correct exposure and position factors.

It will be clear from this lengthy list that the quality of a simple, plain radiograph is affected by many factors, some of which are interactive. Each of them must be carefully controlled if the maximum amount of diagnostic information is to be obtained from the image.

Further reading

Curry T. S., Dowdey J. E. and Murry R. C. Jr. (1984) *Christensen's Introduction to the Physics of Diagnostic Radiology*, 3rd edn. Lea and Febiger, Philadelphia.
Gifford D. (1984) *A Handbook of Physics for Radiologists and Radiographers*. Wiley, Chichester.
Hay G. A. (1982) Traditional X-ray imaging. In *Scientific Basis of Medical Imaging* (ed. P. N. T. Wells), pp. 1–53. Churchill Livingstone, Edinburgh.
Meredith W. J. and Massey J. B. (1977) *Fundamental Physics of Radiology*, 3rd edn. Wright, Bristol.

Exercises

1 What is meant by contrast in a radiograph?
2 Why is contrast reduced by scattered radiation?

3 The definition of contrast used in radiography cannot be used in nuclear medicine. Discuss the reasons for this.

4 What are the advantages and disadvantages of having a radiographic film with a high gamma?

5 A solid bone 7 mm diameter lies embedded in soft tissue. Ignoring the effects of scatter, calculate the contrast between the bone (centre) and neighbouring soft tissue.
Film Gamma = 3.
Linear attenuation coefficient of bone = 0.5 mm^{-1}.
Linear attenuation coefficient of tissue = 0.04 mm^{-1}.

6 How can the effect of scattered radiation on contrast be reduced?

7 Why is a low kVp used to take a mammogram?

8 Give a sketch showing how the relative scatter (scattered radiation as a fraction of the unscattered radiation) emerging from a body varies with X-ray tube kV between 30 kVp and 200 kVp and explain the shape of the curve. What measures can be taken to minimize loss of contrast due to scattered radiation?

9 How can X-ray magnification be used to enhance the detail of small anatomical structures? What are its limitations?

10 What are the advantages and disadvantages of an X-ray tube with a very fine focus?

11 List the factors affecting the sharpness of a radiograph. Draw diagrams illustrating these effects.

12 A radiograph is taken of a patient's chest. Discuss the principal factors that influence the resultant image.

13 What factors, affecting the resolution of a radiograph, are out of the control of the radiologist? (Assuming the radiographer is performing as required).

14 A radiograph is found to lack contrast. Discuss the steps that might be taken to improve contrast.

6

Special Radiographic Techniques

6.1 Introduction

In Chapter 2 the basic principles of X-ray production were presented, and Chapter 3 dealt with the origin of radiographic images in terms of the fundamental interaction processes between X-rays and the body. Chapters 4 and 5 showed how the radiographic image is converted into a form suitable for visual interpretation.

On the basis of the information already presented, it is possible to understand the physics of most simple radiological procedures. However, a number of more specialized techniques are also used in radiology and these will be drawn together in this chapter. These techniques provide an excellent opportunity to illustrate the application to specific problems of principles already introduced, and appropriate references will be made to the relevant sections in earlier chapters.

6.2 Mobile X-ray generators

Clearly there are a number of situations in which it is not practicable to take the patient to the X-ray department and the X-ray set must be taken to the patient. The term mobile X-ray generator applies to machines which can be moved around a hospital but which cannot be dismantled or carried. The latter are strictly called portable X-ray generators and are not considered here. Although mobile units generally take conventional films, there is also a role for mobile image intensifiers both in theatres, for example in orthopaedic practice, and on the wards, for example to monitor the progress of an endoscope as it is inserted into the patient. The manoeuvrability of, say, a C-arm image intensifier allows the radiologist to position the arm so that only the region of interest is irradiated, thereby reducing the dose to adjacent organs. Note, however, that a screening procedure should not be adopted if the same information can be obtained from a few short conventional exposures because the latter will almost certainly result in a lower dose to the patient.

Three basic categories of mobile unit are available and these can be distinguished by the type of X-ray generator used namely; single phase full wave rectified, constant potential and capacitor discharge. The fact that such units must be mobile introduces additional constraints and limitations and these will be the main features of interest in this section.

6.2.1 Single phase full wave rectified generators

These machines operate essentially as ordinary single phase generators powered by the hospital mains with an output wave form as shown in Fig. 6.1. The output and rating limits of a single phase fixed unit described in section 2.5 apply equally to mobile units but the output is dependent in addition on hospital mains supply. Because high currents are drawn, a separate 30 A ring main must be installed for these units. This is expensive and obviously imposes limits on where they can be used. Even with a specially provided circuit, the voltage delivered to the primary of the high voltage transformer, and hence the voltage to the X-ray tube, is dependent on the relative electrical impedances of the input to the transformer and the mains circuit. A mismatch can result in errors of up to 20 kV in the actual kVp produced. Although some units match impedances automatically, most require the operator to make manual adjustments.

Outwith the control of the operator are sudden fluctuations in voltage that can occur in the hospital mains supply following the connection or disconnection of other electrical appliances that draw large currents.

The power output from these machines is limited and a full range of exposures cannot be obtained from them.

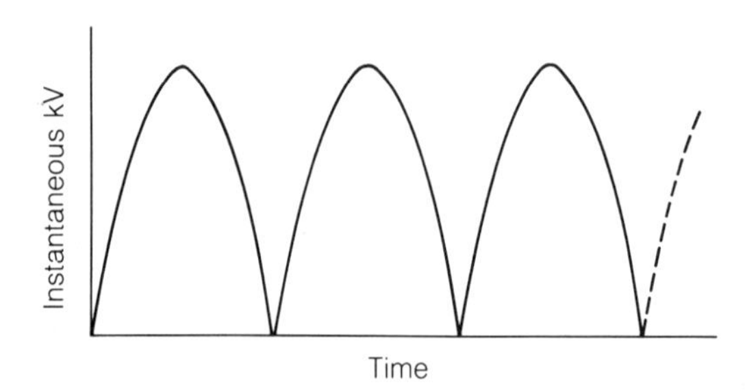

Fig. 6.1. Instantaneous kV of a fully rectified mobile X-ray unit.

6.2.2 Constant potential generators

This category of generator can be further subdivided into two types. The first is truly independent of an electricity supply as it is powered by batteries (which do, however, require charging at a central location). The second type is operated from the normal 13 A hospital mains supply.

Battery powered generators generally use a nickel-cadmium battery which can store a charge equivalent to about 10,000 mA s at normal operating voltages. The direct current voltage of approximately 130 V from this battery must be converted to an alternating voltage before it can be used to supply the generator transformer. This conversion is carried out by an electrical device called an 'invertor'. The AC voltage produced by this invertor differs significantly from the mains AC voltage in that it is at 500 Hz, some 10 times greater than the normal mains frequency. At this high frequency the transformer is more efficient and as a consequence can be much smaller.

Although the transformer output is basically single phase full wave rectified, at 500 Hz the capacitance inherent in the secondary circuit smoothes the output to that shown in Fig. 6.2 so the kV generated is essentially constant. The generator produces a fixed tube current of 100 mA which, as well as simplifying design, allows exposures to be calculated with ease. The battery is depleted and the voltage falls from one exposure to the next so some form of compensation must be applied until the unit is recharged. This compensation can be applied either automatically or manually. Recharging takes place, when necessary, at a low current from any hospital 13 A mains supply plug.

The second type of constant voltage mobile generator is rather sophisticated. The charge for the exposure is stored on a very large capacitor and this charge is then used as a DC supply to an invertor producing a 4.5 kHz supply for the X-ray generator. The large capacitor can be charged using a battery or from the normal hospital 13 A mains supply. During an exposure, the voltage of the X-ray tube is monitored and the output controlled using a microprocessor in such a way that the output is essentially constant as shown in Fig. 6.2.

For both these types of generator the output is continuous, so exposures are shorter than for a single phase mobile unit. Furthermore the output is independent of fluctuations in the hospital mains supply. The output of the battery operated units is low (10 kW) but because of their

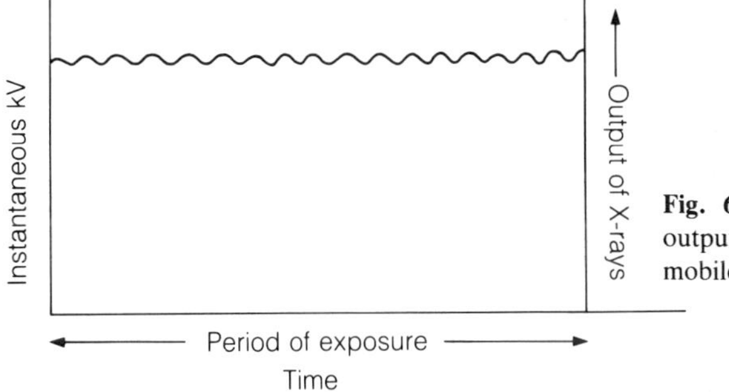

Fig. 6.2. Instantaneous kV and output for constant potential mobile units.

good stability they are particularly suited for chest radiography or premature baby units. Mains driven capacitor units have a higher power (23 kW) and thus can be used in any situation.

6.2.3 Capacitor discharge units

In these units the capacitor is used in a quite different way from that in the constant output units. In the latter, charge is fed from the capacitor through an invertor into what is essentially a conventional X-ray tube generator system. In a capacitor discharge unit, the capacitor, which now always has a value of 1 μF, is used quite differently, being connected directly to a grid-controlled X-ray tube. In this type of X-ray tube there is a third electrode shield or grid placed on the cathode assembly. If this electrode is maintained some 2 kV negative relative to the cathode filament it stops the tube discharging by repelling any electrons emitted from the cathode even when the capacitor is fully charged. This grid control can be turned on and off independently providing instantaneous control of the X-ray tube current and very precisely timed exposures. Note that grid control is useful in any X-ray unit where extremely short exposures (a few milliseconds) are required. It is also useful where rapid repeat exposures are required because the switching mechanism is without inertia.

The capacitor can be charged, at low current, from any hospital 13 A mains supply. Charging will continue until the capacitor reaches a preset kilovoltage. Once an exposure has started the mA s delivered must be monitored and the exposure is terminated as required.

The operating kV of a capacitor discharge unit is high at the start of the exposure and relatively low at the end (Fig. 6.3). This is because the

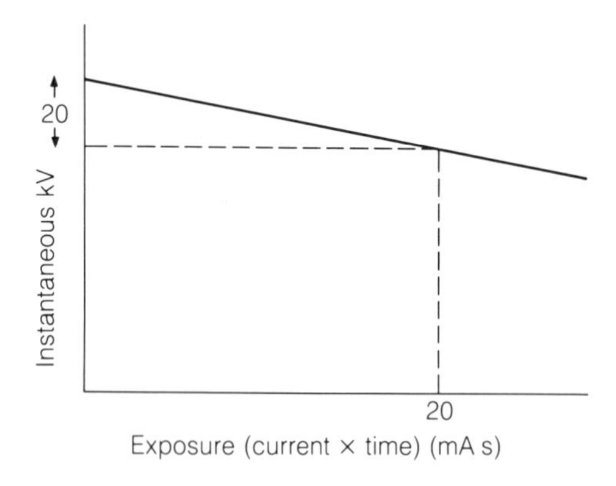

Fig. 6.3. Variation of kV with exposure during exposure for a capacitor discharge mobile unit.

kilovoltage across the tube is reduced as charge is taken off the capacitor. If the capacitor has a value of 1 μF, the reduction in kilovoltage is 1 kV per 1 mA s of exposure. The output falls accordingly during the exposure and failure to realize this, and the cause, has often led to mistakes being made in the setting of exposure factors with these machines. Consequently they have acquired an unjustified reputation of lacking in output.

The equivalent kilovoltage of a capacitor discharge unit, i.e. that setting on a constant potential kilovoltage machine which would produce the equivalent radiographic effect, can be shown to be approximately equal to the starting voltage minus one third of the fall in tube voltage which occurs during the exposure. As the tube voltage drop is numerically equal to the mA s selected for a 1 μF capacitor

$$\text{equivalent kV} = \text{starting kV} - \tfrac{1}{3}\,\text{mA s}$$

If therefore one has a radiographic exposure setting of 85 kV and 30 mA s, the equivalent voltage is 75 kV. If this exposure is insufficient and an under-exposed radiograph is produced, simply increasing the mA s may not increase the blackening on the film sufficiently. For example suppose the exposure is increased to 50 mA s at the same 85 kV. The equivalent kV will now be only 68 kV. Since the equivalent kV is less, much of the increase in mA s will be used to provide soft radiation which does not contribute to the radiograph. The appropriate action is to change the exposure factors to 92 kV and 50 mA s, thereby maintaining the equivalent kV at 75 kV but increasing the exposure to 50 mA s.

If it is desired to reduce the starting kV after the capacitor has been charged or when radiography is finished, the capacitor must be discharged. When the 'discharge' button is depressed, an exposure takes place at a low mA for several seconds until the required charge has been lost. During this exposure the tube produces unwanted X-rays. These are absorbed by a lead shutter which moves across the light beam diaphragm. An automatic interlock ensures that the tube cannot discharge without the lead shutter in place to intercept the beam (Fig. 6.4). It should be noted, however, that this shutter does not absorb all the X-rays produced, especially when the discharge is taking place from a high kV. Neither patient nor film should be underneath the light beam diaphragm during the discharge operation.

Capacitor discharge units require more operator training to ensure optimum performance than other portable units, but when used under optimum conditions they have a sufficiently high output to permit acceptably short exposures for most investigations.

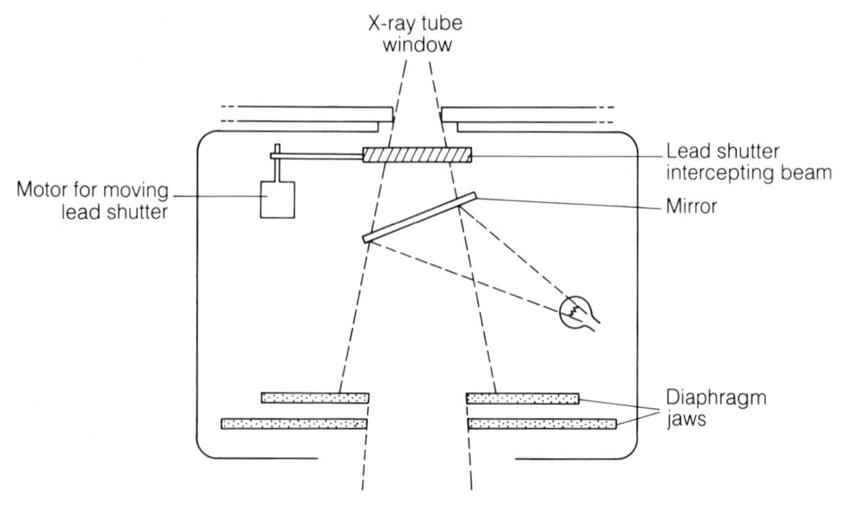

Fig. 6.4. Arrangement of the lead shutter for preventing exposure during capacitor discharge and the light-beam diaphragm assembly on a capacitor discharge mobile [illegible].

6.3 High voltage radiography

Increasing the generator voltage to an X-ray tube has a number of effects, some of which are desirable with respect to the resulting radiograph, some are undesirable. Among the desirable features are increased X-ray output per mA s, more efficient patient penetration, reduced dose to the patient and more efficient film blackening. More scattered radiation reaches the film and this is clearly a disadvantage. Finally, the fact that radiographic contrast falls with increasing tube kilovoltage is generally a disadvantage, except when it is necessary to accommodate a wide patient contrast range. These effects may all be illustrated by considering the technique used for chest radiography. (See, for example, Gayler 1979.)

An operating kilovoltage of 60–70 kVp is a sound choice for small or medium-sized patients. It gives a good balance of contrast, good bone definition and a sharp, clean appearance to the pulmonary vascularity. However, for larger patients, say in excess of 25 cm anterior–posterior diameter, both attenuation and scatter become appreciable. Tube current can only be increased up to the rating limit, then longer exposure times are required. Also if a scatter-reducing grid is used, the dose to the patient is increased appreciably. Finally, use of a grid in this low kV region may enhance contrast excessively, resulting in areas near the chest wall and in the mediastinum being very light and central regions being too dark.

The problems encountered when a high voltage technique (say 125–150 kVp) is adopted are different. Tube output and patient penetration are good, thus for example operating conditions of 140 kVp and 200–500 mA allow short exposure times to be used. Note, however, that the X-ray tube may not tolerate consistent use at high voltage and the manufacturer should be told at the time of installation if the tube is to be used in this way. The kV of the generator output should be checked regularly since the tube will be operating near to its electrical rating limit. High tension cables may develop problems more frequently than at low kV.

Films will be of markedly lower contrast and may have an overall grey appearance. Thus it will be possible to accommodate a much wider range of object contrast but rib detail will not be as distinct. If a patient's anterior–posterior diameter is markedly different over the upper and lower chest, an aluminium wedge filter may be used to compensate for differences in tissue absorption.

Some form of scatter reducing technique must be used. If this is by means of a grid a 10:1 grid ratio is probably a good compromise. A higher grid ratio may improve image quality even more for a few large patients but unless the grid is changed between patients, all patients will receive a lot more radiation.

An alternative method to reduce scatter reaching the film is the air gap technique, in which the film is moved about 15–20 cm from the patient. Increased film contrast is better because

1 scattered radiation generated at an oblique angle misses the film,
2 inverse square law effects are favourable.

It is not clear at first sight why the second factor should reduce the proportion of scatter, since the exposure factors will also have to be increased to allow for the inverse square law. However, the fractional increase in focal spot film distance, which determines the inverse square law effect on the primary beam, is small. Scatter originates at the patient, so this inverse square law effect is determined by the fractional increase in the distance of the mid-plane of the patient to the film which is large. The increase in skin dose with the air gap technique is less than when using a grid. Note the following additional points about the air gap technique:

1 the effect of finite focal spot size (say 0.6 mm or 1.2 mm) on image sharpness decreases, i.e. the image gets sharper as the focus–film distance increases;

2 the effect of shadowing sharpness (penumbra) increases as the patient–film distance increases.

Although the air gap technique would appear to have a number of advantages, it is not widely used, perhaps because the position of the film holder relative to the couch has to be changed.

Finally, although this discussion has been presented in terms of low voltage (60–70 kVp) versus high voltage (125–150 kVp) techniques, intermediate voltages can of course be used, with the consequent mix of advantages and disadvantages.

6.4 Magnification radiography

As discussed in section 5.10.1, an object radiographed onto film is magnified in the ratio

$$\frac{\text{focus–film distance}}{\text{focus–object distance}} = \frac{FFD}{FFD - d} \qquad \text{(equation 5.3)}$$

This geometry is reproduced in Fig. 6.5a for ease of reference. Note that magnification, $M = 1$ only if the object is in contact with the film, i.e. when $d = 0$.

For most routine X-ray examinations, M is kept as small as possible. This is because, as discussed in section 5.9.1, if the focal spot is of finite size, which is always the case in practice, a penumbra proportional to $d/(FFD - d)$ is formed, so the penumbra is also absent only if $d = 0$. Comparison of Fig. 6.5 b and c, shows that for a fixed value of d, the size of the penumbra depends on the size of the focal spot.

In practice structures of interest are never actually in contact with the film. Taking a typical value of $d = 10$ cm, then for a FFD of 100 cm, $M \simeq 1.1$. Sometimes, for example when looking at small bones in the extremities or in some angiographic procedures, it is useful to have a magnified image. One way to achieve this is to magnify, using optical means, a standard radiograph. However, this approach is often not satisfactory as it produces a very grainy image with increased quantum noise (see sections 4.10.2 and 9.6). The alternative is to increase the value of M, and for the purpose of the present discussion it will be assumed that this is effected by increasing d whilst keeping FFD constant. Although increasing d achieves the desired result, this has a number of other consequences as far as the radiographic process is concerned and these will now be considered.

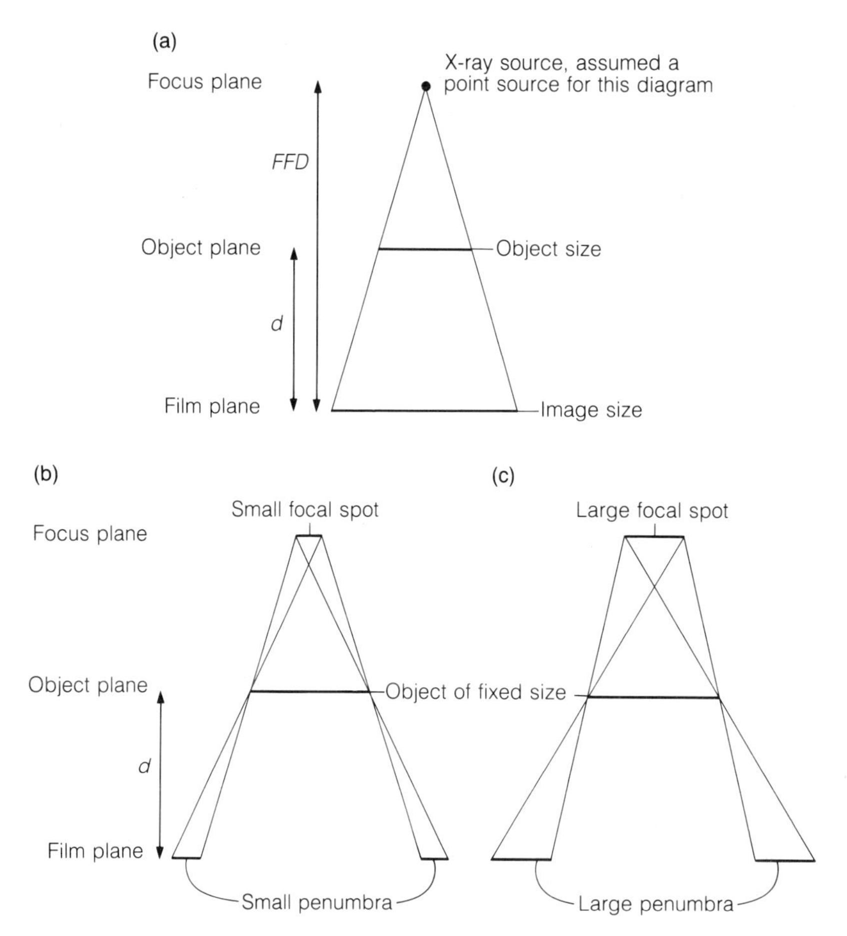

Fig. 6.5. Geometrical arrangements for magnification radiography. (a) Assuming a point source of X-rays, then by similar triangles the magnification $M = FFD/(FFD - d)$. (b) and (c) demonstrate that for an object of fixed size and a fixed magnification, the size of the penumbra increases with the size of focal spot.

Focal spot size

As shown in Fig. 6.5 b and c, the size of the penumbra depends on focal spot size. If it is assumed that the penumbra is part of the magnified image, it may be shown by simple geometry that the true magnification is equal to

$$M + (M - 1)\ F/xy$$

where F is the focal spot size and xy the size of the object. Thus when M is large and F is of the order of xy, the penumbra contributes significantly to the image.

In order that this geometric (penumbral) unsharpness is kept to a minimum, the focal spot used for magnification radiography must be as small as possible. Spots larger than 0.3 mm are little use and 0.1 mm is preferable. This imposes severe rating problems and specially cooled tubes are often required.

A focal spot of 0.3 mm or less is not easy to measure accurately and its size may vary with the tube current by as much as 50% of the expected value (see section 2.3.1). A pin hole may be used to measure the spot size (section 2.7) but to estimate the resolution it is best to use a star test pattern. The performance of a tube used for magnification radiography is very dependent on a good focal spot and careful, regular quality control checks must be carried out. It can be difficult to maintain a uniform X-ray intensity across the X-ray field using a very small focal spot. The intensity distribution may be either greater at the edge than in the centre or, conversely, higher in the middle than at the edge. Such irregularities can cause difficulties in obtaining correct exposure factors.

Film screen unsharpness

Whereas magnification has a deleterious effect on unsharpness due to a finite focal spot size, screen unsharpness is in fact reduced by magnification. To understand the reason for this, consider image formation for a test object that consists of eight line pairs per mm. If the object is in contact with the screen ($M = 1$) the screen must be able to resolve eight line pairs per mm, which is beyond the capability of fast screens. Now suppose the object is moved to a point mid-way between the focal spot and screen ($d = FFD/2$ and $M = 2$). The object is now magnified at the screen to four line pairs per mm, thereby making the imaging task easier.

Movement unsharpness

One further important source of image degradation in magnification radiography is movement unsharpness (see section 5.9.2). If an object is moving at 5 mm s^{-1}, and the exposure is 0.02 s, then the object moves 0.1 mm during the exposure. If the object is in contact with the film ($M = 1$) and the required resolution is four line pairs per mm, corresponding to a separation of 0.25 mm, then movement of this magnitude will not seriously affect image quality. However, as shown in Fig. 5.12, the effect of object movement at the film depends on d. If $d = FFD/2$, i.e. $M = 2$, the shadow of the object at the film will move 0.2 mm and this may cause significant degradation of a system attempting a resolution of 0.25 mm. Hence in magnification radiography exposure times must be kept as short as possible thus placing further demands on tube rating.

Quantum mottle

This is determined by the number of photons per square mm in the image (see sections 4.10.2 and 9.6) which in turn is governed by the required film blackening. As the image is being viewed under normal viewing conditions, the number of photons striking the film per square mm is exactly the same as on a normal radiograph and the quantum mottle is exactly the same.

Patient dose

If the *FFD* is fixed, then exposure factors are unaltered, but if the patient is positioned closer to the focal spot in order to increase magnification then the dose to the patient is increased. Two factors compensate partially for this increased dose. First the irradiated area on the patient can, and must be reduced. This will require careful collimation of the X-ray beam and accurate alignment of the part of the patient to be exposed. Second, an 'air gap' has in effect been introduced (see section 6.3) so it may be possible to dispense with the use of a grid.

Some increase in the *FFD* may also be necessary because if the mid-plane of the patient is positioned 50 cm from the focal spot (to give $M = 2$ for a *FFD* of 100 cm) the upper skin surface of the patient will be very close to the focal spot and the inverse square law may result in an unacceptably high dose. Note that if the *FFD* is increased, exposure factors will have to be adjusted and a higher kVp may be necessary to satisfy rating requirements.

The way in which the effect on image quality of some of these factors may be analysed more quantitatively is outlined in section 9.8.1. Suffice to conclude here that resolution falls off rapidly with magnification and for a 0.3 mm focal spot the maximum usable magnification is approximately 2.0 at the object or about 1.6–1.8 at the skin surface nearest the tube.

6.5 Image subtraction

This is a technique whereby unwanted information is eliminated from an image, thereby making the diagnostically important information much easier to visualize. It is particularly useful where sequential images differ in a small amount of detail which one wishes to highlight. A typical example would be during angiography where one requires to see changes in the position and amount of contrast medium in blood vessels between two images separated by a short time interval.

The basic principle of the technique is quite simple. A radiograph is a negative of the object data. If a negative of this negative is prepared (a positive of the object data) and positive and negative are then superimposed, the transmitted light will be of uniform intensity. This is because regions that were black on the original negative are white on the positive and vice versa, the two compensating exactly. The positive of the original image is often called the 'mask'. When a second radiograph is taken of the patient, with one or two details slightly different, e.g. following the injection of contrast medium, superimposition of the mask and the second radiograph will result in all the unchanged areas transmitting uniformly, but the parts where the first and second radiograph differ will be visualized. The technique can also be used to show the change in position of a structure or a foreign body.

For the technique to be successful, the two images must superimpose exactly and no patient movement must take place between exposures. The exposure factors for the radiograph and the tube output must remain the same to ensure an exact match of optical density which should be in the range 0.3–1.7. The copy film making the mask must have a gamma equal to 1.0. When the films are viewed together the combined optical density is approximately 2.0 and a special viewing box with a high illumination is required.

A much more elegant and slightly more sophisticated technique can be carried out using colour subtraction. For this technique two normal negative radiographs from a series are taken and beams of white light are shone through each of them so that their images superimpose exactly on a screen. One white light beam is then covered with a red filter and the other with a green filter. In regions where the two images are exactly the same the screen will appear white. Lighter areas, corresponding to additional contrast, which are on the film with red light passing through but are not on the other film will appear red and likewise detail only on the 'green filtered' film will appear green.

The use of digital subtraction methods now available with the most sophisticated screening units has made the process very much easier. These techniques are described in section 9.7 and the flexibility associated with their use makes them preferable for subtraction work. Note, however, that the dynamic range is now limited and great care must be taken when setting up the digital mask to be used in the subtraction process. Several attempts are often required to obtain a suitable mask without over-bright or very dark areas. Once a suitable mask has been obtained, however, excellent results can be achieved.

6.6 Mammography

There are several difficulties associated with imaging the breast to determine whether a carcinoma or pre-cancerous condition exists. First, there is no physical density difference between suspect areas and normal breast tissue and only a small difference in atomic number. Any radiological differentiation is therefore very dependent on photoelectric attenuation. Secondly, one of the prime objectives of mammography is to identify areas of microcalcification, even as small as 0.1 mm in diameter. To achieve the necessary geometric resolution, a small focal spot size is required and problems of X-ray tube output and rating must be considered. Finally, breast tissue is very sensitive to the induction of breast cancer by ionizing radiation (especially for women between the ages of 14 years and menopause). High regard must therefore be paid to the radiation dose received during mammographic examinations. This factor has, until relatively recently, prevented the introduction of well-women breast screening programmes using X-rays, as the doses received during mammography were too high. This section will identify a number of technical developments which may make such a programme acceptable in the future.

6.6.1 Optimum kilovoltage and tube design

In order that maximum contrast may be achieved, a low kV must be used (see Chapter 5). The choice of kV is however a compromise. Too low a kV results in insufficient penetration and a very high radiation dose to the breast. X-ray photons below 12–15 keV contribute very little to the radiograph and must, if possible, be excluded. Early experimental studies suggested that the optimum kilovoltage was between 15 and 20 keV. Recent theoretical work, based on signal to noise ratios (see section 9.5), has suggested however that the optimum voltage probably lies somewhere between 21 and 25 keV, especially when the breast thickness exceeds 2.75 cm (Jennings *et al.* 1981). Experimental data supporting this conclusion has also been reported from one or two centres (e.g. Beaman *et al.* 1983).

Although this difference in photon energy is apparently very small, it has a fundamental effect on tube design and construction. Most, if not all mammography units designed to produce the majority of their X-ray photons in the 15–20 keV region, use a molybdenum anode in a tube with a beryllium window. Additional filtration is provided by a molybdenum filter. A tube designed to produce X-rays in the 21–25 keV region uses a tungsten anode tube with special filters.

6.6.2 Molybdenum anode tubes

A typical spectrum for a molybdenum anode tube operating at 35 kV constant potential and using a 0.05 mm molybdenum filter was shown in Fig. 3.15. Nearly all the X-ray photons are in fact Kα (18 keV) and Kβ (20.5 keV) characteristic X-rays from molybdenum. Figure 6.6a shows a similar spectrum at 30 kVp using a slightly different quantity on the *y* axis and showing the characteristic lines resolved. An X-ray tube window of low atomic number (beryllium with $Z = 4$) is used so that wanted X-rays are not attenuated. Note however that the total filtration on such a tube should still be equivalent to at least 0.5 mm of aluminium to remove low energy radiation.

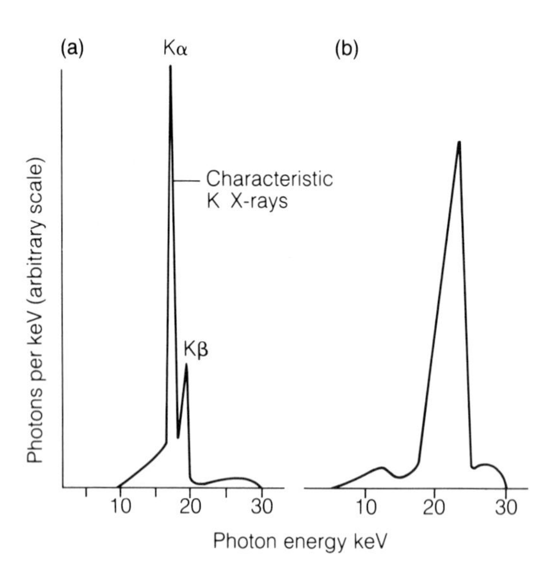

Fig. 6.6. (a) Spectral output of a molybdenum anode X-ray tube operating at 30 kVp with a 0.05 mm molybdenum filter. (b) Spectral output of a tungsten anode X-ray tube operating at 30 kVp with a 0.05 mm palladium filter.

For reasons of geometric resolution, discussed in the general remarks on mammography, a small focal spot size (0.6 mm or less) and a focus–film distance of at least 60 cm should be used. A stationary anode is used to minimize extra-focal radiation and the tube operates typically at about 40 mA. Because a stationary anode of low atomic number material (for Mo, $Z = 42$) is being used, X-ray output is low and rather long exposure times, of the order of 3 s, are required.

The spectral advantage of using a molybdenum anode starts to be lost for breast thicknesses above about 2.75 cm because higher kVps have to be used to achieve sufficient penetration. The additional high energy photons are bremsstrahlung radiation and are not associated with the Kα and Kβ characteristic X-rays. As these higher energy photons have a range of energies up to the maximum keV, the contrast falls.

To counteract the effect of body attenuation as much as possible, whilst retaining a low kV, mammography units always use special applicators to compress the breast to as small a thickness as possible. Applying compression also minimizes or even eliminates movement unsharpness during the relatively long exposures, reduces geometric unsharpness because the breast is closer to the receptor and improves contrast by reducing scatter (see section 6.6.6).

6.6.3 Tungsten anode tubes

The effect of a 0.05 mm palladium filter on the output spectrum of a tungsten anode tube operating at 30 kVp is shown in Fig. 6.6b. The K absorption edge of palladium is at 24.3 keV, so below this energy the attenuation of the filter is very much lower than at higher energies (see section 3.6). Thus a 'window' of energies is transmitted rather readily and such a filter is sometimes known as a K-edge filter. The spectrum transmitted by palladium most closely matches that which theory predicts will be most suitable for mammography.

Another situation in which there is a need for low kV work at low dose is paediatric radiology, and other K-edge filters such as rhodium, silver and cadmium, which have K-edges at 23.2, 25.5 and 26.9 keV respectively, have been used to good advantage. The dose reduction achieved with these filters can be considerable.

Since a tungsten anode tube gives a good output, a small focal spot (0.2 mm) can be used and this can be reduced to an effective focal spot of 0.1 mm by mounting the tube at an angle of 5°. A further advantage of the tungsten tube is that it may be used quite easily for xeroradiography (see section 6.7), thus eliminating the need for a mammography unit that wishes to use both techniques to acquire two tubes.

Note that greater care is required in setting and checking the generator kilovoltage since emission is not 'locked in' to characteristic line spectra as in the case of molybdenum.

6.6.4 Film–screen combinations

Many different films and intensifying screens are offered by manufacturers exclusively for mammography. The films are single-sided thus eliminating parallax and are used with a single screen, which is often much thinner than standard screens. The screen is positioned behind the film as this causes slightly less loss in resolution and screen and film are often pulled into very close contact by using a flexible plastic cassette which can be vacuum evacuated. The screens are generally rare earth, e.g. lanthanium bromide or gadolinium oxysulphide or occasionally calcium tungstate. They are not as fast as normal intensifying screens (about one-tenth as fast) thereby ensuring that resolution is maintained as quantum mottle is eliminated. The optimum film screen combination for a given mammography unit is generally found by trial and error and may involve combining the film of one manufacturer with the screen of another. To enhance contrast differences, the film has a high gamma and hence a small film latitude (see section 4.4.4). Therefore to achieve film blackening in the optimum range for diagnosis, the exposure must always be controlled by using a photo timer or other form of automatic control. The optical density of the film may be in excess of 3.0 and a high intensity viewing box is required.

6.6.5 Patient doses

As mentioned in the introduction to this section, skin doses must be kept very low because the breast is very radiosensitive. As techniques have improved, the mean dose to glandular tissue has fallen from several tens of mGy per exposure 15 years ago to less than 1 mGy using the best current techniques with molybdenum anodes. The patient dose using a tungsten anode tube and a K-edge filter is even less than that using a molybdenum anode. At operating voltages of about 30 kVp the dose reduction factor for a thin breast is approximately 2. This dose advantage is partially offset by a very slight reduction in image quality with the tungsten tube as the molybdenum tube is particularly suitable for radiographing thin breasts. For thicker breasts not only can the dose reduction factor rise to as much as 5 but the image quality produced by the tungsten tube is also superior. The mean dose for a two view examination of a 5 cm thick compressed breast, without using a grid, might be between 0.8 and 1.5 mGy depending on the film–screen combination chosen.

6.6.6 Contrast improvement

In mammography, contrast is reduced by scatter by as much as 50% for a thicker breast. The use of compression to reduce scatter has already been mentioned. Contrast may also be improved by using either a stationary grid with a high strip density or a moving grid. The grid must have high transmission for low energy photons and may improve contrast by a factor of between 1.2 (2 cm breast) and 1.7 (8 cm breast). The dose to the breast will of course be increased—by a factor of between two (stationary grid) and three (moving grid).

As an alternative to using a grid, an air gap technique with a magnification of about 1.5 will reduce scatter sufficiently to improve contrast by about 25% for a 6 cm breast. Overall image sharpness may also improve provided that the focal spot size is sufficiently small ($\sim$ 0.1 mm) for geometrical unsharpness not to increase significantly. The dose to the breast will of course be increased (see section 6.4) and greater demands will be placed on the X-ray tube generator.

6.7 Xeroradiography

This method of recording X-ray images is particularly suited to mammography but has also been used for other high definition work particularly imaging bones and extremities. For an early account of the method see Boag *et al.* (1972).

Instead of using a film as the recording medium, xeroradiography uses a photo conductor, which is a special class of semiconductor, as the surface on which the image is recorded. As discussed in section 4.1, a semiconductor can be converted from an insulator into a conductor under certain conditions. In a photoconductor the energy gap between the valence and conduction bands is so small that even the energy provided by the absorption of a photon of visible light (for green light of wavelength 540 nm, the quantum energy is about 2.3 eV) is sufficient to cause this change. Thus if a photoconducting surface is covered uniformly with charge and then exposed to visible or X-ray photons, exposed regions will become conducting and the charge will flow away. Hence the pattern of charge-free regions after exposure is an image of the irradiated area.

The photoconductor used in xeroradiography is amorphous selenium and the construction of a xeroradiography plate is shown in Fig. 6.7. The aluminium plate must be very smooth and the selenium layer laid down upon it must be very uniform in thickness. The cellulose acetate layer, as well as protecting and thus extending the life of the selenium layer, also prevents the lateral conduction of charge across the surface.

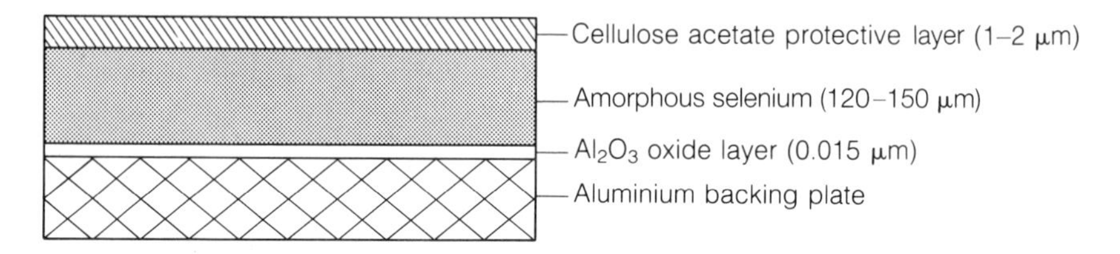

Fig. 6.7. Construction of a 'selenium plate' for xeroradiography.

6.7.1 Mode of use

The selenium is charged to a potential difference of about 1500 V. The charge, which is normally positive, must be very uniformly distributed on the plate surface. The plate acts as a capacitor so a negative charge is induced on the aluminium. This charge is prevented from leaking into the selenium by the aluminium oxide layer. The plate is then placed in a light-tight cassette, since the photoconductor is sensitive to light. During an exposure, X-ray photons interact with the selenium plate and these interactions differ from the interaction of light photons in two ways. First, the X-ray photons may penetrate some distance into the selenium plate before they interact. Secondly, each primary interaction results in the release of secondary electrons that have the capability of causing many more conduction centres in the selenium plate. Where there are many X-ray photon interactions most of the charge leaks away, and where there are few only a small amount of charge leaks away. The X-ray photon image is thus converted to a charge image on the plate.

This charge image is converted to a visual image by first coating the surface of the plate with a very fine (< 3 μm diameter) charged powder. This powder is attracted to the surface by the residual charge. Excess powder is removed. The plate is then covered with a special paper and the powder image is transferred into it electrostatically. Finally the paper is heated to fuse the powder into it creating a permanent record. By altering the charge on the powder, either a positive or negative can be produced.

Any powder remaining on the selenium plate can be cleaned off and the plate is then stored, uncharged, ready for reuse.

6.7.2 Formation of powder image

The electric field across a charge distribution is shown in Fig. 6.8. This figure illustrates the unique advantage of xeroradiography over conventional radiography. At the edge of a body of charge the field does not fall to zero simply. Instead, on moving away from the charged region, there is first a sharp rise in the field strength, immediately followed by a sharp negative field pulse, before the zero field strength position is reached. Charged powder near the edge is either very strongly attracted or

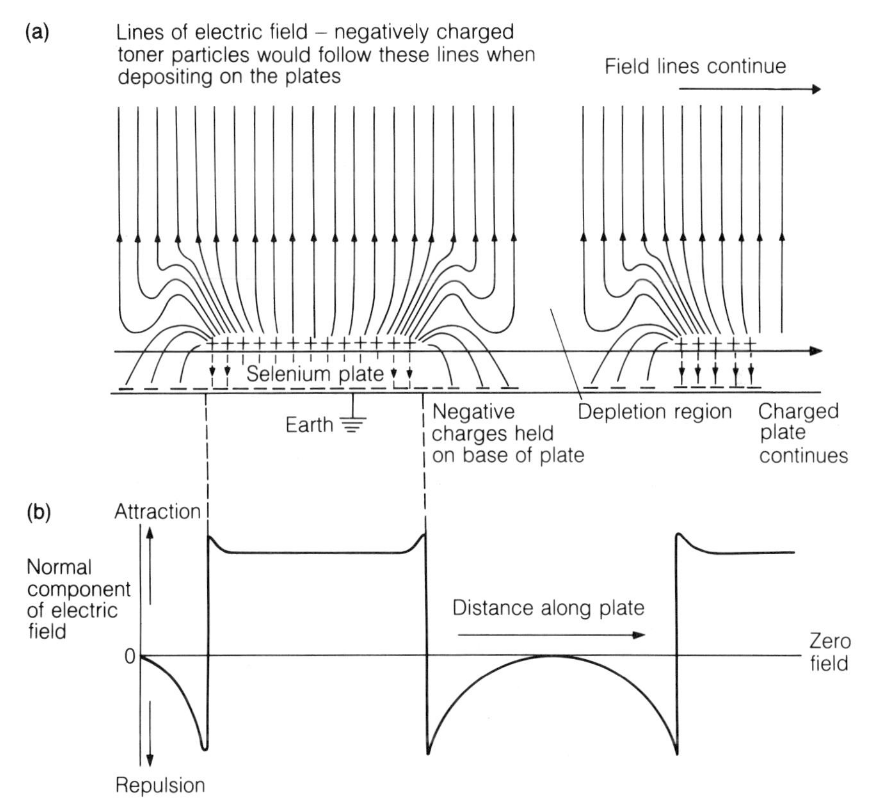

Fig. 6.8. The effect of charge distribution on (a) the electric field lines, (b) the component of electric field normal to the charged surface.

repulsed. This gives the characteristic 'outlining' effect of a xeroradiograph making it particularly suitable for visualizing edges where there may be only a small contrast difference. As shown in Fig. 6.9, a difference in potential of only some 10–20 V at an edge is enough to cause differences in electric field that will cause the edge to be sharply outlined in the resulting image.

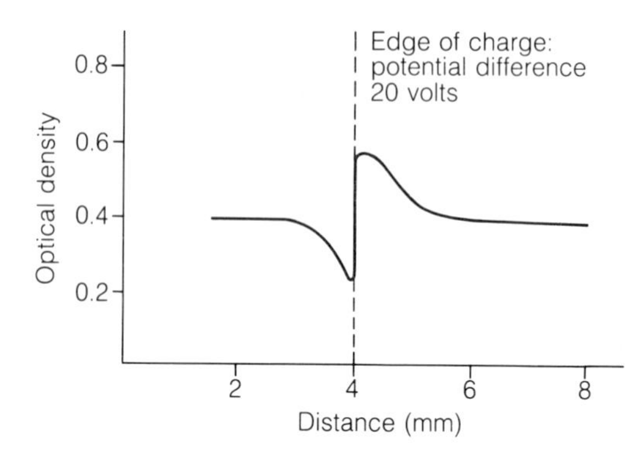

Fig. 6.9. Variation in optical density on a xerographic plate across an edge that represents a potential difference of only 20 V.

Conversely xeroradiography produces poor images of large, homogeneous structures, because the field lines are weak and uniformly spaced in these regions, and has poor broad area contrast. For a more quantitative treatment in terms of modulation transfer functions of these effects on image formation see section 9.8.1.

If very fine particles (< 3 μm in diameter) are used, resolution of approximately 50 line pairs per mm can be achieved on a xeroradiograph. However, this is better than the resolution limit imposed by the tube focal spot size and/or patient movement. Thus in practice the resolution limit of the system is closer to 10 line pairs per mm. However, this does mean that micro calcifications can be seen easily, cysts of 1–2 mm can be identified and carcinomas of 2–5 mm can be found.

6.7.3 Sensitivity and dynamic range

Sensitivity, defined as the radiation exposure required to reduce the charge on the plate to a half, is a function of the thickness of selenium and kVp. For thin layers of selenium, sensitivity increases with increasing thickness because a greater proportion of X-ray photons is stopped. At greater thicknesses charge migration within the selenium becomes more difficult and the sensitivity falls. As shown in Fig. 6.10, the optimum thickness increases with increasing kVp. It is also affected to some extent by the X-

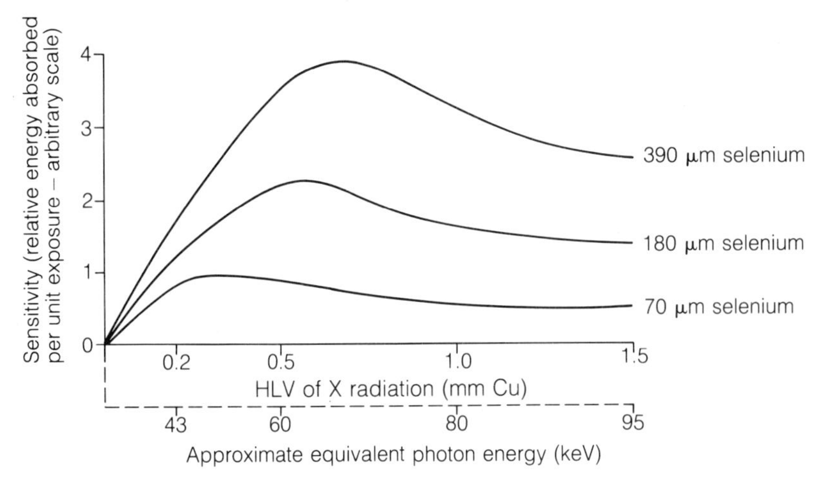

Fig. 6.10. Variation in sensitivity of a selenium plate with thickness of selenium and photon energy.

ray spectrum and is therefore dependent on the generator wave form and particularly the anode material and filtration. A tungsten anode is normally used for xeroradiography.

When a plate is uniformly exposed, charge will be lost uniformly and the potential of the plate will fall uniformly. An alternative way to consider that fact that a xeroradiograph has poor broad area contrast is that, when the exposure is uniform, the optical density changes very little with plate voltage and hence radiation exposure. This has the advantage that a xeroradiograph exhibits a very wide latitude, with large and small exposures producing almost the same background optical density. Conversely, changes in optical density associated with edges are almost equally sharp whatever the exposure. One practical example of this advantage is that blood vessels are visualized equally well under a thick layer or a thin layer of breast tissue. With conventional radiographs it might be necessary to increase the kVp for the thicker tissue with consequent loss of contrast.

6.7.4 Advantages and disadvantages of xeroradiography

In addition to the imaging advantages already discussed, xeroradiography has the cost advantage of not being based on an imaging system involving silver and, in theory at least, a plate is reusable indefinitely. A xeroradiograph has a short processing time and can be read quickly. The nature of the edge enhancement process also means that there is a very broad exposure latitude in the system allowing good results to be achieved with less than optimum generator settings.

Possible disadvantages of the technique are that very great care must be taken to ensure a uniform charge distribution prior to exposure. Mechanical damage to the selenium surface and exposure to light must be avoided. The charging process must be performed immediately prior to use because a certain amount of charge leakage is inevitable if charged plates are stored. A further limitation is that the selenium plate cannot be reused too quickly as it carries a 'memory' and may produce a 'ghost' of the original image.

The major disadvantage, however, is that most reports to date indicate that the dose to the breast required to take a xeroradiograph is some three to 10 times higher than for a conventional radiograph. Given the sensitivity of the breast to induction of cancer, this is sufficient reason for use of the technique to be very carefully controlled. It should be noted, however, that one or two centres are currently reporting reduced doses during xeroradiography using tungsten anode tubes with special filters and the situation with respect to patient dose may change in the future.

6.8 Ionography

As described in section 4.8 the isotropic production of light following interaction of an X-ray photon with an intensifying screen degrades the image. Although the distance travelled by the light photons from their point of origin is quite large, the electrons released in the initial ionization processes travel only short distances (see Table 1.2). Ionography attempts to increase the resolution of the imaging process by using these electrons to form a direct image rather than via the production of light. For this reason it is sometimes called Electron Radiography (Johns 1976).

In ionography the X-ray photons interact with a high atomic number gas, usually either Xenon or Freon at pressures of up to 10 atmospheres. The number of ion pairs produced at any point in the gas is proportional to the X-ray photon intensity at that point. The gas is contained in a chamber between two electrodes as shown in Fig. 6.11. The lower aluminium electrode has a fixed shape but the top electrode which is made of plastic coated Melinex foil is very thin and is held in position and shape by the perspex top plate. The size of the chamber depends on the pressure of the gas in it. At atmospheric pressure the distance between the electrodes is approximately 4 cm, but at high pressures the distance falls to approxi-

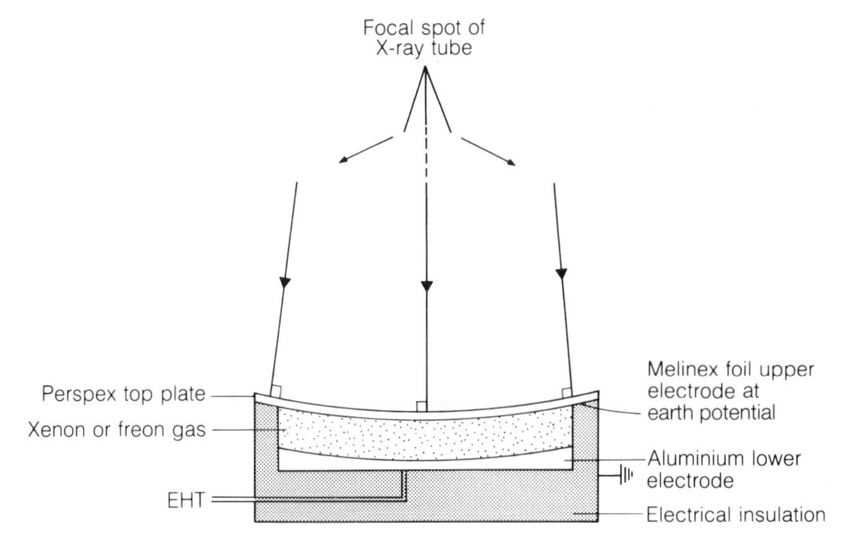

Fig. 6.11. Cross-section through an ionography chamber.

mately 1 cm. One practical problem of using a chamber at a high pressure is that the Melinex foil on which the final image is formed can be more difficult to change. Both electrodes are curved with a radius of curvature appropriate to a particular electrode–tube focal spot distance. The distance is critical and if the chamber is used with any other electrode–focal spot distance, image quality is degraded.

When the ion chamber is at the correct distance from the focal spot, X-ray photons strike the chamber normally. The electric field between the two electrodes is also normal to the surface (Fig. 6.12). Thus ions produced during an ionizing event move, under the influence of the field, along the same line as the original X-ray photon. The voltage applied between the

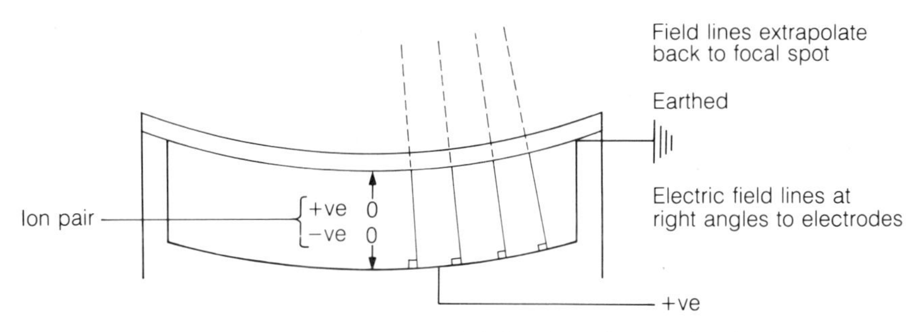

Fig. 6.12. Distribution of electric field lines in an ionography chamber.

two electrodes must be sufficient to ensure that no recombination takes place. Its value depends on the gas used but is typically sufficient to produce an electric field of 1 kV mm^{-1} for Freon. Charge of one polarity is deposited on the plastic surface of the upper electrode which is kept at earth potential for safety reasons. Depending on the polarity of the lower electrode with respect to earth, either a positive or a negative charge image can be produced. In practice, it is found that, despite the original name electron radiography, setting the bottom electrode at a positive potential relative to earth so that the positive ions are collected on the plastic produces slightly better resolution. This is because the positive ions are heavier and hence tend to diffuse apart less than electrons. The plastic/ Melinex foil can be removed after exposure and the charge distribution can be imaged using a charged powder. As in xeroradiography there is some edge enhancement.

The current resolution limit of ionography is about 10 line pairs per mm with some deterioration towards the edge of the chamber. The maximum chamber size used at present is limited to 30 cm diameter. The dose required to produce an image, although greater than for a normal radiograph is less than that required to produce a xeroradiograph. It is anticipated that these factors will be improved as the technique is developed. One possible improvement is to replace the gas by a high atomic number liquid. A much smaller electrode gap could then be used with a thinner front face and this might allow much higher energy photons to be imaged thereby extending use of ionography into the field of nuclear medicine.

6.9 Thermography

Thermography is the imaging of infra-red radiation emitted from the skin and the evaluation of these images. According to Stefan's law, the rate of emission per unit area per second $E = \varepsilon\sigma T^4$ where ε is the emissivity of skin, σ the Stefan constant and T the Kelvin temperature. ε is close to 1.0 for skin.

If two areas differ in temperature by a small amount δT, the difference in rate of emission $\delta E = 4\varepsilon\sigma T^3\delta T$. Thus the fractional change in energy emission $\delta E/E = 4\delta T/T$. For example if $T = 300$ K and $\delta T = 1$ K, $\delta T/T$ is about 0.3% but $\delta E/E = 1.2\%$. Thus measurement of infra-red emission is four times as sensitive as any method based on direct measurement of change in temperature.

As tissue is virtually opaque to infra-red radiation, it is essentially the variation in cutaneous temperature that is being measured. Although skin emits infra-red radiation with wavelengths varying from 4 to 40 μm at 30°C, the peak radiation occurs at 9 μm. Detectors are normally semiconductor devices in which electrons are freed following the absorption of infra-red photons. Two types can be used. The more common, indium antimonide, detects radiation between 2 and 5.6 μm, the other, mercury cadmium telluride, is sensitive to wavelengths between 3 and 14 μm. Both detectors have to be operated at liquid nitrogen temperature to avoid noise.

The image of a point on the body surface is focused onto the detector crystal by a concave mirror or lens system. Glass will not transmit infra-red in the region of interest so materials such as quartz or germanium have to be used. All mirrors must be front surface reflecting.

An image is built up by scanning the point from which radiation is detected across the body in a raster fashion. Some systems use rotating mirrors to scan, others a rotating pair of prisms. The signal produced by the detector is then used to modulate the signal on the display.

Diagnosis often depends on the comparison of the thermogram of one area of the body with that of another thought to be normal. In purely physical terms spatial resolution is good (better than 0.5 mm) and temperature resolution is very good, probably 0.1–0.2°C. Most of the problems with the technique are biological in origin with temperature differences of up to 2°C occurring naturally in local areas and, in the case of the breast, to the breast as a whole. Differences of 2°C or greater are considered significant of changes produced in the vascular system by a neoplasm. A distortion of the isothermal contours is also considered suspicious.

Very few centres now use thermography even for well-women surveys for breast cancer. Although it has the obvious advantage of being completely non-invasive, most centres found that too many false positives and negatives were being diagnosed. The few centres that persevered however do report better than average results. If used in a special temperature-regulated room by a skilled operator, thermography may be worth considering for screening purposes. The skill, however, appears to be as much an art as a science.

Other applications of thermography are in the study of arthritis and inflammation in peripheral joints, tissue viability following plastic surgery, vascular studies and scrotal thermography, but none of these has developed to the point of being an important diagnostic tool for the radiologist.

6.10 Diaphanography

The use of visible light to investigate lesions in the breast was first suggested as early as 1929. This pioneer work was, however, not developed until the 1970s, when, particularly in Sweden, an increased awareness of the contribution the technique could make to investigations of the breast was realized.

Light scattering in skin causes preferential removal of the shorter wavelengths from a white-light beam, but in the red and near infra-red part of the spectrum sufficient light penetrates the skin and tissues to allow an observer to visualize variations in brightness over the skin. The technique involves placing a bright tungsten light under the breast and viewing the illuminated breast in a darkened room. Dark areas are taken to indicate that a malignant lesion may be present, bright areas imply the presence of a fluid-filled cyst. Blood-filled cysts and bruises can also give dark areas. Note that radiation that is further into the infra-red than a wavelength of about 1000 nm must be filtered out to prevent burning the skin.

Although sufficient light penetrates for a visual diagnosis to be made using a tungsten lamp, the percentage transmission is so low that a much higher intensity light source is required for the duration of the exposure to produce a permanent record on film. This light source is generally a xenon flash tube which is synchronized to the camera to obtain optimum exposures. The recording film is infra-red colour film which requires special developing procedures. If this film cannot be developed locally, problems do arise because it takes an unacceptably long time for the developed film to be returned.

An alternative method of producing an image using television techniques has now been developed. Using a Silicon Vidicon, sensitive up to a wavelength of 1000 nm, it is possible to image using a tungsten filament lamp, thus removing the need for the xenon flash tube (Watmough 1982).

Although the resolution of the technique cannot match that of X-ray mammography, this approach appears to be better than standard thermography and may have future potential for examination of the breast.

References and further reading

Bassett L. W. and Gold R. H. (1982) *Mammography, Thermography and Ultrasound in Breast Cancer Detection*. Grune & Stratton, New York.

Beaman S., Lillicrap S. C. and Price J. L. (1983) Tungsten anode tubes with K-edge filters for mammography. *Br. J. Radiol.* **56**, 721–727.

Boag J. W., Stacey A. J. and Davis R. (1972) Zerographic recording of mammograms. *Br. J. Radiol.* **45**, 633–640.
Chesney D. N. and Chesney M. O. (1984) *X-ray Equipment for Student Radiographers*, 3rd edn. Blackwell Scientific, Oxford.
Curry T. S., Dowdey J. E. and Murry R. C. Jr. (1984) *Christensen's Introduction to the Physics of Diagnostic Radiology*. Lea & Febiger, Philadelphia.
Gayler B. W. (1979) Chest technique. In *Multiple Imaging Procedures*, Vol. 1 'Pulmonary system', eds S. S. Siegelman, F. P. Stitik and W. R. Summer, pp. 1–18. Grune & Stratton, New York.
Jennings R. J., Eastgate R. J., Siedband M. P. and Ergun D. L. (1981) Optimal X-ray spectra for screen-film mammography. *Med. Phys.* **8**, 629–639.
Johns H. E. (1976) New methods of imaging in diagnostic radiology. *Br. J. Radiol.* **49**, 745–764.
Johns H. E. and Cunningham J. R. (1983) *The Physics of Radiology*, 4th edn, pp. 624–630. Thomas, Springfield.
Jones C. H. (1982) Thermographic imaging. In *Scientific Basis of Medical Imaging,* ed. P. N. T. Wells. Churchill Livingstone, New York.
Pyke A. H. (1975) Mobile, portable and dental units. In *Principles of Diagnostic X-ray Apparatus*, ed. D. R. Hill, pp. 189–196. Macmillan, London.
Watmough D. J. (1982) A light torch for transillumination of female breast tissues. *Br. J. Radiol.* **55**, 142–146.

Exercises

1 Compare and contrast a condenser discharge mobile X-ray unit with a constant potential mobile X-ray unit.
2 Since a low kVp is required for high contrast, why should the use of a high kVp sometimes be advantageous?
3 Outline the principles of magnification radiography and indicate its limitations.
4 Suggest some situations in which the subtraction technique might be used and indicate how the required information would be obtained.
5 What are the basic requirements of an X-ray tube that is to be used for mammography? How can these be achieved with
(a) a tungsten anode tube,
(b) a molybdenum anode tube.
6 What steps would you take to obtain the best results when using rare earth screens for mammography?
7 Discuss the factors that control the contrast on a radiograph, illustrating your answer by considering the design of equipment for mammography.
8 Discuss the relative advantages and disadvantages of magnification mammography.

9 Explain why the images produced in xeroradiography have a dark line round the outside of them.

10 What are the relative merits of xeroradiography and normal mammography techniques for investigation of the female breast?

11 What advantages does ionography have over conventional radiography?

12 Compare and contrast the use of diaphanography with thermography for the imaging of breasts.

7

Radiation Measurement

7.1 Introduction

Lord Kelvin (1824–1907), who is probably best remembered for the absolute thermodynamic scale of temperature, is reported to have stated on one occasion: 'Anything that cannot be expressed in numbers is valueless.' In view of the potentially harmful effect of X-rays it is particularly important that methods should be available to 'express in numbers' the 'strength' or intensity of X-ray beams.

With respect to measurement, three separate features of an X-ray beam must be identified. The first consideration is the flux of photons travelling through air from the anode towards the patient. The ionization produced by this flux is a measure of the **radiation exposure**. If expressed per unit area per second it is the **intensity**. Of more fundamental importance as far as the biological risk is concerned is the **radiation dose**. This is a measure of the amount of energy deposited as a result of ionization processes. Finally, it may be important to know about the energy of the individual photons. Because of the mechanism of production, an X-ray beam will contain photons with a wide range of energies. A complete specification of the beam would require determination of the full spectral distribution as shown in Fig. 2.2. This represents information about the **quality** of the X-ray beam.

Clearly the intensity of an X-ray beam must be measured in terms of observable physical, chemical or biological changes that the beam may cause, so it will be useful to review briefly relevant properties of X-rays. Two of them are sufficiently fundamental to be classified as **primary properties**—that is to say measurements can be made without reference to a standard beam.

1 *Heating effect*. X-rays are a form of energy which can be measured by direct conversion into heat. Unfortunately, the energy associated with X-ray beams used in diagnostic radiology is so low that the temperature rise could scarcely be measured (see section 11.2).

2 *Ionization*. X-rays cause ionization by photoelectric, Compton and pair production processes in any material through which they pass. The number of ions produced in a fixed volume under standard conditions of temperature and pressure will be fixed.

A number of other properties of X-rays can, and often are, used for dosimetry. In all these situations, however, it is necessary for the system to be calibrated by first measuring its response to beams of X-rays of known intensity so these are usually called **secondary properties**.

3 *Physical effects*. When X-rays interact with certain materials, visible light is emitted. The light may either be emitted immediately following the interaction (**fluorescence**); after a time interval (**phosphorescence**); or, for some materials, only upon heating (**thermoluminescence**).

4 *Physico-chemical effects*. The action of X-rays on photographic film is well-known and widely used.

5 *Chemical changes*. X-rays have oxidizing properties, so if a chemical such as ferrous sulphate is irradiated, the free ions that are produced oxidize some Fe^{++} to Fe^{+++}. This change can readily be detected by shining ultraviolet light through the solution. This light is absorbed by Fe^{+++} but not by Fe^{++}.

6 *Biochemical changes*. Enzymes rely for their action on the very precise shape associated with their secondary and tertiary structure. This is critically dependent on the exact distribution of electrons, so enzymes are readily inactivated if excess free electrons are introduced by ionizing radiation.

7 *Biological changes*. X-rays can kill cells and bacteria, so, in theory at least, irradiation of a suspension of bacteria followed by an assay of survival could provide a form of biological dosemeter.

Unless specifically stated otherwise, in the remainder of this chapter references to X-rays apply equally to gamma rays of the same energy.

7.2 Ionization in air as the primary radiation standard

There are a number of important prerequisites for the property chosen as the basis for radiation measurement.

1 It must be accurate and unequivocal, i.e. personal, subjective judgment must play no part.

2 It must be very sensitive producing a large response for a small amount of radiation energy.

3 It must be reproducible.

4 The measurement should be independent of intensity, i.e. an intensity I for time t must give the same answer as an intensity $2I$ for time $t/2$. This is the **Law of Reciprocity**.

5 The method must apply equally well to very large and very small doses.

6 It must be reliable at all radiation energies.

7 The answer must convert readily into a value for the absorbed energy in biological tissues or 'dose' since this is the single most important reason for wishing to make radiation measurements.

None of the properties of ionizing radiation satisfies all these requirements perfectly but ionization in air comes closest and has been internationally accepted as the basis for radiation dosimetry. There are two good reasons for choosing the property of ionization:

1 Ionization is an extremely sensitive process in terms of energy deposition. Only about 35 eV is required to form an ion pair, so if a 100 keV photon is completely absorbed, almost 3000 ion pairs will have been formed when all the secondary ionization has taken place.

2 As shown in Chapter 11, the extreme sensitivity of biological tissues to radiation is directly related to the process of ionization so it has the merit of relevance as well as sensitivity.

There are also good reasons for choosing to make measurements in air:

1 It is readily available.

2 Its composition is close to being universally constant.

3 More important, for medical applications, the mean atomic number of air (Z = 7.6) is very close to that of muscle/soft tissue (Z = 7.4). Thus, provided ionization and the associated process of energy absorption is expressed per unit mass by using mass attenuation coefficients rather than linear attenuation coefficients, results in air will be closely similar to those in tissue.

7.3 The unit of radiation exposure

The unit of radiation exposure (X) is defined as that amount of radiation which produces in air ions of either sign equal to 1 C (coulomb) kg^{-1}. Expressed in simple mathematical terms

$$\mathrm{X} = \frac{\Delta Q}{\Delta m}$$

where ΔQ is the sum of all the electrical charges on all the ions of one sign produced in air when all the electrons liberated in a volume of air whose mass is Δm are completely stopped in air. The last few words ('are completely stopped in air') are extremely important. They mean that if the electron generated by say a primary photoelectric interaction is sufficiently energetic to form further ionizations (normally it will be), all the associated ionizations must occur within the collection volume and all the electrons contribute to ΔQ.

The older, obsolescent unit of radiation exposure is the roentgen (R). One roentgen is that exposure to X-rays which will release one electrostatic unit of charge in one cubic centimetre of air at standard temperature and pressure (STP). Hence 1 R = 2.58×10^{-4} C kg^{-1}.

7.4 The ionization chamber

In this section measurement of radiation exposure will be considered. Figure 7.1 shows a direct experimental interpretation of the definition of radiation exposure. The diaphragm, constructed of a heavy metal such as

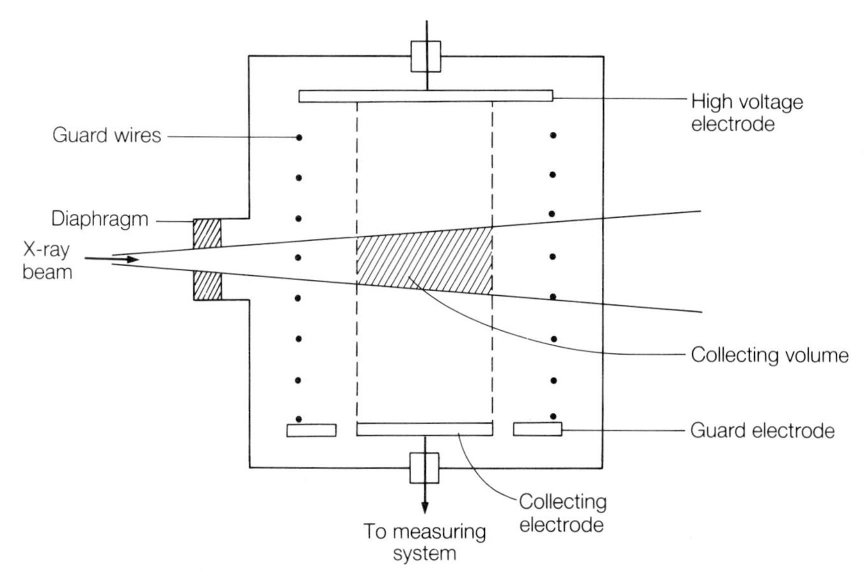

Fig. 7.1. The free air ionization chamber. From Whyte (1959).

tungsten or gold, defines an X-ray beam of accurately known cross-section A. This beam passes between a pair of parallel plates in an air-filled enclosure. The upper plate is maintained at a high potential relative to the lower and, in the electric field arising from the potential difference between the plates, all the ions of one sign produced in the region between the dashed lines move to the collecting electrode.

Either the current flow (exposure rate) or the total charge (exposure) may be measured using the simplified electrical circuits shown in Fig. 7.2 a and b.

Since 1 ml of air weighs 1.3×10^{-6} kg at STP, a chamber of capacity 100 ml contains 1.3×10^{-4} kg of air. A typical exposure rate might be 2.5 μC kg^{-1} h^{-1} (a dose rate of approximately 0.1 mGy h^{-1}, see section 7.10) which corresponds to a current flow of $(2.5 \times 10^{-6})/(60 \times 60) \times 1.3 \times 10^{-4}$ or about 10^{-13} A.

If $R = 10^{10}\Omega$, then $V = 1$ mV which is not too difficult to measure. However, the voltmeter must have an internal resistance of at least $10^{13}\Omega$ so that no current flows through it and this is quite difficult to achieve.

Since the free air ionization chamber is a primary standard for radiation measurement, accuracy better than 1% (i.e. more precision than the figures quoted here) is required. Although it is a simple instrument in principle, great care is required to achieve such precision and a number of

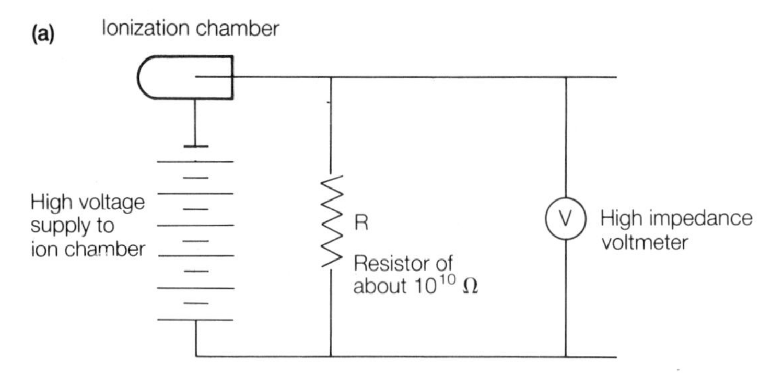

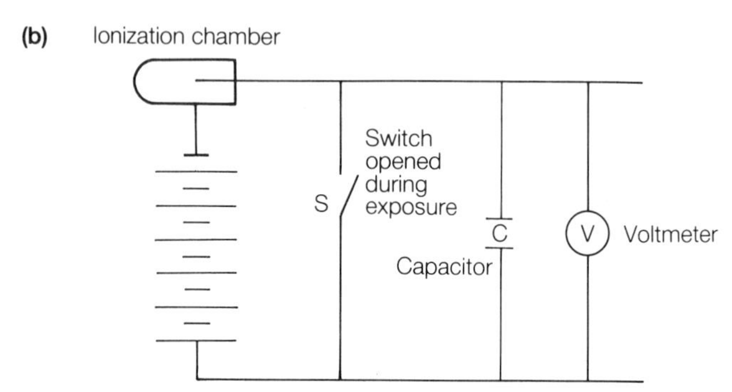

Fig. 7.2. Simplified electrical circuits for measuring (a) current flow (exposure rate), (b) total charge (exposure).

corrections have to be applied to the raw data. For example a correction must be made if the air in the chamber is not at STP. For air at pressure P and temperature T, the true reading R_T is related to the observed reading by

$$R_T = R_0 \cdot \left(\frac{P_0}{P}\right) \cdot \left(\frac{T}{T_0}\right)$$

where P_0, T_0 are STP values.

Rather than trying to memorize this equation, the reader is advised to refer to first principles. Ion pairs are created because X-ray photons interact with air molecules. If the air pressure increases above normal atmospheric, the number of air molecules will increase, the number of interactions will increase, and the reading will be artificially high. Changes in temperature may be considered similarly.

The requirement for precision also creates design difficulties. For example, great care must be taken, using guard rings and guard wires (see Fig. 7.1), to ensure that the electric field is always precisely normal to the plates. Otherwise, electrons from within the defined volume may miss the collecting plate or, conversely, may reach the collecting plate after being produced outside the defined volume.

Major difficulties arise as the X-ray photon energy increases, especially above about 300 keV, because of the ranges of the secondary electrons (see Table 1.2). Recall that all the secondary ionization must occur within the air volume. If the collecting volume is increased, eventually it becomes impossible to maintain field uniformity.

Thus the free air ionization chamber is very sensitive in the sense that one ion pair is created for the deposition of a very small amount of energy. However, it is insensitive when compared to solid detectors that work on the ionization principle because air is a poor stopping material for X-rays. It is also bulky and operates over only a limited range of X-ray energies. However, it is a primary measuring device and all other devices must be calibrated against it.

7.5 Secondary instruments—air equivalent wall

Fortunately, there is a technique which, to an acceptable level of accuracy for laboratory instruments, eliminates the requirement for large volume air chambers.

Imagine a large volume of ethylene gas with dimensions much bigger than the range of secondary electrons. Now compress the gas to solid polyethylene, leaving only a small volume of gas at the centre (Fig. 7.3). The radiation exposure will be determined by the density of electrons within the gas and if the gas volume is small, the number of secondary electrons either being created in the gas or coming to the end of their range there will be negligible compared to the electron density in the solid.

However, the electron density in the solid is the same as it would be at the centre of the large gas volume. This is because the electron flux across any plane is the product of rate of production per unit thickness and range. The rate of production depends on the number of interactions and is higher in the solid than in the gas by $\varrho_{solid}/\varrho_{gas}$ (the ratio of the densities). But once the electrons are formed they have a fixed energy and lose that energy more rapidly in the solid due to collisions. Thus the ratio range of electrons in solid/range of electrons in gas is in the inverse ratio $\varrho_{gas}/\varrho_{solid}$.

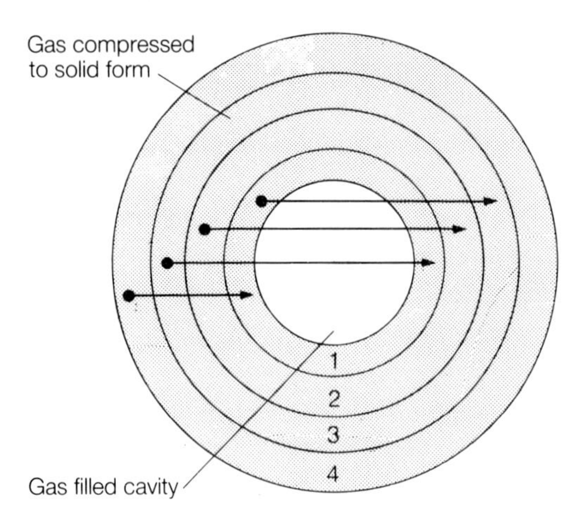

Fig. 7.3. Secondary electron flux in a gas-filled cavity. Consider the solid as a series of layers starting at the edge of the cavity. Layers 1, 2 and 3 contribute to the ionization in the cavity but layer 4 does not.

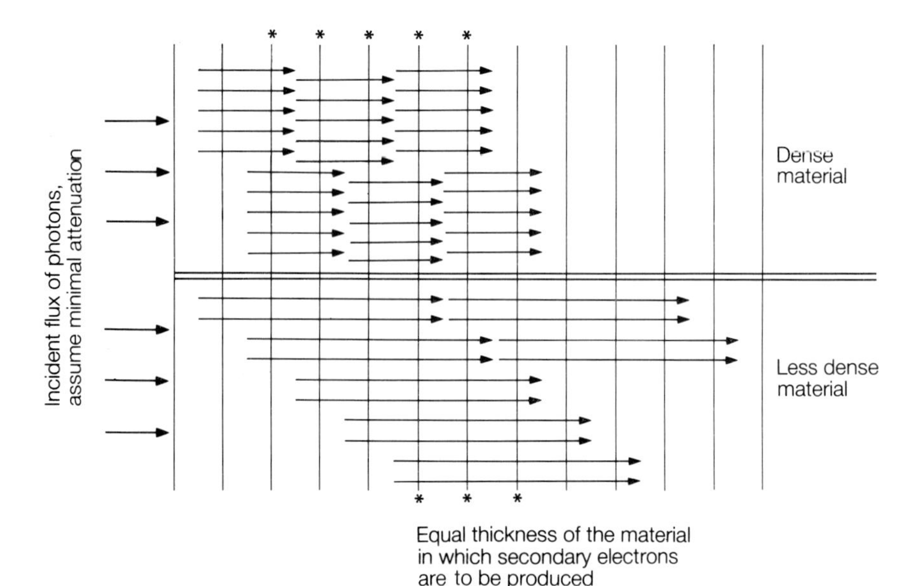

Fig. 7.4. A schematic demonstration that the flux of secondary electrons at equilibrium is independent of the density of the stopping material. The upper material has 2.5 × the density of the lower. It therefore produces 2.5 × more electrons per slice. However, the range of these electrons is reduced by exactly the same factor.

When equilibrium is established, shown by a * for each material, 10 secondary electrons are crossing each vertical slice (the electron density) in each case. Note that equilibrium establishes quicker in the more dense material.

Hence the product (rate of production of secondary electrons × range of secondary electrons) is independent of density as shown schematically in Fig. 7.4.

An alternative way to view this situation is that, provided the atomic composition is the same, the gas in the cavity does not know if it is surrounded by a big volume of gas or by a much smaller volume of solid resulting from compression of the gas.

The result is precise for polyethylene and ethylene gas because the materials differ only in density. No solid material is exactly like air in terms of its interaction with X-ray photons by photoelectric and Compton processes at all photon energies but good approximations have been constructed and a correction can be made for the discrepancy.

A simple, compact, secondary instrument, suitable for exposure rate measurements around an X-ray set, is shown in Fig. 7.5. The dimensions now only require that the wall thickness should be greater than the range of secondary electrons in the solid medium. Note that a correction must also be made for attenuation of the primary beam in the wall surrounding the measurement cavity.

An important modern detector that uses the principle of ionization in air is the high pressure xenon gas chamber. Xenon is chosen because it is an inert gas and its high atomic number ($Z = 54$) ensures a large cross-section for photoelectric interactions. It is sometimes mixed with krypton ($Z = 82$). The pressure is increased to about 25 atmospheres to improve sensitivity. Although the latter is still poor by comparison with solid detectors (see section 7.12ff.), it is possible to pack a large number of xenon gas detectors of very uniform sensitivity into a small space. The use of Xe/Kr high pressure gas detectors in computed tomography is discussed in section 10.3.2.

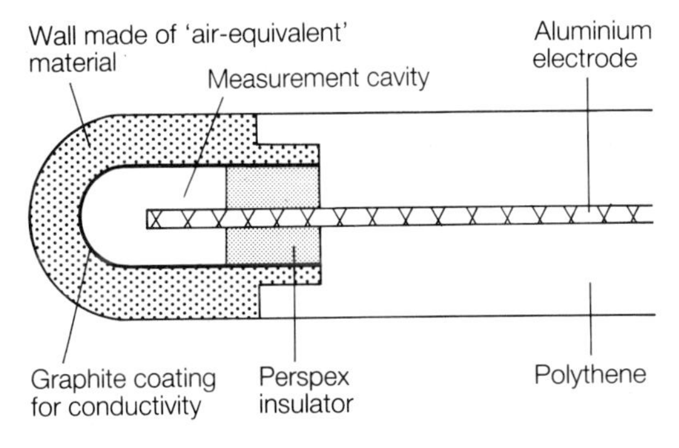

Fig. 7.5. A simple, compact secondary ionization chamber that makes use of the 'air equivalent wall' principle.

7.6 Pocket exposure meters for personnel monitoring

Before leaving the ionization principle for radiation monitoring, the so-called 'fountain pen' dosemeter should be considered briefly. This is illustrated in Fig. 7.6 and is based on the principle of a gold leaf electroscope. When the electroscope is charged, the leaves will diverge because of electrostatic repulsion. If the gas around the leaves is then ionized, the instrument will be discharged.

The dosemeter is charged by an auxiliary charger to a specific potential as shown by the position of the needle on a calibrated scale which may be viewed through the telemicroscope. When placed in a radiation field, the device will be discharged by an amount proportional to the ionization provided by the radiation and the pointer will move across the scale.

The instrument is not very accurate (±20%) and since it must be robust, the casing is usually too thick to allow low energy radiation to penetrate. However, it is a compact, portable device which fits in the top pocket and can provide a useful safeguard in an area where radiation levels may be high.

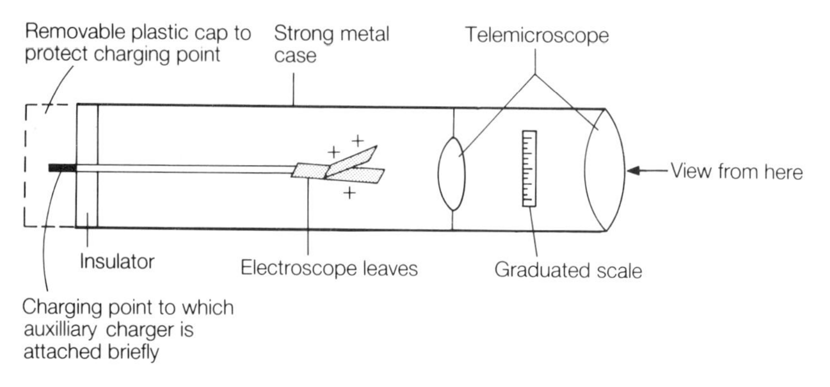

Fig. 7.6. The fountain pen dosemeter.

7.7 Electrical conductivity of a gas

If, when using the equipment shown in Fig. 7.1, the X-ray beam intensity were fixed but the potential difference between the plates were gradually increased from zero, the current flowing from the collector plate would vary as shown in Fig. 7.7. Initially, all ion pairs recombine and no current is

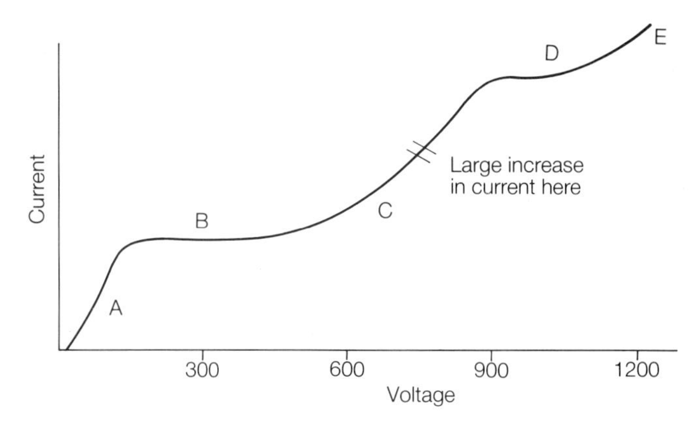

Fig 7.7. Variation of current appearing across capacitor-plates with applied potential difference for a fixed X-ray beam intensity. (A) Loss of ions by recombinations. (B) Ionization chamber region. (C) Proportional counting. (D) Geiger–Müller region. (E) Continuous discharge. The voltage axis shows typical values only.

registered. As the potential difference increases (region A) more and more electrons are drawn to the collector, until, at the first plateau B, all the ion pairs are being collected. This is the region in which the ionization chamber operates and its potential difference must be in the region B.

At C, the current increases again. This is because secondary electrons gain energy from the electric field between the plates and eventually acquire enough energy to cause further ionizations (see Fig. 7.8). **Proportional counters** operate in the region of C. They have the advantage of increased sensitivity, the extra energy having been drawn from the electric field, and, as the name implies, the strength of signal is still

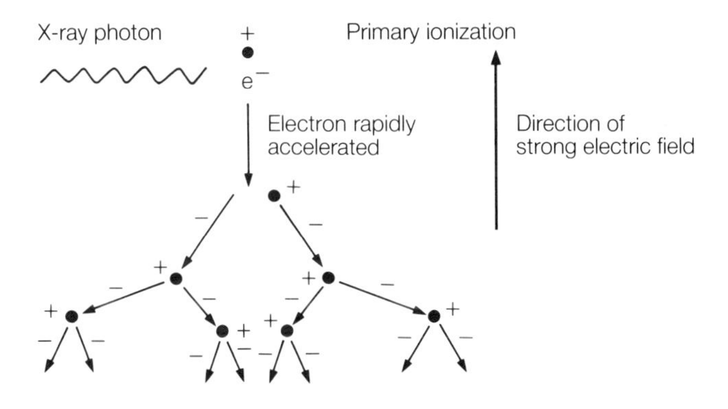

Fig. 7.8. Amplification of ionization by the electric field.

proportional to the amount of primary and secondary ionization. Hence, proportional counters can be used to measure radiation exposure. However, very precise voltage stabilization is required, since the amplification factor is changing rapidly with small voltage changes, and such a device is unsuitable for precision work with a portable measuring device.

Beyond C the amplification increases rapidly until the so-called Geiger–Müller (GM) plateau is reached at D. Continuous discharge takes place at E.

7.8 The Geiger–Müller tube

Essential features of a GM tube are shown in Fig. 7.9 and the most important details of its design and operation are as follows.

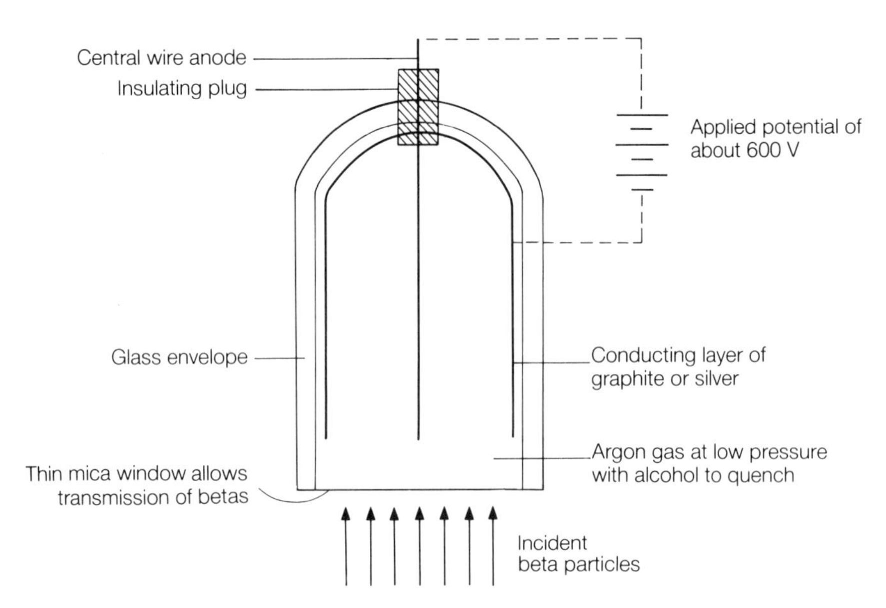

Fig. 7.9. Essential features of an end-window Geiger–Müller tube suitable for detecting beta particles.

1 In Fig. 7.7, the GM plateau was attained by applying a high voltage between parallel plates. In fact it is the electric field $E = V/d$, where d is the distance between the plates, that accelerates electrons. High fields can be achieved more readily using a wire anode since near the wire E varies as

$1/r$, where r is the radius of the wire. Thus the electric field is very high close to a wire anode even for a working voltage of 300–400 V. When working on the GM plateau, D, count rate changes only slowly with applied voltage so very precise voltage stabilization is not necessary.

2 The primary electrons are accelerated to produce an avalanche as in the proportional counter, but in the avalanche discharge excited atoms as well as ions are formed. They lose this excitation energy by emitting X-ray and ultraviolet photons which liberate outer electrons from other gas atoms creating further ion pairs by a process of photoionization. As these events may occur some distance from the initial avalanche, the discharge is spread over the whole of the wire. Because of the high electric fields in a GM tube, the positive ions reach the cathode in sufficient numbers and with sufficient energies to eject electrons. These electrons initiate other pulses which recycle in the counter thus producing a continuous discharge.

3 The continuous discharge must be stopped before another pulse can be detected. This is done by adding a little alcohol or bromine to the counting gas which is either helium or argon at reduced pressure. The alcohol or bromine 'quenches' the discharge because their ionization potentials are substantially less than those of the counting gas. During collisions between the counting gas ions and the quenching gas molecules, the ionization is transferred to the latter. When these reach the cathode, they are neutralized by electrons extracted by field emission from the cathode. The electron energy is used up in dissociating the molecule instead of causing further ionization. The alcohol or bromine also has a small effect in quenching some of the ultraviolet photons.

4 The discharge is also quenched because a space charge of positive ions develops round the anode, thereby reducing the force on the electrons.

5 Finally quenching can be achieved by reducing the external anode voltage using an external resistor and this is triggered by the early part of the discharge.

Once a discharge has been initiated, and during the time it is being quenched, any further primary ionization will not be recorded as a separate count. The instrument is effectively 'dead' until the externally applied voltage is restored to its full value, typically after about 300 μs. This is known as the **dead time**.

Thus the true count is always higher than the real count. The difference is minimal at 10 counts per second but at 1000 counts per second the monitor is dead for $1000 \times 300 \times 10^{-6} = 0.3$ s in every second and losses become appreciable.

7.9 Comparison of ionization chambers and Geiger–Müller tubes

Both instruments have important but well defined roles in radiological monitoring so their strengths and weaknesses must be clearly understood.

Type of radiation
Both respond to X and gamma rays and to fast beta particles. By using a thin window the response of the GM tube can be extended to low energy beta particles, but not the very low energy beta particles from H-3 (18 keV max).

Sensitivity
Because of internal amplification, the GM tube is much more sensitive than the ionization chamber and may be used to **detect** low levels of contamination (but see section 7.11).

Nature of reading
The ionization chamber is designed to collect all primary and secondary radiation and hence to give a reading of exposure or exposure rate. With the GM tube there is no proportional relationship between the count rate and the number of primary and secondary ionizations so it is not a radiation **monitor**.

Size
The ionization chamber must be big enough to collect all secondary electron ionizations. Since the GM tube does not have the property of proportionality, there is no point in making it large and it can be much more compact.

Robustness and simplicity
Generally favour the GM tube.

In conclusion a GM tube is an excellent **detector** of radiation, but it must not be used to measure radiation exposures for which an appropriate **monitor** is required.*

*A 'Compensated Geiger' is sometimes used as a radiation monitor. However it is not very suitable for diagnostic radiology because of poor sensitivity at low energies.

7.10 Relationship between exposure and dose

The second aspect of radiation measurement is to obtain the dose or amount of energy deposited in matter. In respect of the biological damage caused by ionizing radiation, this is more relevant than radiation exposure.

The unit of radiation dose is the gray (Gy), where 1 Gy = 1 J kg^{-1}, and this has replaced the older unit, rad. One rad was 100 erg g^{-1} and thus 1 Gy = 100 rad.

Whereas radiation exposure, by definition, refers to ionization in air, the dose, or energy absorbed from the radiation, may be expressed in any material. Calculation of the dose in say soft tissue, when the radiation exposure in air is known, may be treated as a two stage problem as follows:

1 conversion of exposure in air to dose in air,

2 conversion of dose in air to dose in tissue.

When treated in this way, both parts of the calculation become fairly easy.

Conversion of exposure in air to dose in air

A term that is being used increasingly in radiation dosimetry is KERMA. This stands for Kinetic Energy Released per unit Mass and must specify the material concerned. Note that KERMA places the emphasis on removal of energy from the beam of indirectly ionizing particles (X or gamma photons) in order to create secondary electrons. Radiation dose relates to where those electrons deposit their energy in the medium.

There are two reasons why KERMA in air (K_A) may differ from Dose in air (D_A). First, some secondary electron energy may be reradiated as bremsstrahlung. Second, the point of energy deposition in the medium is not the same as the point of removal of energy from the beam because of the range of the secondary electrons. However, at diagnostic energies bremsstrahlung is negligible and the ranges of secondary electrons are so short that $K_A = D_A$ to a very good approximation.

Now the number of ion pairs generated in each kilogram of air multiplied by the energy required to form one ion pair (W) is equal to the energy removed from the beam. But the first term is the definition of radiation exposure, say E, and the third term is the definition of KERMA in air K_A. Thus

$$E(\mathrm{C\ kg^{-1}}) \times W(\mathrm{J\ C^{-1}}) = K_A(\mathrm{J\ kg^{-1}})$$

Or, since $K_A = D_A$, expressing dose as the subject

$$D_A(\mathrm{J\ kg^{-1}}) = E(\mathrm{C\ kg^{-1}}) \times W(\mathrm{J\ C^{-1}})$$

The energy to form one ion pair, W, is close to 34 J C^{-1} (34 electron volts per ion pair) for all types of radiation of interest to radiologists and, coincidentally, over a wide range of materials of biological importance. By definition 1 Gy = 1 J kg^{-1}. Thus

$$D_A(\mathrm{Gy}) = 34\ E(\mathrm{C\ kg^{-1}})^*$$

In this book, dose in air will be used in preference to exposure. 'Skin dose' will be used to describe the dose in air at the patient.

Conversion of dose in air to dose in tissue

To convert a dose in air to dose in any other material, recall that for a given incident flux of photons, the energy absorbed per unit mass depends only on the mass absorption coefficient of the medium. Hence

$$\frac{D_M}{D_A} = \frac{(\mu_a/\varrho)_M}{(\mu_a/\varrho)_A}$$

Note the use of a subscript 'a' to distinguish the mass absorption coefficient from the mass attenuation coefficient.

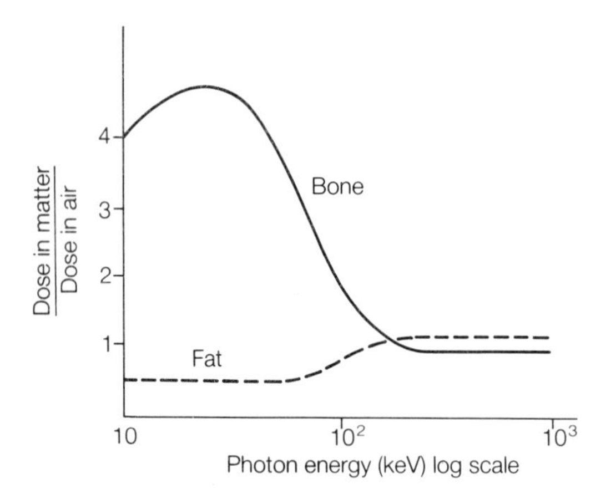

Fig. 7.10. The ratio (dose in matter/dose in air) plotted as a function of radiation energy for bone and fat.

*Many older textbooks use the units roentgen for exposure in air and rad for dose. Since 1 R = 2.58×10^{-4} C kg^{-1}, and 1 Gy = 100 rads, D_A(rad) = $100 \times 2.58 \times 10^{-4} \times 34\ E$ (roentgen). Hence, D_A (rad) = $0.88E$ (roentgen).

It follows that only a knowledge of the relative values of μ_a/ϱ is required to convert a known dose in air to the corresponding dose in any other material.

The ratio D_M/D_A is plotted as a function of different radiation energies for different materials in Fig. 7.10. It is left as an exercise to the reader to justify the shapes of these curves from a knowledge of:

1 the mean atomic number for each material,

2 the relative importance of the Compton and photoelectric effects at each photon energy.

7.11 Radiation doses and dose rates in diagnostic radiology

It is important to have an appreciation of the doses or dose rates involved in some typical radiological situtions and this information is given in Table 7.1. Unless otherwise stated, all figures represent doses or dose rates in air at the skin surface or 'skin doses'. Note the change of scale in the dose rate column for screening.

Values quoted are considered typical. The actual exposure settings used for similar examinations in fact vary widely from one centre to another and it is a matter of concern that some X-ray departments find it is necessary to give a much bigger dose than others to obtain the requisite diagnostic information. Note that because of significant attenuation in body tissues, the doses received by organs that are particularly at radiation risk may be appreciably less than the figures given here.

7.12 Scintillation detectors and photomultiplier tubes

The fundamental interaction process in a scintillation detector is fluorescence which was discussed in section 4.2. One material widely used for such detectors is sodium iodide to which about 0.1% by weight of thallium has been added—NaI (Tl). The traps generated by thallium in the NaI lattice are about 3 eV above the band of valency electrons so the emitted photon is in the visible range. The detector is carefully designed and manufactured to optimize light yield. Note that whereas the **number** of photons emitted is a function of the energy imparted by the X-ray or gamma ray interaction, the energy or wavelength of the photons depends only on the positions of the energy levels in the scintillation crystal.

When a scintillation crystal is used as a monitor, its advantages over the detectors discussed so far are as follows:

Table 7.1. Typical doses or dose rates associated with radiographic and nuclear medicine procedures

Outputs		
Normal X-ray set (80 kVp 75 cm *FFD*) — single phase (full wave)		0.1 mGy mAs^{-1}
Normal X-ray set (80 kVp 75 cm *FFD*) — three phase		0.15 mGy mAs^{-1}
Fluoroscopy (100 kVp 1 mA 50 cm *FFD*)		20 mGy min^{-1}
*Skin doses for planar films**		Dose (mGy)
Abdomen (80 kVp 50 mAs 100 cm *FFD*)		6
Lateral lumbar spine (90 kVp 200 mAs 100 cm *FFD*)		30
Chest (65 kVp 12 mAs 200 cm *FFD*)		0.2
Neonate chest or abdomen (60 kVp 1 mAs 90 cm *FFD*)		0.05
*At image receptor**		Dose (μGy)
To produce unit optical density on screen film		10
At image intensifier for a single image in digital angiography		3
(dose rate at image intensifier surface in fluoroscopy~0.5 μGy.s^{-1})		
CT skin doses†		Dose (mGy)
Chest (140 kVp 40 mA 15 slices)		15
Abdomen		5
Nuclear medicine	Typical activity (MBq)	Effective dose equivalent to patient (mSv)‡
Bone scan (Tc-99m)	600	4
Lung scan (Tc-99m)	80	1
Myocardial imaging (Tl-201)	80	7
Functional renal imaging (I-123)	20	0.3
Whole body—abscess (In-111)	40	1.5

* Note that the skin dose to the patient is determined to a large extent by the air dose required at the front surface of the image receptor. The major factor relating these two is attenuation within the patient, and grid if any, which, as shown above, will vary from about 5 in a neonate to perhaps 3000 for a lateral lumbar spine.
† Corresponding doses to the bone marrow are, respectively, 5 mGy and 0.3 mGy.
‡ For the purposes of this table, 1 mSv=1 mGy—see section 11.4.

1 Since it is a high density solid, its efficiency, especially for stopping higher energy gamma photons, is greatly increased. A 2.5 cm thick NaI (Tl) crystal is almost 100% efficient in the diagnostic X-ray energy range. Contamination monitors that must be capable of detecting of the order of 30 counts per second from an area of 1000 mm^2 invariably contain scintillation crystals.

2 It has a rapid response time, in contrast to an ionization chamber which responds only slowly owing to the need to build up charge on the capacitor plates.

3 Different scintillation crystals can be constructed that are particularly sensitive to low energy X-rays or even to neutrons. Beta particles and alpha particles can be detected using plastic phosphors.

NaI (Tl) detectors are used extensively in nuclear medicine and the properties that render them particularly appropriate for *in vivo* imaging will be discussed in Chapter 8. Alternative scintillation detectors are caesium iodide doped with thallium, and bismuth germanate. Like NaI (Tl), the latter has a high detection efficiency, and is preferable at high counting rates (e.g. for CT) because it has little 'after glow'—persistence of the light associated with the scintillation process. Bismuth germanate detectors also exhibit a good dynamic range and long-term stability.

The light signal produced by a scintillation crystal is too small to be used until it has been amplified and this is almost invariably achieved by using a photomultiplier tube (PMT).

The main features of the PMT coupled to a scintillation crystal (Fig. 7.11) are as follows:

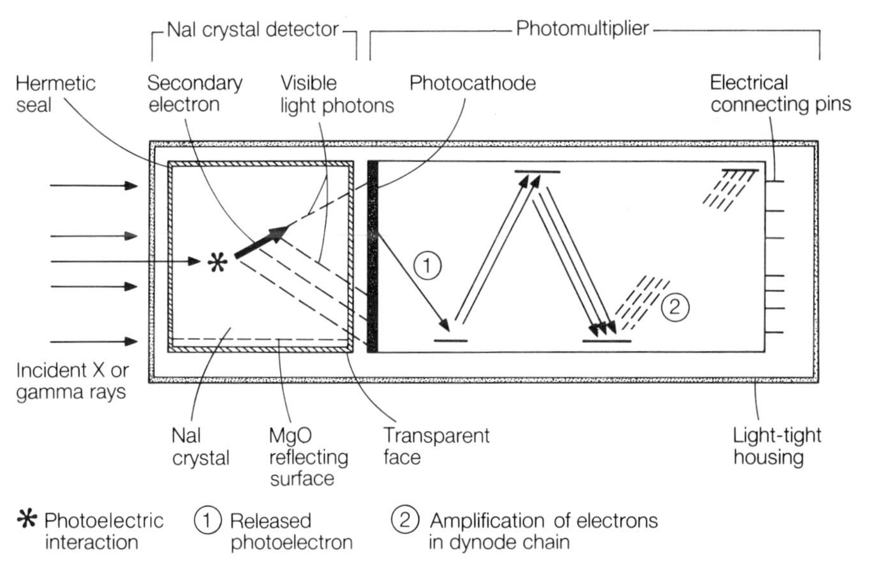

Fig. 7.11. The main features of a PMT coupled to a scintillation crystal for radiation detection.

1 An evacuated glass envelope, one end of which has an optically flat surface. Since photon losses must be minimized, the scintillation crystal must either be placed in contact with this surface, or if it is impracticable, must be optically coupled using a piece of optically transparent plastic—frequently referred to as a 'light guide' or 'light pipe'.

2 A layer of photoelectric material such as caesium–antimony. The characteristic of such materials is that their work function, i.e. the energy required to release an electron, is very low. Thus electrons are emitted when visible or ultraviolet photons fall on the photocathode, although the efficiency is low with only one electron emitted for every 10 incident photons.
3 An electrode system to provide further amplification. This system consists of a set of plates, each maintained at a potential difference of about 100 V with respect to its neighbours and coated with a metal alloy, say of magnesium–silver, designed to release several electrons for every one incident on it. Each plate is known as a dynode and there may be 12 dynodes in all so the potential difference across the PMT will be in the region of 1200 V. If each dynode releases four electrons for each incident electron and there are 12 dynodes, the amplification in the PM tube is 4^{12} or about 10^7. Further more, this figure will be constant to within about 1% provided the voltage can be stabilized to 0.1%.

Note that during the complete detection process in the crystal and PMT, the signal twice takes the form of photons, once as X-ray photons and once as visible light photons, and twice takes the form of electrons. Since a PMT is an extremely sensitive light detector, great care must be taken to ensure that no stray light enters the system.

7.13 Spectral distribution of radiation

Thus far nothing has been said about the third factor in a complete specification of a beam of ionizing radiation, namely the energy spectrum of the photons. Under appropriate conditions, a scintillation crystal used in conjunction with a PMT may be used to give such information.

If the crystal is fairly thick, and made of high density material, preferably of high atomic number, most X or gamma rays that interact with it will be completely stopped within it and each photon will give up all its energy to a single photoelectron. Note that for this discussion it is unimportant whether there is a single photoelectric interaction or a combination of Compton and photoelectric interactions (contrast with the discussion in section 8.2.2).

Since the number of visible light photons is proportional to the photoelectron energy, and the amplification by the PMT is constant, the strength of the final signal will be proportional to the energy of the interacting X or gamma ray photon. A pulse height analyser may now be used to determine the proportion of signals in each predetermined range of strengths and if the pulse height analyser is calibrated against a monoenergetic beam of gamma rays of known energy, the results may be converted into a spectrum of incident photon energies (Fig. 7.12).

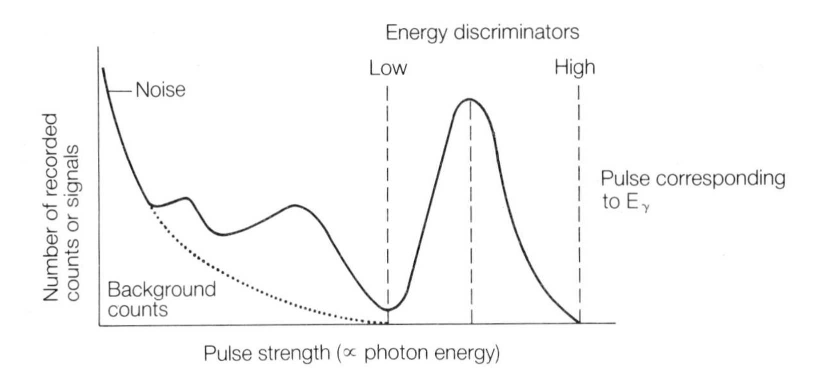

Fig. 7.12. Use of a pulse height analyser to determine the spectrum of photon energies in an X-ray beam. A typical pulse height spectrum obtained after monoenergetic gamma rays, E_γ, have passed through scattering material, and the use of energy discriminators to select the peak are shown. The tail of pulses of lower strength is due to Compton scattering. The sharp rise for very low pulses is due to noise.

Unfortunately, because of statistical problems, this method of determining the spectral distribution of a beam of X or gamma rays is not as precise as one would wish. One limitation of the scintillation crystal and PMT combination as a detector of ionizing radiation is that the number of electrons entering the PMT per primary X or gamma ray photon interaction is rather small. There are two reasons:

1 about 30 eV of energy must be dissipated in the crystal for the production of each visible or ultraviolet photon, and

2 even assuming no loss of these photons, only about one photoelectron is produced for every 10 photons on the PMT photocathode.

Thus to generate one electron at the photocathode requires about 300 eV and a 140 keV photon will produce only about 400 electrons at the photocathode. This number is subject to considerable statistical fluctuation ($N^{\frac{1}{2}} = 20$ or 5%). The result is that a monoenergetic beam of gamma rays will produce a range of pulses and will appear to contain a range of energies (Fig. 7.13a). This is a particular problem in the gamma camera and will be considered further in Chapter 8.

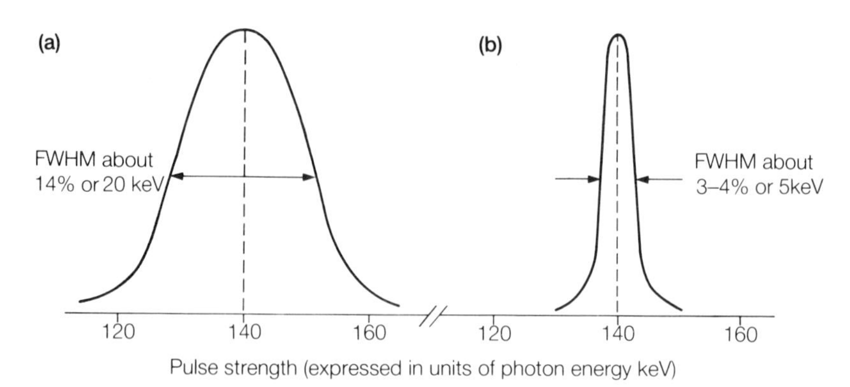

Fig. 7.13. Typical spread in the strength of signals from a monoenergetic beam of gamma rays when using as the primary detector (a) a NaI (Tl) crystal, (b) a solid-state device.

7.14 Semi conductor detectors

In a semi conductor, the forbidden band of energy levels is very narrow and therefore only small quantities of energy, sometimes as little as 2.5–3 eV, are required to release electrons from the filled band into the conduction band. As a result, a 140 keV photon is capable of releasing very large numbers of electrons and the statistical fluctuations are considerably reduced.

In theory, a very pure semi conductor such as silicon or germanium could be used as a detector. If electrodes were attached across a slice of the material, the electrons set free by ionization events would be collected at the anode and electrons from the cathode would drift into the material to neutralize the positive ions. The detector would have a high sensitivity per unit volume because it is solid and, for the reason given above, would have a high energy resolution.

Until quite recently it has not been possible to manufacture pure enough material to prevent electrons becoming 'lost' in the lattice so more complex structures have been adopted. One possibility is illustrated in Fig. 7.14a. Very small quantities of impurity, say a few parts per million are added deliberately to the purest obtainable silicon. The material on the left contains an n-type impurity, e.g. antimony or arsenic which has five valence electrons compared to the four in silicon and germanium, and is readily able to make free electrons available to the lattice. The material on the right contains a p-type impurity, e.g. gallium with only three valence electrons. It provides a 'hole' where spare electrons may reside. At the

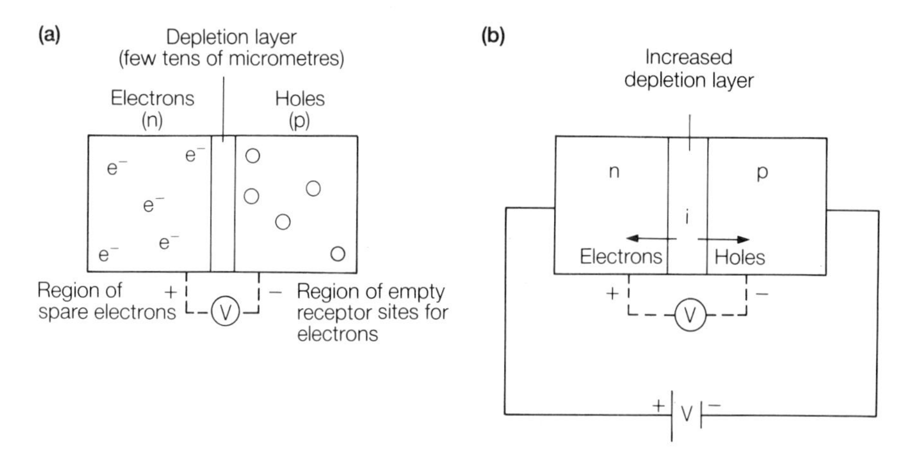

Fig. 7.14. Schematic arrangement of the disposition of mobile electrons and holes in an n–p silicon semi conductor (a) when no voltage is applied, (b) when the depletion layer is increased by applying a potential difference such that the bulk of the n-type material is at a higher potential than the bulk of the p-type material.

interface between the n-type and p-type materials, a certain amount of diffusion takes place with electrons occupying vacant holes. This process ceases when the electron imbalance has established a potential difference that is sufficient to prevent further flow. A non-conducting depletion zone results as shown in Fig. 7.14a.

Such a structure has rectifying properties (see section 2.3.4) because if a potential is applied such that the p material is positive with respect to the n material, the internal potential barrier is reduced and both electrons and holes flow freely. However, if a reverse potential is applied, the electrons and holes are drawn away from the junction, increasing the depletion layer until the internal potential across it is equal and opposite to the applied potential (Fig. 7.14b). No current then flows.

The depletion layer provides an excellent radiation detector because if any electrons are generated as a result of ionizing interactions, they can migrate to the anode and be registered as a current. The thickness of the depletion layer is determined by the magnitude of V.

In practice only very thin silicon-based detectors can be constructed to this design so they are not very suitable for direct detection of X or gamma rays. However, they do respond very well to visible light and near infra red (Fig. 7.15) so they can be used as silicon photodiodes in conjunction with a scintillation crystal. This detector is 10^4 times more sensitive than a gas filled detector of similar volume operating at atmospheric pressure. A factor of 10^3 comes from the ratio of densities, the other factor of 10

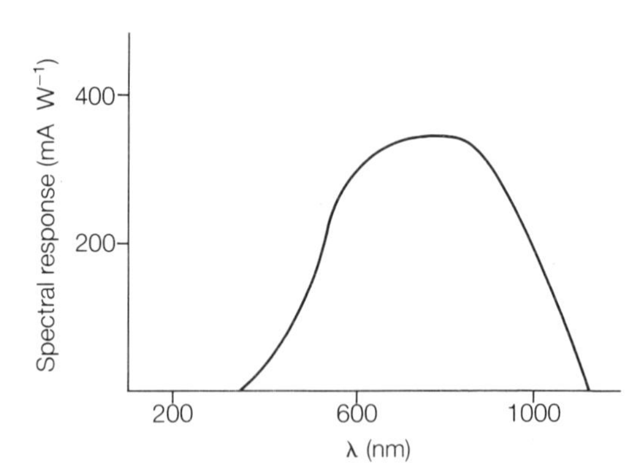

Fig. 7.15. Spectral response of a silicon photodiode.

because only 3 eV is now required to form an ion pair, not 30 eV. Further advantages of the solid state photodiode are linearity over a wide range of intensities (9–10 orders of magnitude), a response time generally better than 1 μs and an operating temperature range of −20 to +80°C.

For direct detection of X and gamma radiations, thicker crystals of germanium with lithium diffused into them are used. They have adequate efficiency but, unlike silicon-based detectors which can operate at room temperature, they must be cooled to liquid nitrogen temperature (−190°C) during the whole of their working life. Very high purity germanium detectors (only about 10^9–10^{11} electrical impurities per cubic centimetre), which only need to be cooled during operation to reduce noise, are now available and have replaced lithium drifted detectors for most purposes.

Solid state detectors dispense with the need for relatively bulky PMTs and the requirement for stabilized high voltage supplies. As previously intimated, they have a high ionization yield so the energies of photoelectrons generated by X or gamma ray absorption can be measured with very high precision (Fig. 7.13b).

7.15 Thermoluminescence dosemeters

In a thermoluminescent dosemeter (TLD), which also depends upon electrons being trapped in the forbidden energy band (see section 4.2), the radiation message is not emitted spontaneously in the form of visible light but is stored almost indefinitely and only released when the TLD is heated, generally to about 300–400°C. TLDs have wide application for personnel dosimetry and will be discussed in detail in Chapter 11. Here their use as general-purpose dosemeters is considered briefly.

Among their advantages are the following:

1 response is linear with dose over a wide range;
2 sensitivity is almost energy independent (see section 7.17);
3 adequate sensitivity is achieved in a very small volume.

Disadvantages of TLDs are:

1 they must be calibrated against standard radiation sources;
2 careful annealing is required after read-out to ensure that the TLD material returns to the same condition in respect of the number of available traps otherwise the sensitivity and hence calibration factor of the TLD may change.

7.16 Photographic film

The properties of photographic film as a recording medium for X-rays were discussed in Chapter 4. Here it is sufficient to note that as a dosemeter, apart from its specialized use as a personnel monitor (Chapter 11), film has a number of disadvantages:

1 Because of the shape of its characteristic curve, it is a highly non-linear device. Therefore calibration is required at a large number of radiation exposures.
2 Again because of the shape of the characteristic curve, a given film may only be used over a limited range of dose.
3 Sensitivity expressed as film blackening per gray in air is very dependent on energy, especially in the diagnostic range (see section 7.17).

7.17 Variation of detector sensitivity with photon energy

An important question to consider for any radiation monitor is whether or not its sensitivity will change with photon energy. A good example of a detector which shows marked variation in sensitivity is photographic film (see Fig. 7.16).

Since, in general, the response of a detector is proportional to the amount of energy it absorbs, the question can be answered in terms of absorbed energy. However, as shown on the left hand axis of Fig. 7.16, 'sensitivity' means detector response per gray, or per unit absorbed dose in air. The latter depends on the mass absorption coefficient of air. Thus

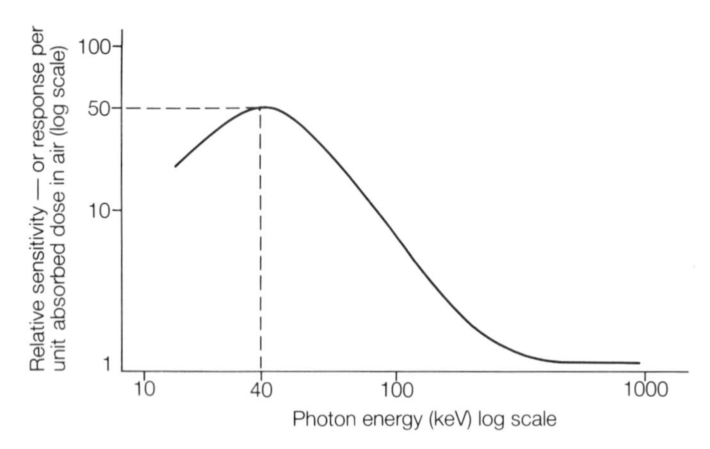

Fig. 7.16. Variation in sensitivity of photographic film as a function of photon energy.

sensitivity is determined by the relative mass absorption coefficients of the detector and air. If both vary with photon energy in a similar manner, detector sensitivity will not change, but if the variation of the two mass absorption coefficients with energy is dissimilar, detector sensitivity will change.

With this information, it is possible to interpret Fig. 7.16. In the vicinity of 1 MeV, interactions are by Compton processes. The mass absorption coefficient is therefore independent of atomic number and sensitivity is independent of photon energy. However as the photon energy decreases and approaches 100 keV, photoelectric absorption becomes important. This effect is much greater in film, which has a high mean atomic number (Z for silver = 47, Z for bromine = 35), than it is in air (Z = 7.6). Thus the mass absorption coefficient for film $(\mu/\varrho)_F$ increases much more rapidly than $(\mu/\varrho)_{air}$ and film sensitivity increases.

Note that below about 40 keV film sensitivity decreases. This is because of absorption edges for silver and bromine at 26 keV and 13 keV respectively. Near an absorption edge, although the incidence of photoelectric interactions is high, conditions are very favourable for the generation of characteristic radiation. Since a material is relatively transparent to its own characteristic radiation, a high proportion of this energy is reradiated and is therefore not available for film blackening (Fig. 7.17).

Detectors that have a mean atomic number close to that of air or soft tissue, for example lithium fluoride, show little variation in radiation sensitivity with photon energy, and are sometimes said to be 'tissue equivalent'. Note that lithium fluoride does show some variation in sensitivity at

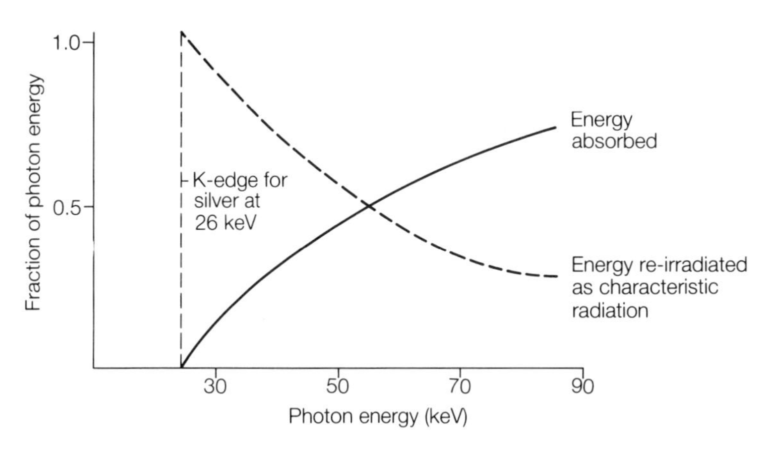

Fig. 7.17. Fraction of photon energy finally absorbed as a function of photon energy in the vicinity of a K absorption edge.

very low energies where even small differences in atomic number can be important.

7.18 Conclusions

A wide range of properties of ionizing radiations is available for radiation measurement. Ionization in air is taken as the reference standard, partly because it can be related directly to a fundamental physical process and partly because it is very sensitive in terms of the number of ion pairs created per unit energy deposition.

The choice of instrument in a given situation will depend on a variety of factors including sensitivity, linearity, dynamic range, speed (at high counting rates), variation in response with photon energy, uniformity of response between detectors, long-term stability, size, operating conditions such as temperature or requirements for established voltage supplies, and cost.

It is important to distinguish carefully between radiation measurement with for example an ionization chamber and radiation detection with a Geiger–Müller counter. The latter is very sensitive and may be very suitable for detecting radiation leakage or contamination from spilled radioactivity. It should not normally be used as a radiation measuring device.

Three aspects of radiation measurement have been identified—exposure, dose and the quality or spectral distribution of the radiation. If an absolute measure of absorbed dose is required, reference back to the ionization process and cross-calibration will be required. However, for many applications, for example in digital radiology and nuclear medicine, there is no need to convert the numerical data into dose routinely so uniformity of response and long-term stability are more important. One area in which quantitative dose measurements are required is personnel monitoring and this aspect will be discussed in Chapter 11.

If information is required on spectral distributions, a detector whose response is proportional to the energy of an individual photon must be used. A NaI (Tl) scintillation detector will normally be chosen and might be used for example to check the homogeneity of an X-ray beam or to identify (and reject) low energy scattered radiation from a monochromatic gamma ray beam. If very fine energy resolution is necessary a cooled solid-state detector may be required.

References and further reading

Greening J. R. (1985) *Fundamentals of Radiation Dosimetry*, 2nd edn. (Medical Physics Handbooks 15) Adam Hilger Ltd, Bristol, and the Hospital Physicists' Association.
Lovell S. (1979) *An Introduction to Radiation Dosimetry*. Cambridge University Press.
McAlister J. M. (1979) *Radionuclide Techniques in Medicine*. Cambridge University Press.
Tait W. H. (1980) *Radiation Detection*. Butterworths, London.
Whyte G. N. (1959) *Principles of Radiation Dosimetry*. Wiley, New York.

Exercises

1 Suggest reasons why ionization in air should be chosen as the basis for radiation measurement.
2 Draw a labelled diagram of a (free air) ionization chamber and explain the principle of its operation.
3 Explain how the following problems are overcome in an ionization chamber:
(a) recombination of ions,
(b) definition of the precise volume from which ions are collected.
4 Explain how a (free air) ionization chamber might be used to measure the exposure at a given point in an X-ray beam. How would you expect the exposure to change if thin aluminium filters were inserted in the beam about half way between the source and chamber?

5 How would you expect the dose rate in air to vary with distance from an X-ray set and why?
6 Outline briefly the important features of an experimental arrangement for measuring exposure in air and discuss the factors which limit the maximum energy of the radiation that can be measured in this way.
7 Explain from first principles why the reading on a free air ionization chamber will decrease if the temperature increases.
8 Explain the importance of the concept of electron equilibrium in radiation dosimetry.
9 Describe a small cavity chamber for measurement of radiation exposure. Discuss the choice of material for the chamber wall and its thickness.
10 Describe the operation of a Geiger–Müller tube and explain what is meant by 'dead time'.
11 Define the gray and show how absorbed dose is related to exposure.
12 Show graphically how (dose in matter/dose in air) varies with radiation energy for water, bone and muscle between 10 keV and 1 MeV, and explain the results.
13 Explan how a scintillation detector works.
14 Show that the energy of a photon in the visible range is about 3 eV.
15 Describe the sequence of events that leads to a pulse of electrons at the anode of a photomultiplier tube if the tube is directed at a NaI scintillation crystal placed in a beam of photons.
16 Explain how a scintillation detector may be used to measure
(a) the energy and
(b) the intensity of a beam of radiation.
17 Explain in as much detail as possible the shape of the curve in Fig. 7.16.
18 Describe the process of thermoluminescence and explain how a thermoluminescent material may be calibrated for dosimetry.

8

Diagnostic Imaging with Radioactive Materials

8.1 Introduction

Radioactive materials are used in diagnosis in many ways and a useful subdivision is in terms of the type of procedure involved. For example some studies are carried out entirely *in vitro*. A specimen, usually blood, is taken from the patient and is incubated with radioactive precursors of the metabolite of interest. Various constituents can then be isolated and their gamma or beta activity counted in a well counter or liquid scintillation counter respectively.

Other studies are carried out entirely *in vivo* and these may be further subdivided into those which are primarily concerned with counting and those which involve imaging. Counting studies are normally designed to measure very small quantities of radioactivity in the body, for example natural potassium-40 or trace elements like copper and zinc. For such work the principal requirement is high sensitivity. Alternatively the emphasis may be on imaging, with the primary requirement being to obtain information concerning the spatial distribution of activity. This chapter deals with the physical principles involved in obtaining diagnostic quality images after a small quantity of radioactive material has been administered to the patient in a suitable form.

The basic requirements of a good imaging system are:

1 a device that is able to use the radiation emitted from the body to produce high resolution images, supported by electronics, computing facilities and displays that will permit the resulting image to be presented to the clinician in the manner most suitable for interpretation;

2 a radionuclide that can be administered to the patient at sufficiently high activity to give an acceptable number of counts in the image without delivering an unacceptably high dose of radiation to the patient;

3 a radiopharmaceutical, i.e. a radionuclide firmly attached to a pharmaceutical, that shows high specificity for the organ or region of interest in the body.

Each of these features will be considered in this chapter.

It is important to recognize that, when detecting *in vivo* radioactivity, sensitivity and spatial resolution are mutually exclusive (see Fig. 8.1). The arrangement on the left (Fig. 8.1a) has high sensitivity because a large amount of radioactivity is in the field of view of the detector, but poor resolution. The arrangement on the right (Fig. 8.1b) has better resolution but correspondingly lower sensitivity. Since gamma rays are emitted in all directions, sensitivity can be increased by increasing the detector area and,

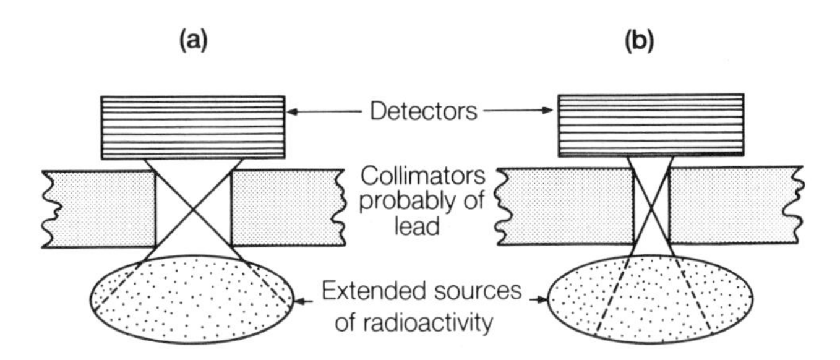

Fig. 8.1. Collimator design showing conflicting requirements of sensitivity and resolution. Arrangment (a) where the detector has a wide acceptance angle will have high sensitivity but poor resolution, whereas arrangement (b) will have much better resolution but greatly reduced sensitivity.

when this area is maximized in the so-called 'whole body counter', resolution has been effectively reduced to zero. For a review of whole body counting see Andrews *et al.* (1973). In diagnostic imaging spatial resolution is important and sensitivity must be sacrificed. A modern gamma camera (see section 8.2.2) is at least 100 times less sensitive than a typical whole body counter and records no more than 1 in 10^4 of the gamma rays emitted from that part of the patient within the field of view of the camera. Furthermore, any additional loss of sensitivity in the complete system will result in an image of inferior quality unless the imaging time is extended to compensate. Therefore this chapter also considers the factors that limit image quality and the precautions that must be taken to optimize the images obtained using strictly controlled amounts of administered activity and realistic imaging times.

8.2 Principles of imaging

The radiations that are most suitable for *in vivo* imaging are medium energy gamma rays in the range 100–200 keV. Lower energy gammas, alpha particles and negative beta particles are stopped in the body, whilst higher energy gammas are difficult to stop in the detector. Further discussion of this point and its influence on the choice of radionuclide and radiopharmaceutical is deferred until section 8.3 where factors affecting the quality of radionuclide images are considered.

In all commercial equipment currently available, the radiation detector is a crystal of sodium iodide doped with about 0.1% thallium—NaI (Tl). The sodium iodide has a high density (3.7×10^3 kg m^{-3}) and since iodine has a high atomic number ($Z = 53$) the material has a high stopping efficiency for gamma rays. Furthermore, provided the gamma ray energy is not too high, most of the interactions are by the photoelectric effect (see section 3.5) and result in a light pulse proportional to the gamma ray energy. This is important for discriminating against scatter (see section 8.2.1). The thallium increases the light output from the scintillant, about 10% of the gamma ray energy being converted into light. This yields about 4000 visible photons from a 140 keV gamma ray. Finally, the light flashes have a short decay time, of the order of 0.2 μs. Thus the crystal has only a short dead time and can be used for quite high counting rates. One disadvantage of the NaI (Tl) detector is that it is hygroscopic and thus must be placed in a hermetically sealed container.

Two fundamentally different methods have been developed for obtaining information about the spatial distribution of radioactivity in the body.

The older and conceptually simpler approach is to 'scan' the patient. Positional information is obtained by the purely mechanical method of moving two relatively small scintillation crystal detectors positioned one above and one below the patient, in a systematic raster motion. The field of view is determined by the scan limits.

A more widely used method now is the gamma camera. This is a stationary device incorporating a much larger scintillation crystal. An array of photomultiplier tubes (PMTs) and appropriate electronic circuits are used to identify the exact position of a scintillation in the crystal and to convert this information into an image. For a parallel hole collimator (see section 8.2.2) the field of view is now defined by the crystal area.

Scanners and cameras and the display systems used with them are discussed in more detail in the following sections.

8.2.1 The scanner

Essential features of a typical scanner are shown in Fig. 8.2. They may be considered under three main headings:

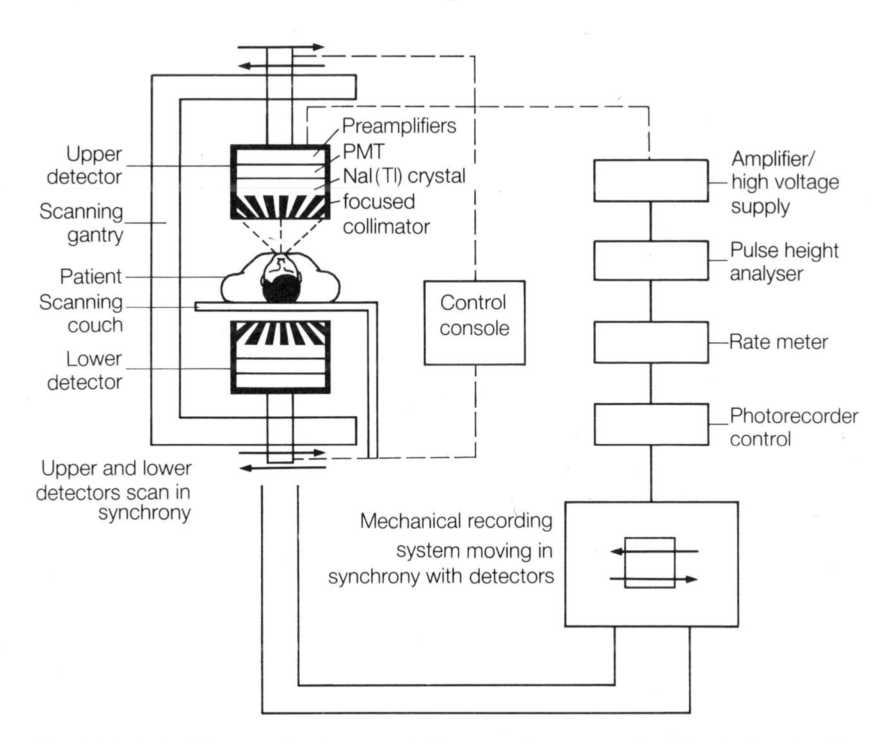

Fig. 8.2. Labelled diagram showing essential features of a scanner. For details of the functions of various parts see text.

The detector system

Two detectors are arranged, one above and one below the patient. Each detector is a scintillation crystal which, nowadays, is invariably sodium iodide doped with thallium, coupled optically to a PMT. The crystals are typically 10 cm in diameter and 5 cm thick. As discussed in section 7.13, when a gamma ray interacts with such a crystal by the photoelectric effect, a light pulse is produced which is proportional to the gamma ray energy. This is subsequently converted into an electronic pulse with proportionality maintained.

The field of view is defined by a focused collimator (Fig. 8.3). Typically, the holes are tapered so that the collimator is 'focused' at a point 10–15 cm below its lower surface. This design ensures that the system has a resolution comparable with that of a small hole but sensitivity is increased by a factor N, the number of holes. Note that the collimator is only 'focused' in the sense that all channels are directed to the same point. No refraction of gamma rays takes place, as when light is focused.

An electronic system amplifies the pulses generated by the PMTs in the detector head and **pulse height analysis** is used to select those pulses corresponding to the gamma ray energy of the radionuclide being emitted. For a radionuclide emitting monoenergetic rays, the technique of pulse height analysis ought, in principle, to discriminate completely between scattered and unscattered rays. Consider for example a 140 keV gamma ray from technetium-99m. When it interacts with an NaI (Tl) crystal, it does so primarily by the photoelectric effect, producing a signal that is proportional to the gamma ray energy. Any ray that has been scattered in

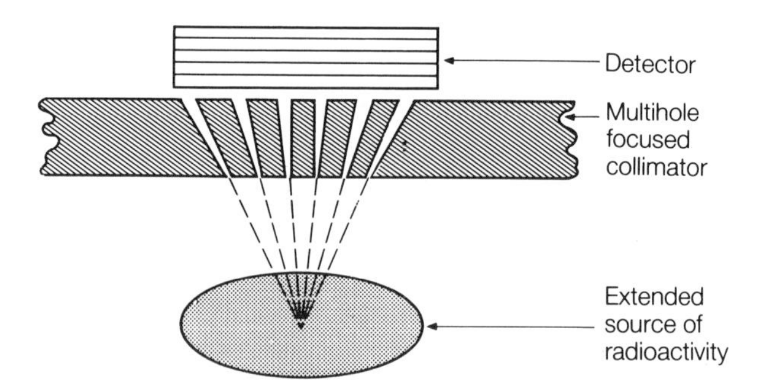

Fig. 8.3. Design of a multihole focused collimator which combines some of the advantages of good resolution with increased sensitivity.

the patient by the Compton effect will be of lower energy and will produce a smaller signal that can be identified and rejected (but see section 8.3.4).

Motor drives move the detectors in a raster pattern over the region of interest with typical sampled areas varying from 1 mm × 1 mm for a small organ such as the thyroid, to 7 mm × 7 mm when scanning the whole body.

2 *The display system*

Pulses selected from the amplifier are used as input signals to a display unit. The density of the final record must correspond to the density of information gathered by the detectors, and this can be achieved by using an analogue rate meter to monitor the input pulse rate. The continuously variably voltage output of the analogue rate meter is used to drive a voltage sensitive pulse generator.

To obtain a permanent record, a recorder head with a built-in source of light can be driven in a raster pattern, which matches that of the detectors, over a standard X-ray film. The light intensity is controlled by the voltage sensitive pulse generator. Alternatively, a printing head can be driven

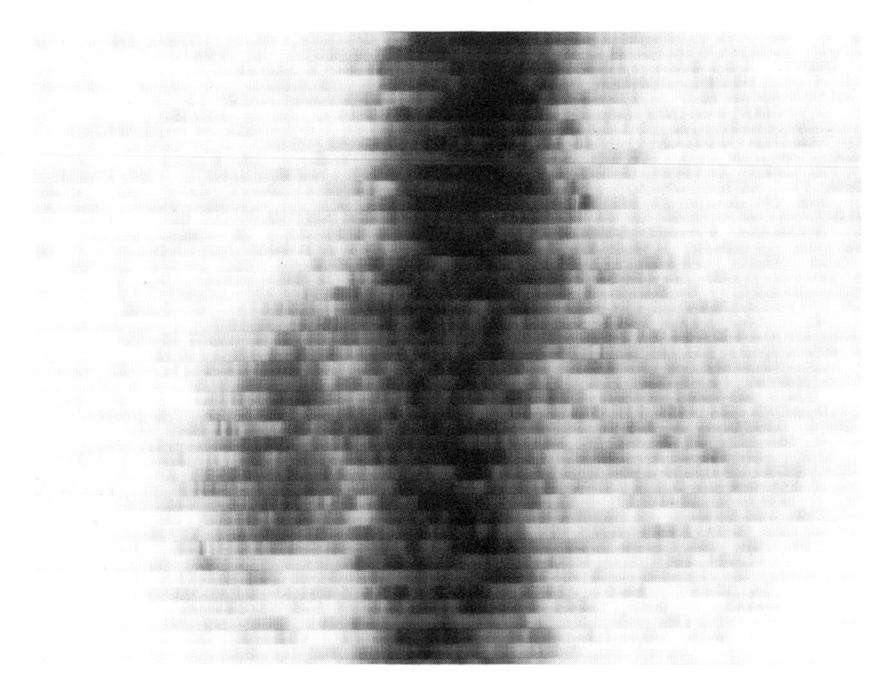

Fig. 8.4. An example of part of an early bone scan produced by a rectinlinear scanner. The region shown is the lower dorsal and upper lumbar spine.

across a sheet of paper, producing an image in which either the density of dots represents the count rate, or, using multicoloured ribbons, ranges of count rate may be printed in different colours. This is an elementary form of digitized image. Part of an early bone scan produced by a scanner is shown in Fig. 8.4. The raster pattern is clearly visualized.

Mechanical features

The control console contains adjustments for setting energy windows for the photoelectric peak of the radionuclide, changing the speed of movement of the detector heads in accordance with the count rate observed on an auxiliary rate meter and changing scan limits. A useful accessory is an anatomical marker which allows intense additional spots to be superimposed on the image to mark the position of reference points on the patient, e.g. suprasternal notch, or to indicate left and right where confusion might exist. The examination couch must be comfortable since the patient will have to remain still for perhaps 15–20 min.

8.2.2 The gamma camera

A gamma camera may consist of two units, the collimated detector mounted on a stand to allow it to manoeuvre around the patient, and the console containing pulse processing electronics and displays. Compared with a scanner, a much bigger but thinner crystal is now used, backed by a large number of PMTs. Modern gamma cameras contain at least 37 PMTs and some contain as many as 91.

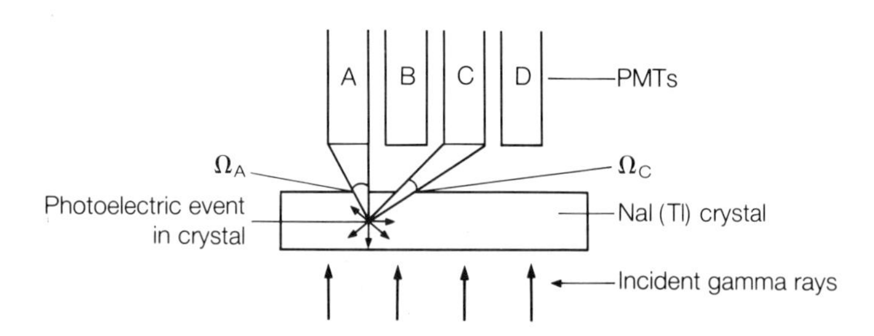

Fig. 8.5. Use of an array of PMTs to obtain spatial information about an event in an NaI (Tl) crystal. Light photons spread out in all directions from an interaction and the signal from each PMT is proportional to the solid angle subtended by the PMT at the event. The signal from PMT A is proportional to Ω_A and much greater for the event shown than the signal from PMT C which is proportional to Ω_C.

The principles of operation are as follows. First, collimation is used to establish a direct spatial relationship between the point of emission of a gamma ray in the patient and the point at which it strikes the crystal. When an incident gamma ray interacts with the scintillation crystal (an event), visible light photons are produced. These spread out into the crystal and some go to each PMT. As shown in Fig. 8.5, the number of photons reaching each PMT, and hence the strength of the signal, will be determined by the solid angle subtended by the event at that PMT. Hence, by analysing all the PMT signals, it is possible to determine the position of the gamma ray interaction in the crystal. Essential features of the gamma camera may be considered under four headings.

The detector system

Components of the detector system are shown in Fig. 8.6. In the gamma camera, crystal thickness must be a compromise. A very thin crystal reduces sensitivity whereas a very thick crystal degrades resolution (Fig. 8.7). A camera crystal is typically 9–12 mm thick. As shown in Table 8.1, it stops most of the 140 keV photons from Tc-99m but is less well suited to higher energies.

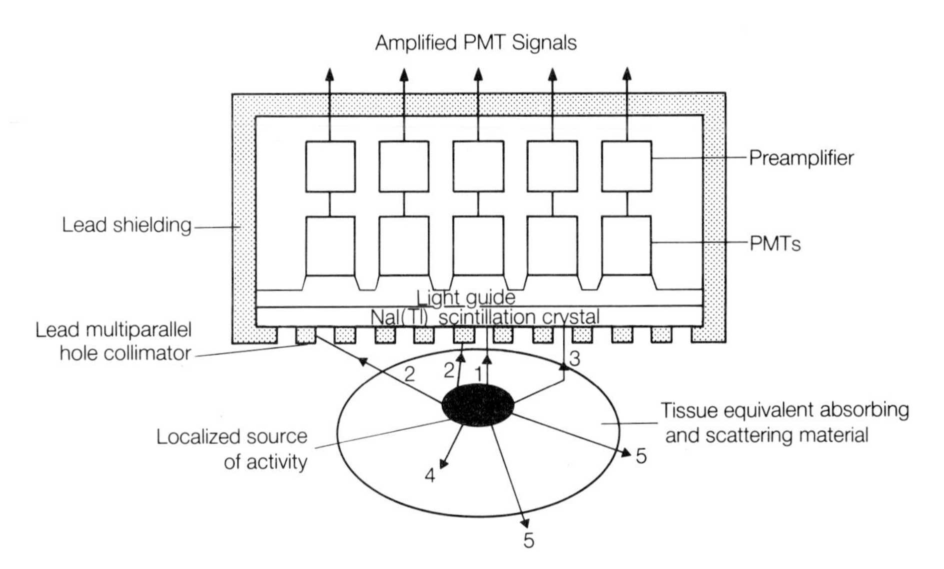

Fig. 8.6. Basic components of a gamma camera detector system. The fates of photons emitted from the source may be classified as follows: **1** useful photon, **2** oblique photon removed by collimator, **3** scattered photon removed by pulse height analysis, **4** absorbed photon contributing to patient dose but giving no information, **5** wasted photons emitted in the wrong direction.

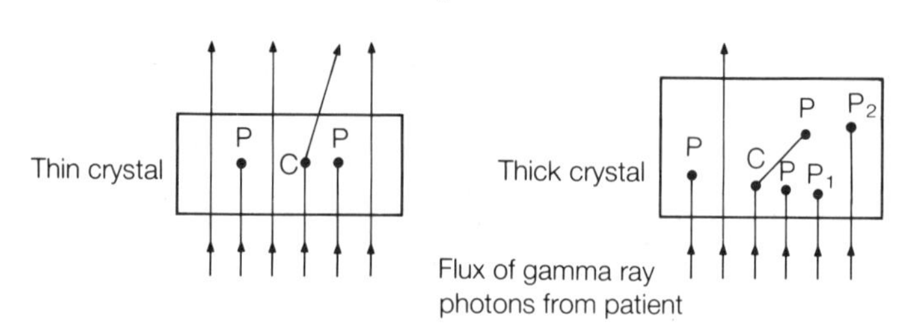

Fig. 8.7. Interactions of gamma rays with thin and thick NaI (Tl) crystals. P = photoelectric absorption. C = Compton scattering. With a thin crystal, many photons may pass through undetected, thereby reducing sensitivity. With a thick crystal the image is degraded for two reasons. First, the distribution of light photons to the PMTs for an event at the front of the crystal such as P_1 will be different from the distribution for an event at the rear of the crystal such as P_2. Second, scatter in the crystal degrades image quality since the electronics willposition 'the event' somewhere between the two points of interaction in the crystal.

Table 8.1. Stopping capability of a 12.5 mm thick NaI (Tl) crystal for photons of different energy

Photon energy (keV)	Interactions (%)
80	100
140	89
200	60
350	23
500	15

The function of the collimator is to establish a direct spatial relationship between the point of origin of the radiation in the patient and a point of contact in the crystal (see Figs. 8.6 and 8.8a). Fig. 8.6 also illustrates two fundamental problems. The wanted gamma ray, ray 1, must be distinguished from unwanted rays which might reach the same point in the crystal either directly (ray 2) or after scattering (ray 3). The purpose of the collimator is to remove rays such as 2. For the 140 keV gamma rays from Tc-99m, the half value thickness for lead is 0.3 mm, but as shown in Fig. 8.9, oblique gamma rays usually either pass through several strips or pass through a single strip at an oblique angle.

In some applications the camera may be fitted with a diverging collimator in which the holes diverge towards the patient. This produces a scintillation image that is smaller than the object and might be used when

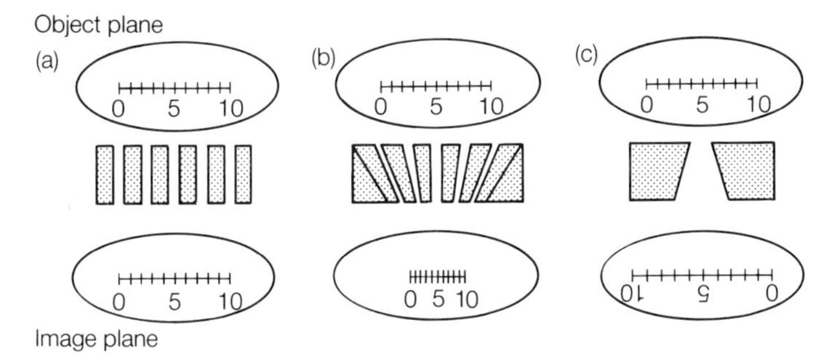

Fig. 8.8. The effect of different collimator designs on image appearance. (a) The parallel hole collimator produces the most faithful reproduction of the object. (b) The diverging collimator produces a minified image but is useful when the required field of view is larger than the detector area. (c) The pinhole collimator produces an enlarged inverted image and is useful for very small fields of view.

the required field of view is larger than the available crystal area (Fig. 8.8b). Alternatively, for small organs such as the thyroid, a single pinhole collimator may be used to produce a larger but inverted image (Fig. 8.8c).

Note that both diverging and pinhole collimators introduce distortion because the magnification or minification factor depends on the distance from the object plane to the collimator and is therefore different for activity in different planes in the object.

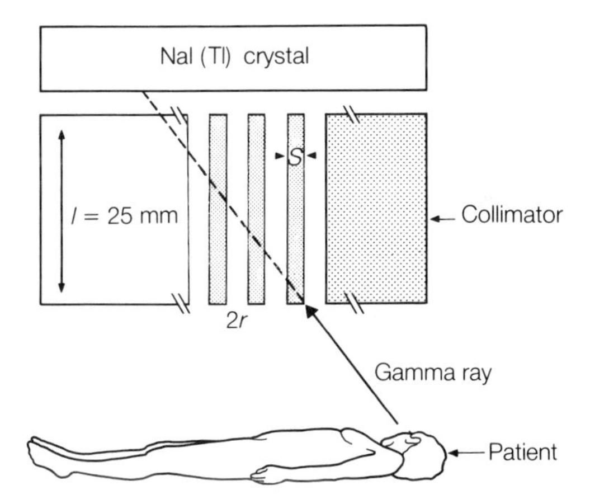

Fig. 8.9. Diagram showing that oblique gamma rays will pass through many lead strips, or septa, before reaching the detector. Typical dimensions for a low energy collimator are l = 25 mm, $2r$ = 3 mm, s = 0.2 mm. The number of holes will be approximately 15,000.

Pulse positioning

Pulse arithmetic circuits convert the outputs from the PMTs into three signals, two of which give the spatial coordinates of the scintillation, usually denoted by X and Y, and the third the energy of the event Z.

Each PMT has two weighting factors applied to its output signal, one producing its contribution to the X coordinate, the other to the Y coordinate. Several different mathematical expressions have been suggested for the shape of the weighting factors. Those which give the greatest weight to PMTs nearest the event are to be preferred since these will be the largest signals and hence least susceptible to statistical fluctuations due to noise (for a fuller discussion see Sharp *et al.* 1985). The final X and Y signals are obtained by summing the contributions from all tubes.

The energy signal Z is produced by summing all the unweighted PMT signals. This signal is then subjected to pulse height analysis as described in section 8.2.1 and the XY signal is only allowed to pass to the display system if the Z signal falls within the preselected energy window.

Data display and image recording

The display unit will contain a persistence oscilloscope on which each scintillation flash is displayed for about 2 s. For each successive gamma ray the oscilloscope beam shifts to the new position just before the signal arrives and an image of low count density is quickly built up. This image assists patient positioning. The display unit also contains a high quality screen from which a permanent record may be obtained by exposing Polaroid or nuclear medicine X-ray type transparency film. The quality of the final image is very dependent on the total counts collected (see section 8.3.1).

If the data are encoded and recorded on magnetic tape or disc, under computer control, various operations such as background subtraction, smoothing and correction for image distortions may be performed. The result is displayed on a video unit and this so-called digitized image will be discussed later.

Mechanical aspects

The control console enables the operator to set limits on energy selection circuits, control overall intensity on the display and choose either time for collection or total count limit for the image. As for the scanner, an anatomical marker is a useful accessory. The camera head can be raised, lowered or tilted at any angle and a comfortable examination couch minimizes patient movement.

For reasons discussed in section 8.4, the gamma camera is now the instrument of choice for a wide range of imaging investigations (Table 8.2), so further remarks will be limited to this type of device.

Table 8.2. Some important or potentially important nuclear medicine imaging examinations

Examination	Radiopharmaceutical	Static/ Dynamic	Principal application
Bone	Tc-99m methylene diphosphonate	S	Secondary spread of malignancy
Liver	Tc-99m sulphur colloid	S	Secondary spread of malignancy/cirrhotic changes
Brain	Tc-99m pertechnetate	S/D	Occult metastases/brain damage/vascular problems
Lung (perfusion)	Tc-99m human albumin microspheres-macroaggregates	S	Pulmonary embolism
Lung (ventilation)	Xe-133 gas	D	Pulmonary emphysema
	Tc-99m aerosol	D	Pulmonary emphysema
	Kr-81m gas	D	Regional ventilation rate
Kidneys	I-123 hippuran/ Tc-99m diethylenetriamine pentacetic acid (DTPA)	D	Renal function
Hepatobiliary system	Tc-99m iminodiacetic acid	D	Patency of biliary tree/liver function
Thyroid	I-123 iodide Tc-99m pertechnetate	S	Thyroid function
Heart (perfusion)	Tl-201 chloride	S	Cardiac infarction and ischaemia
Heart/blood pool	Tc-99m pertechnetate/ Au-195m colloid Tc-99m red cells	D	Intra-cardiac shunts/ cardiac wall motion/ ventricular volumes
Blood pool	Tc-99m red cells	S/D	Intra-abdominal bleeding
Blood products	In-111 platelets	S	Thrombus formation
	In-111 monoclonal antibodies	S	Tumour localization
	In-111 white cells	S	Abscess
Whole body	Ga-67 citrate	S	Soft tissue tumour/sepsis

8.2.3 Variations in the standard gamma camera

The mobile camera

The standard gamma camera is not readily moved from one room to another so with the increasing importance of cardiac work, the need for a mobile camera that could be used on the ward became apparent. Mobile cameras are designed primarily for cardiac work so they have a small field of view, and since they are only used with radionuclides that emit low energy gamma rays, for example in addition to Tc-99m, currently mainly thallium-201 which emits gamma rays at 80 keV, the NaI (Tl) crystal can be thinner, typically 6–9 mm. Performance is comparable with that of a standard camera.

The scanning camera

This combines the advantages of a camera with some of those of a scanner to obtain a whole body scan, especially a bone scan, in a single image. Design details vary somewhat but one technique is to use a collimator that diverges in one direction, a so-called fish-tail collimator, so that the full width of the patient is in the field of view which is increased to 50 cm by this technique. Some distortion is introduced (see section 8.2.2).

The detector moves on rails along the length of the patient and the *Y* position signals have an offset DC voltage signal applied to them which is a function of detector position. All the data are collected in a single pass of the camera over the patient thereby producing a non-overlapping image, hence facilitating interpretation, in the shortest possible scan time.

8.3 Factors affecting the quality of radionuclide images

In respect of radionuclide images, the radiologist should always ask 'Are the pictures of good quality—and if not, why not?' The numerous factors that affect image quality will now be considered.

8.3.1 Information in the image

It is well-known that the quality of an image depends on the number of photons it contains. Figure 8.10 shows three images of a simple phantom frequently used in nuclear medicine with different numbers of counts in each image. Unfortunately, because injected radioactivity spreads to all parts of the body and is retained for several hours (in contrast to X-rays which can be confined to the region of interest and 'switched off' after the study), *in vivo* nuclear medicine investigations are always photon-limited by the requirement to minimize radiation dose to the patient. The number of useful gamma rays is further reduced by the heavy collimation that has to be employed. Hence a typical photon density in radionuclide imaging is of the order of 1 mm^{-2} compared with about 10^5 mm^{-2} in radiography and 10^{10} mm^{-2} in conventional photography.

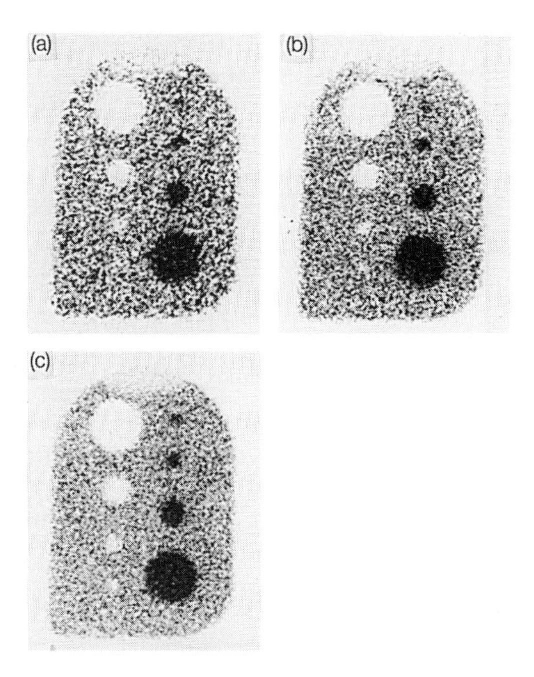

Fig. 8.10. Images of a phantom at different count densities. (a) 45 kilocounts, (b) 105 kilocounts, (c) 173 kilocounts. The cold areas on the left and the hot areas on the right become sharper as the number of counts in the image increases.

The information in the image can, in theory, always be increased by increasing the time of data collection. However this time will be limited by the length of time the patient can lie still, and the work load on the camera. In some situations physiological factors, e.g. heart movement, may negate the potential gain in image quality from a long data collection time. The primary objective is therefore to obtain the maximum number of counts in the image in a given time subject to the limitation of an acceptable radiation dose to the patient. In achieving this objective, choice of radionuclide and choice of radiopharmaceutical are the two main factors to be considered.

8.3.2 Choice of radionuclide

It is important to use a short half life radionuclide so that, for a given injected activity, the radiation dose to the patient *after* the examination is as low as possible. Note, however, that the half-life should not be too short compared with the planned duration of the study, and very short half-life materials may create problems of availability.

For the same reason a radionuclide which decays to a non-radioactive or very long half-life daughter should be chosen. If both parent and daughter are radioactive, the ratio of their activities is the inverse ratio of their half-lives. Thus a long half-life daughter is excreted before any significant dose can arise from its decay.

The radionuclide selected should emit no beta particles or, even worse, alpha particles. These would be stopped in the body, adding to the radiation dose, but contributing nothing to the image.

Only gamma ray energies within a limited range are really well suited to *in vivo* imaging. For example they must be sufficiently energetic not to be absorbed in the patient. For this and other reasons gamma rays of energy

Table 8.3. Properties of some radionuclides used for *in vivo* imaging

Nuclide	Half-life	Type of emission	Example of use
Carbon-11	20 min*	β^+ giving 511 keV γ rays	Cerebral glucose metabolism §
Nitrogen-13	10 min*	β^+ giving 511 keV γ rays	Amino acids for myocardial metabolism §
Oxygen-15	2 min*	β^+ giving 511 keV γ rays	Gaseous studies with labelled O_2, CO_2 and CO§
Fluorine-18	110 min*	β^+ giving 511 keV γ rays	Identification of receptor sites §
Gallium-67	72 h	92 keV, 182 keV and 300 keV γ rays	Soft tissue malignancy and infection
Technetium-99m	6 h†	140 keV γ rays	Numerous
Indium-111	2.8 days	173 keV and 247 keV γ rays	Labelling blood products
Iodine-123	13 h	160 keV γ rays	Renal function
Iodine-131	8.0 days	360 keV γ rays *and* β^- particles	Metastases from carcinoma of thyroid
Xenon-133	5.3 days‡	81 keV γ rays *and* β^- particles	Pulmonary emphysema
Gold-195m	3.5 s†	262 keV γ rays	Radionuclide angiocardiography
Thallium-201	73 h	80 keV X-rays and Auger electrons	Cardiac infarction and ischaemia

* Cyclotron produced.
† Generator produced. Note that short half-life radionuclides that cannot be produced on site are of limited value for *in vivo* imaging.
‡ Since Xe-133 is used in gaseous form, the biological half-life is very short so the β^- particle dose is small.
§ Not used widely at present.

less than 100 keV should be avoided if possible. Conversely, the gamma rays must be stopped in the detector or they will be wasted. The 12 mm thick crystals used in gamma cameras become inefficient above about 300 keV (Table 8.1). For equipment based on the scanning principle, where the scintillation crystal can be made quite thick, gamma ray energies up to 500 keV may be used.

A range of radionuclides is used in diagnostic imaging (Table 8.3) but well over 90% of routine investigations are performed with Tc-99m. In addition to its short half-life and near monoenergetic gamma ray at 140 keV, Tc-99m emits no particulate radiations and decays to a long half-life daughter (Tc-99, $T_{\frac{1}{2}} = 2 \times 10^5$ years).

Availability of the 6 h half-life material is not a problem because it is possible to establish a generator system. As explained in section 1.7, equilibrium activity in the decay series

$$^{99}_{42}\text{Mo} \xrightarrow[67\text{ h}]{\beta^-} {}^{99\text{m}}_{43}\text{Tc} \xrightarrow[6\text{ h}]{\gamma} {}^{99}_{43}\text{Tc} \xrightarrow[2\times 10^5\text{ yr}]{\beta^-}$$

is governed by the activity of Mo-99 which has a half-life of 67 h.

A Mo-Tc generator consists of Mo-99 adsorbed onto the upper part of a small chromatographic column filled with high grade alumina (Al_2O_3). When 0.9% saline solution is passed down the column, the Mo-99 remains firmly bound to the alumina but the Tc-99m, which is chemically different, is eluted. Since the Tc-99m activity builds up fairly rapidly (see section 1.7), it is possible to elute the column daily to obtain a ready supply of Tc-99m (Fig. 8.11). The generator can be replaced weekly, by which time the Mo-99 activity will have decreased significantly.

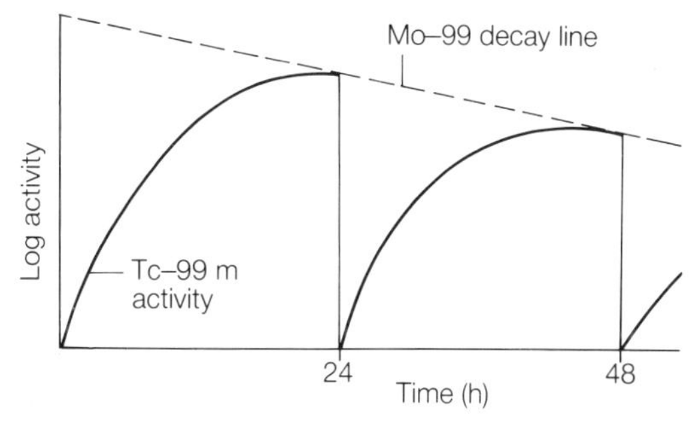

Fig. 8.11. Curve showing the Tc-99m activity in a Mo-99/Tc-99m generator as a function of time, assuming the column is eluted every 24 h.

8.3.3 Choice of radiopharmaceutical

For good counting statistics, the radionuclide must be firmly bound to an appropriate pharmaceutical and the resulting radiopharmaceutical must achieve a high target:non-target ratio. In addition it must satisfy criteria that are not generally relevant for non-radioactive drugs. It must be easy to produce, inexpensive, readily available for all interested users, have a short effective half-life and be of low toxicity. Very short half-life material may constitute a radiation hazard to the radiopharmacist if it is necessary to start the preparation with a high activity.

Radiopharmaceuticals concentrate in organs of interest by a variety of mechanisms including capillary blockage, phagocytosis, cell sequestration, active transport, compartmental localization, ion exchange and pharmacological localization. The reader is referred to a more specialized text (e.g. Rhodes & Croft 1978, or Saha 1979) for further details.

The impact that a radiopharmaceutical of high specificity may have on a particular diagnostic study is well illustrated by the increase in bone scanning that resulted from the introduction of Tc-99m labelled polyphosphates, especially methylene diphosphonate (Fig. 8.12). However, one disadvantage of Tc-99m is that it is not easily bound to biologically relevant molecules and its chemistry is complex (for a very readable account see the

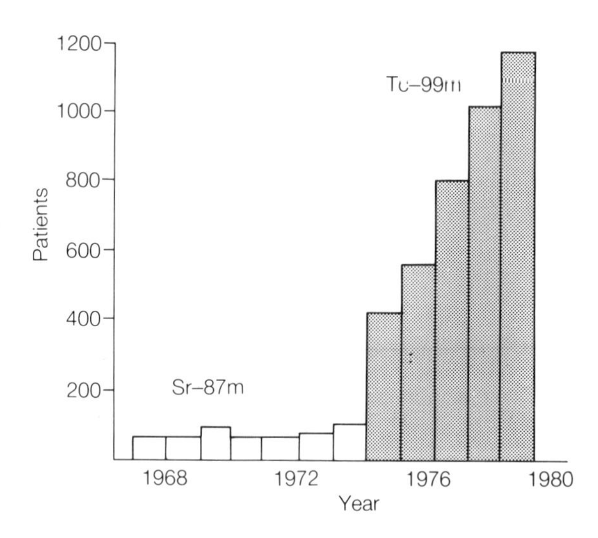

Fig. 8.12. Histogram showing the increase in bone scanning in one nuclear medicine centre over the period 1968–1980 as a result of replacing Sr-87m by Tc-99m polyphosphates.

article by Bremer 1984). I-123 is preferable from this point of view but no generator system is available and its 13 h half-life causes problems of availability.

Poor specificity of radiopharmaceuticals for their target organs remains a weak point in nuclear medicine imaging, with most commonly employed radiopharmaceuticals showing very poor selectivity, generally less than 20% in the organ of interest. Note that the obvious elements to choose for synthesizing specific physiological markers, hydrogen, carbon, nitrogen and oxygen, have no gamma emitting isotopes. Pharmaceuticals containing radionuclides of some of these elements can be used for positron emission tomography as discussed in section 10.5.

8.3.4 Performance of the imaging device

Much has been written about the performance of the gamma camera and only the most salient features will be summarized here.

Collimator design

As already explained, resolution and sensitivity are mutually exclusive. Collimator design must also take into consideration the energy of the gamma ray. Thicker septa will be required for I-131 (364 keV) than for Tc-99m (140 keV), and even so some radiation will probably penetrate the septa at the higher gamma ray energies.

Intrinsic resolution

This is determined primarily by the performance of the scintillation detector crystal. Although a complex problem to treat rigorously, the following simplified explanation contains the essential physics. In principle, by arranging a large number of very small PMTs behind the crystal, one might expect to localize the position of a gamma ray event in the crystal to any required degree of accuracy. However, each 140 keV gamma ray only releases about 4000 light photons and if these are shared between 40 PMTs, the average number, N, reaching each tube is only 100. The process is random, so variations in the signal due to Poisson statistics of $\pm N^{\frac{1}{2}}$ or ± 10 will ensue.

Some PMTs will get more than 100 photons, some will get a lot less, but the 'error' on the signal from each PMT, which will contribute to the error in positioning the event, will increase rapidly if one attempts to subdivide the original signal indefinitely.

Spatial linearity and non-uniformity

The outputs from the PMT array must be converted into the X and Y signals that give the spatial coordinates of the scintillation. Any error in this process, caused perhaps by a change in the amplification factor in one PMT, will result in counts being misplaced in the ensuing image. This will result in distortion (or non-linearity) if a narrow line source of radioactivity is imaged, or in non-uniformity for a uniform extended source.

Several methods of expressing non-uniformity have been suggested. For example, integral non-uniformity examines maximum and minimum variations from the mean in counts over predefined areas

$$U_+ = \frac{C_{\text{max}} - C_{\text{mean}}}{C_{\text{mean}}} \times 100\ (\%)$$

$$U_- = \frac{C_{\text{min}} - C_{\text{mean}}}{C_{\text{mean}}} \times 100\ (\%)$$

where C_{max}, C_{min} and C_{mean} are the maximum, minimum and mean counts per pixel over a predefined area, whilst differential non-uniformity U_{D} is based on the maximum rate of change of count density,

$$U_D = \frac{\Delta C}{C_{\text{mean}} \times \Delta x} \times 100\ (\%)$$

where ΔC is the maximum difference in counts between any two adjacent image elements and Δx is the length of the side of an element. From the viewpoint of accurate diagnosis, camera non-uniformities must be minimized or they may be wrongly interpreted as real variations in the image count density. With appropriate corrections, non-uniformity of the latest cameras can be as low as 1%.

Effect of scattered radiation

Although pulse height analysis ought, in principle, to reject all scattered radiation, discrimination is far from perfect. This is primarily because of the statistical nature of the process that converts a gamma photon into visible light in the crystal. The result, as shown in Fig. 8.13, is that even monoenergetic gamma rays produce light signals with a range of intensities. This spread, expressed as the ratio of the full width at half maximum height (FWHM) of the photopeak spectrum to the photopeak energy, is a measure of the energy resolution of the system and is about 12% for a gamma camera at 140 keV.

Unscattered photons contribute information about the image so a wide

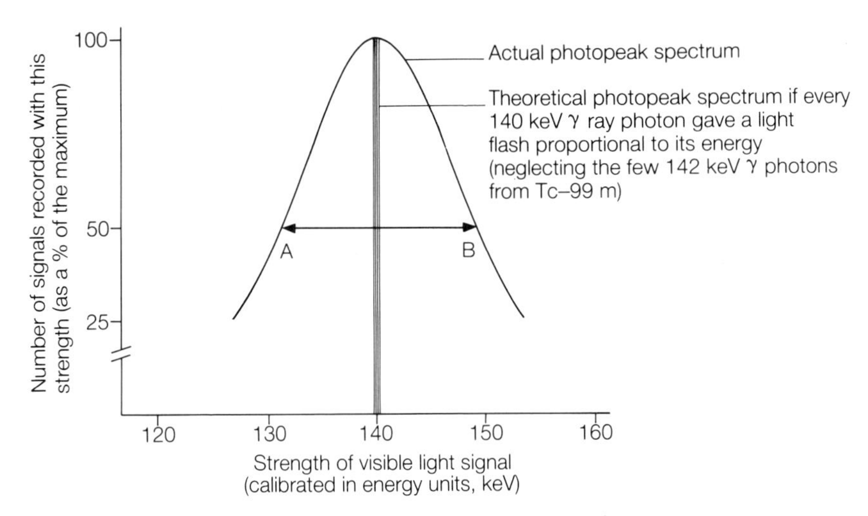

Fig. 8.13. Graph demonstrating the energy resolution of the NaI (Tl) crystal in a gamma camera. The FWHM (AB) is about 16 keV or 12% of the peak energy.

energy window (typically about 20%) must be used. Unfortunately a wide energy window permits some gamma photons that have been Compton scattered through quite large angles, and may have lost as much as 20 keV, to be accepted by the pulse height analyser. The problem is greater for low energy gamma photons, for two reasons.

1 fewer light photons are produced, so statistical variations are greater,

2 the energy lost during a Compton interaction, for fixed scattering angle, is smaller.

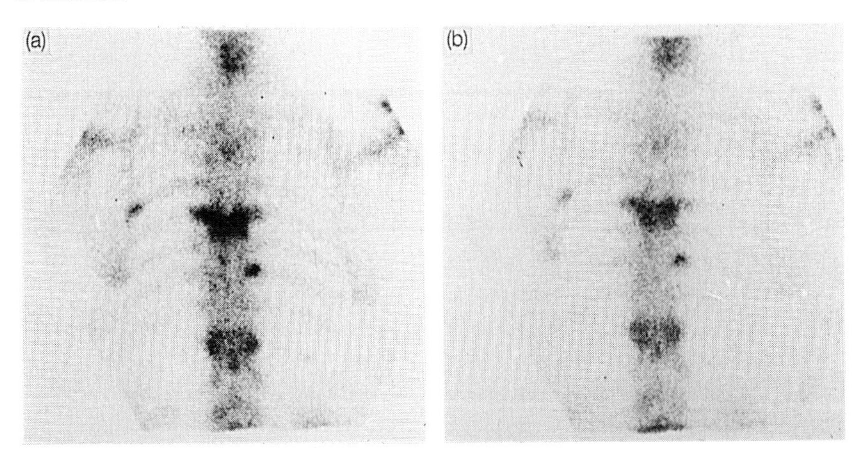

Fig. 8.14. Appearances of a bone scan: (a) normal, (b) with additional scattering material interposed between the patient and camera.

As shown in Fig. 8.14, scattered radiation causes a marked deterioration in image quality. The patient is the major source of scattering material and there is an obvious difference in image quality for say a bone scan of a very thin person when compared with that of an obese person. Careful setting up of the pulse height analyser is essential for good image quality.

System resolution

The intrinsic resolution of the crystal and PMT detector system is further degraded by the presence of a collimator, by the fact that the radioactivity is some distance from the collimator face, and because the intervening space is filled with scattering medium.

The resolution of the complete system can be obtained by imaging a narrow line source of radioactivity—for example, nylon tubing of 1 mm internal diameter filled with an aqueous solution of Tc-99m pertechnetate. A result such as that shown in Fig. 8.15 would be obtained and the spread of the image can be expressed in terms of the full width at half maximum height (FWHM) calculated as shown. For the arrangement suggested, the FWHM is 9.7 mm which is substantially greater than the intrinsic resolution.

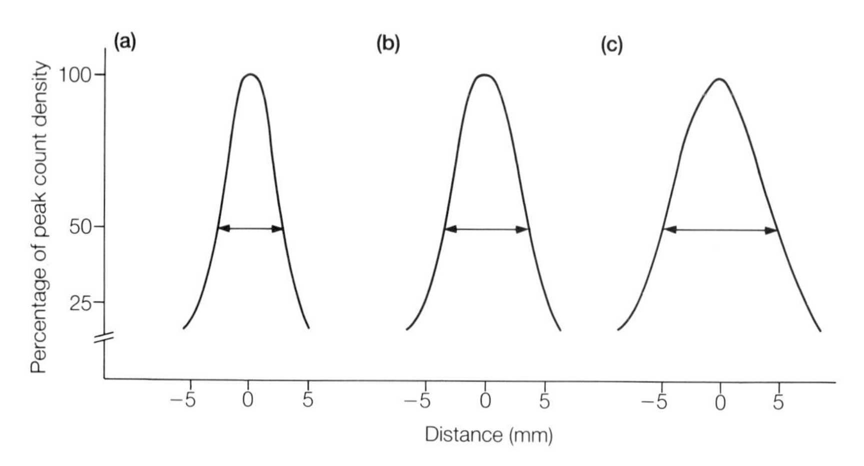

Fig. 8.15. Derivation of system resolution from line spread function measurements and the effect of scattering material on system resolution. The traces are typical of images of a 1 mm line source of 99m-Tc obtained under different conditions: (a) no scattering material FWHM = 5.7 mm, (b) 5 cm tissue equivalent scattering material, FWHM = 7.1 mm, (c) 10 cm of tissue equivalent scattering material FWHM = 9.7 mm.

High count rates
The reasons for loss of counts at high count rates and implications for quantitative work will be discussed in section 8.4.2. Some degradation of image quality also occurs. The main reason is thought to be that the system fails to separate in time two scattered photons whose summed energy falls within the pulse height analyser window. Thus an event is recorded but in a position unrelated to the activity in the patient.

8.3.5 Data display

The quality of the final image is also influenced by the performance of the recording medium. Methods of displaying data will be discussed briefly.

Persistence oscilloscope
When setting up a study, an image of the distribution of activity may greatly assist patient positioning. This may be achieved using a video monitor on which each flash of light representing a collected gamma ray persists long enough for a transient image to be formed. This image is ideal for positioning but totally unsuitable for diagnosis.

Hard copy
Two types of hard copy have been widely used. Black and white Polaroid film has the advantage of high film speed with high resolution and the picture is available quickly so the view can be repeated if necessary. Disadvantages are that it is expensive, has poor contrast with a film gamma of only 1.3, and limited dynamic range. Transparency film, which is singled-sided because only light photons are being recorded by this stage in the detection process, is now the preferred method of display. Multiformatting permits several views to be displayed on a piece of film measuring 360 mm × 280 mm and the most sophisticated equipment can produce up to 81 images on a sheet and simultaneously record images at two different intensities. Such a system is particularly useful for recording sequential images in a dynamic study.

Analogue versus digitized images
The images discussed so far have been mainly 'analogue' images in which one pulse of light will be allowed to fall on the film at each point corresponding to the event in the sodium iodide crystal. This method of display is normally adequate for static images. If image manipulation or numerical information is required, the data must be 'digitized'. That is to say, the

image space is sub-divided into a matrix of pixels, usually 64 × 64 or 128 × 128 or 256 × 256 and the total number of counts in each pixel is recorded. A digital display has many advantages. For example the image can be manipulated prior to viewing by altering the contrast or the background count can be suppressed by raising the threshold. It also permits quantitative information to be produced and sophisticated forms of image processing are possible. For further information on some of these techniques see Sharp *et al.* (1985).

A 128 × 128 matrix over a typical gamma camera face corresponds to a pixel size of about 3 mm and many observers find the underlying matrix pattern visually intrusive when viewing images. However, both theoretical calculations and experimental observations indicate that the inherent resolving capability of the camera does not warrant finer pixellation (see Chapter 9).

If the images collected in a dynamic study are to be analysed quantitatively, they must of course be digitized. This aspect will be discussed in more detail in section 8.4. A block diagram showing the essential components of a gamma camera equipped with reasonably comprehensive data processing facilities is shown in Fig. 8.16.

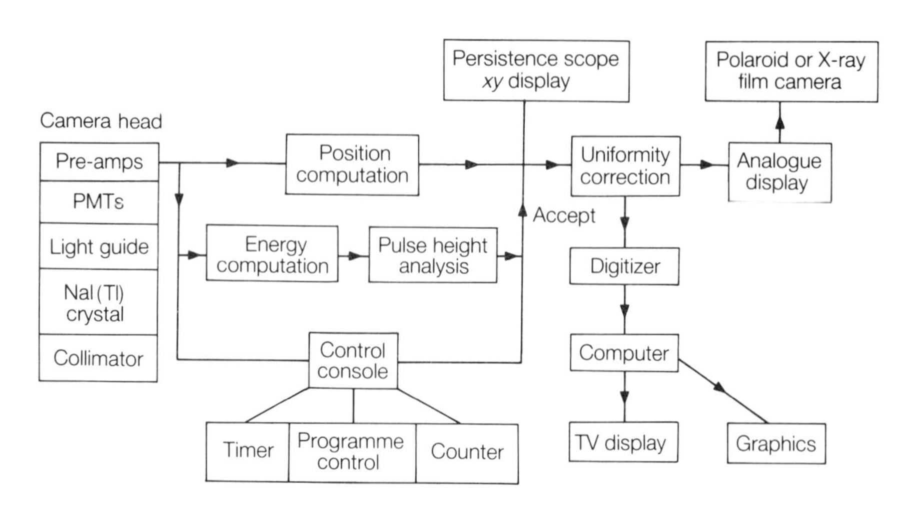

Fig. 8.16. Block diagram of a gamma camera with fairly comprehensive data processing, display and storage facilities.

Grey scale versus colour images

When an image has been digitized, a range of count densities may be assigned to either a shade of grey or a spectral colour. Much has been written about the relative merits of grey scale and colour images and this controversial subject cannot be discussed fully here. The following simple philosophy suggests an approach to each type of display. The sharp visual transition from one colour to another may alert the eye to the possibility of an abnormal amount of uptake of radioactivity and colour images can be useful for this purpose. However, by the same token, this colour change may represent an increase or decrease of only one or two counts per pixel and may not be significant statistically. Therefore it is prudent, if not essential, to examine grey scale images to verify the presence of an abnormality and minimize the risk of over-reporting.

8.4 Dynamic investigations

The potential for performing dynamic studies, in which changes in distribution of the radiopharmaceutical are monitored throughout the investigation, was recognized at an early stage. However, two developments were essential before dynamic imaging became feasible on a routine basis. The first was an imaging device with a reasonably large field of view, sufficiently sensitive to give statistically reliable counts in short time intervals. The gamma camera satisfies these requirements although for dynamic studies it is not uncommon to choose a collimator design that increases sensitivity at the expense of some loss of resolution. The dominance of gamma cameras over scanners in nuclear medicine departments is very largely due to their capability for dynamic studies.

The second development was the availability of reasonably priced data handling hardware and software powerful enough to handle the large amount of data collected. Hence dynamic imaging has only been widely available in general hospitals since the late 1970s.

Important features of dynamic imaging will now be considered under two general headings, but with specific reference to some frequent dynamic investigations.

8.4.1 Data analysis

Consider as an example the study of kidney function—nowadays probably best performed with I-123 hippuran which is actively secreted by the renal tubules. Historically, fixed detectors of small area or 'probes' were used for such studies but the results obtained were critically dependent on probe positioning. Using a gamma camera, a set of images could be obtained and regions of interest for study could be selected retrospectively. Thus there was a need to develop methods of data display that would show both the spatial distribution of radiopharmaceutical and temporal changes in the distribution.

Cine mode
A useful starting point is to examine individual image frames looking for aspects that require further detailed study. This can be done by running a 'cine-film' of the frames using a continuous loop so that the display automatically returns to the start of the study and continues until interrupted.

Time–activity curves
The system will plot activity as a function of time for regions of interest selected by the operator. Examples of such regions of interest and the resulting curves, again taken from renography, are shown in Fig. 8.17. Some other features of this apparently simple procedure must be men-

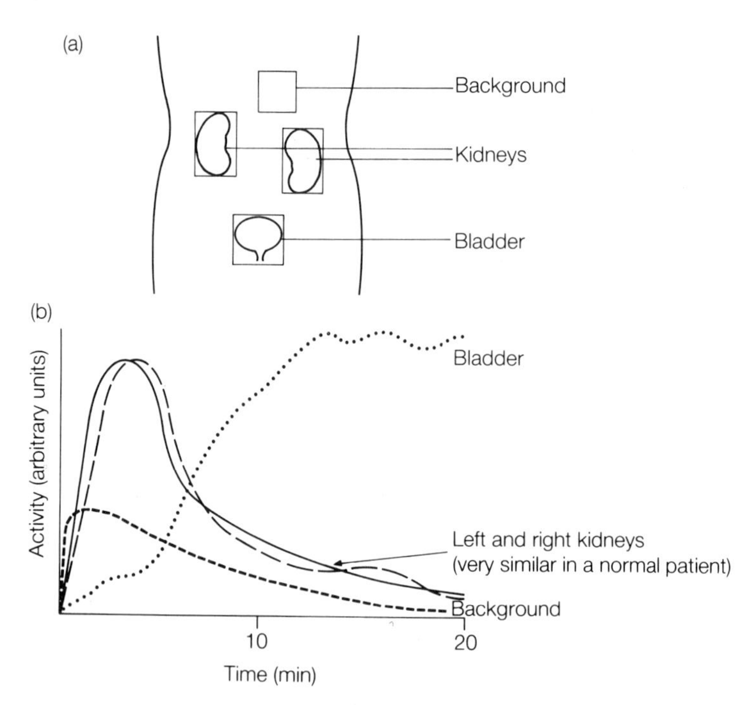

Fig. 8.17. (a) Schematic drawings of regions of interest around kidneys and bladder for a renogram study. A background region is also shown. (b) Typical activity–time curves for such regions of interest.

tioned. First, to achieve a better image (improved counting statistics) on which the regions of interest can be drawn, it may be necessary to add several sequential frames. Data smoothing will also help to keep noise to a minimum. Finally, it is important to subtract background counts arising from activity in overlying and underlying tissue. This is usually done by defining a region of interest between the two kidneys, avoiding major blood pools.

Deconvolution

Although radioactivity is injected intravenously as a bolus, after mixing with blood and passing through the heart and lungs, it arrives at the kidneys over a period of time. Also some activity may recycle. Thus the measured activity–time curve is a combination (convolution) of a variable amount of activity and the rate of handling by the organ. The requirement is to measure the mean transit time for the organ and deconvolution is a mathematical technique that offers the possibility for removing arrival time effects. In nuclear medicine the presence of noise limits the power of deconvolution methods.

Functional imaging

One way to present spatial information on temporal data is to plot the spatial distribution of a mathematical index that characterizes the behaviour of activity with time. Consider for example a lung ventilation study in which the patient breathes a mixture of air and a radioactive gas or aerosol in a closed system until equilibrium is established. The circuit is then switched so that the patient breathes air alone and data collection is continued as the activity is 'washed out' with inactive air. The whole investigation is recorded as a dynamic study, the patient sitting with his back against the gamma camera.

The complete time–activity curve shows wash in, equilibrium and wash out phases but the index of greatest clinical interest is the rate of clearance of gas from the lungs. A functional image showing either the time constant of the wash out phase, if it is a simple exponential, or the mean transit time, defined as the area under the washout curve divided by the equilibrium count rate, is an effective way to display the data. Images such as those shown in Fig. 8.18 are obtained. A high value for the index reflects poor ventilation.

A major difficulty with functional imaging is the choice of a reasonably simple mathematical index that is relevant to the physiological condition being studied.

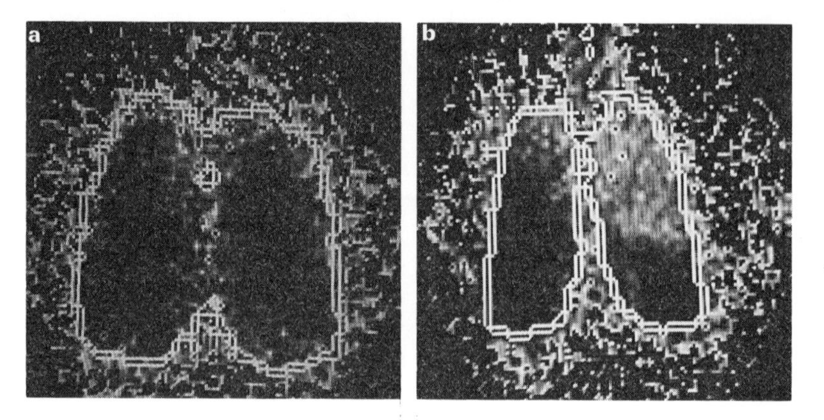

Fig. 8.18. Functional images displaying the mean transit time for clearance of Xe-133 gas from the lungs. In the normal patient on the left the mean transit time is uniform in both lungs. In the emphysematous patient imaged on the right, the mean transit time is elevated in the upper right lobe and, to a lesser extent in the upper left lobe demonstrating slow wash out. Both images are posterior views with the lung outline shown. (Reproduced with permission from Sharp *et al.* 1985.)

8.4.2 Camera performance at high count rates

When a dynamic study requires rapid, sequential imaging, the gamma camera may have to function at high count rates to achieve adequate statistics. However, the system imposes a number of constraints on maximum count rate. For example, the decay time of the scintillation in the crystal has a time constant of 0.2 μs and about 0.8 μs is required for maximum light collection. The electronic signal processing time is also a major limitation. The signal from the camera pre-amplifier has a sharp rise but a tail of about 50 μs so pulses have to be truncated or 'shaped' to last no longer than about 1 μs. The pulse height analyser and pulse arithmetic circuits have minimum processing times and it may be an advantage to bypass the circuits that correct for spatial non-linearity, if any, to reduce processing time.

For all these reasons, if sources of known, increasing activity are placed in front of a gamma camera under typical imaging conditions, a graph of observed against expected count rate might be as shown in Fig. 8.19. The exact shape of curve is very dependent on the thickness of scattering material used and on the width of the pulse height analyser window because the system has to handle counts that will eventually be rejected by the pulse height analyser. Loss of counts is likely to occur above 2×10^4

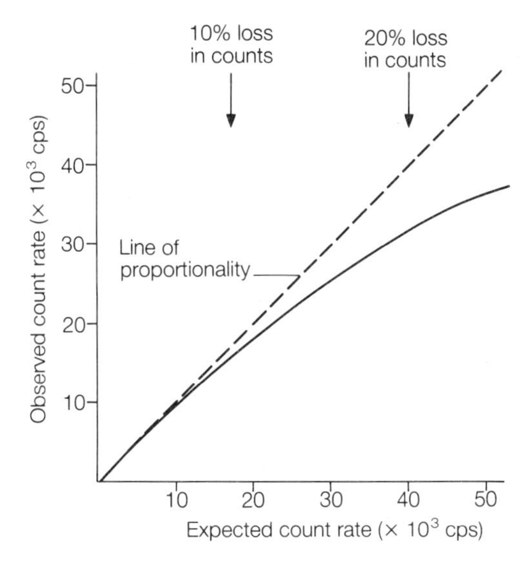

Fig. 8.19. Curve showing how the various dead times in the camera system result in count rate losses. In this example 10% and 20% losses in counts occur at about 17×10^3 and 40×10^3 cps respectively. The maximum observed count rate would be about 75×10^3 cps after which the recorded count rate would actually fall with increasing activity.

counts per second and this can be a serious problem when making quantitative measurements at high count rates.

The highest count rates and hence most stringent demands on both camera and radiopharmaceutical are encountered in cardiac studies. In a first-pass study a bolus injection is given at the highest possible radioactive concentration. One-second frames are taken and the study is complete in about 30 s. Ultrashort half-life radionuclides, for example gold-195m, which is generator produced and has a $T_{\frac{1}{2}}$ of 30.5 s, are ideal because rapid physical decay means high activity during the study but not a big patient dose. Also residual background activity rapidly decays to zero so repeat studies from a different angle or during exercise may be performed within a few minutes. The count rate performance of cameras needs to be improved for such studies.

Cardiac ventricular function and wall motion can also be studied over a longer time-scale provided the frames are gated to the patient's electrocardiogram. Each image then represents a discrete fraction of the cardiac cycle and images may be summed to give adequate statistics. Edge detection techniques similar to those discussed in section 9.7 may be used to estimate ventricular volumes.

8.5 Summary

The primary detector of gamma rays in nuclear medicine is invariably a sodium iodide crystal, doped with about 0.1% of thallium. The advantages of this detector are:

1 A high density and high atomic number ensure a good gamma ray stopping efficiency for a given crystal thickness.

2 The high atomic number favours a photoelectric interaction, thus a pulse is generated which represents the full energy of the gamma ray.

3 Thallium gives a high conversion efficiency of the order of 10%.

4 A short 'dead time' in the crystal generally permits acceptable counting rates except for very rapid dynamic studies, when dead times both in the crystal and elsewhere in the system can be important.

The instrument of choice for a wide range of static and dynamic examinations is the gamma camera. Its mode of operation may be considered in two parts:

1 A direct spatial relationship between the point of emission of a gamma ray in the patient and the point at which it strikes a large Na (Tl) crystal is established by collimation.

2 An array of PMTs backed by appropriate electronic circuits is used to identify the position at which the gamma ray interacts with the crystal.

Over 90% of all nuclear medicine examinations are carried out with Tc-99m. The advantages of Tc-99m for radionuclide imaging are:

1 A monoenergetic gamma ray is emitted—this facilitates pulse height analysis;

2 The gamma ray energy is high enough not to be heavily absorbed in the patient, hence minimizing patient dose, but low enough to be stopped in a thin sodium iodide crystal;

3 No high LET radiations are emitted;

4 A decay product that delivers negligible dose;

5 A half-life that is long enough for most examinations but short enough to minimize dose to the patient;

6 Ready availability as an eluate from a Mo-99/Tc-99m generator.

The quality of radionuclide images is influenced by

1 The number of counts that can be collected for given limits on radiation dose to the patient, required resolution, and time of examination;

2 The presence of, and ability to discriminate against, scattered radiation;

3 Overall performance of the imaging device including spatial and temporal linearity, uniformity and system resolution (Table 8.4).

Table 8.4. Typical performance figures for a gamma camera

Parameter	Value	Conditions
Intrinsic spatial resolution	3.6 mm	FWHM over the useful field view
System spatial resolution	9.7 mm	FWHM with 10 cm scattering material and high resolution collimator
Intrinsic energy resolution	12.4%	FWHM at 140 keV
	14.3%	FWHM at 140 keV with collimator and scatter
Integral uniformity	3.4%	Central region
Differential uniformity	0.8%	Central region
Count rate performance	40,000 cps	20% loss of counts
	75,000 cps	Deterioration of intrinsic spatial resolution to 3.9 mm
System sensitivity	10 cpm per kBq	Tc-99m and a general purpose collimator

Analogue images are produced by transmitting the signal from the positioning circuits as a light flash directly to a hard copy recording device such as X-ray film. If the signals are fed to a computer and digitized, all counts within one pixel are summed. There is little to be gained by digitizing data for visual interpretation, but digitized images can be manipulated before display and can be used to extract, under computer control, functional data for specified regions of interest. The gamma camera is being used increasingly for dynamic studies where important information is obtained by numerical analysis of digitized images on a frame-by-frame basis.

References and further reading

Andrews G. A., Gibbs W. D., Morris A. C. Jr. and Ross D. A. (1973) Whole body counting. *Semin. Nucl. Med.* **3**, 367–388.

Bremer P. O. (1984) The principles and practice of radiopharmaceutical production. In *Technical advances in Biomedical Physics*, eds P. P. Dendy, D. W. Ernst and A. Şengün, pp. 287–318 and other relevant chapters. (NATO ASI series E, Applied Sciences 77) Martinus Nijhoff, the Hague.

Dendy P. P., Sharp P. F., Keyes W. I. and Mallard J. R. (1980) Radionuclide emission imaging—single photon techniques including radiopharmaceutical developments. *Br. Med. Bull.* **36**, 223–30.

Goodwin P. N. and Rao D. V. (1977) *An Introduction to the Physics of Nuclear Medicine.* Thomas, Springfield.

McAlister J. M. (1979) *Radionuclide Techniques in Medicine*. Cambridge University Press.
Parker R. P., Smith P. H. S. and Taylor D. M. (1984) *Basic Science of Nuclear Medicine*, 2nd edn. Churchill Livingstone, Edinburgh.
Rhodes B. A. and Croft B. Y. (1978) *Basics of Radiopharmacy*. Mosby, St Louis.
Saha G. B. (1979) *Fundamentals of Nuclear Pharmacy*. Springer, New York.
Sharp P. F., Dendy P. P. and Keyes W. I. (1985) *Radionuclide Imaging Techniques*. Academic Press, New York.
Sodee D. B. and Early P. J., eds (1975) *Technology and Interpretation of Nuclear Medicine Procedures*, 2nd edn. Mosby, St Louis.

Exercises

1 What is a radionuclide generator?

2 What mass of iodine-123 will give 80 MBq of activity? (Half-life of I-123 = 13 h, the Avogadro constant = 6.02×10^{23} mole^{-1}.)

3 A dose of Tc-99m macroaggregated human serum albumin for a lung scan had an activity of 180 MBq in a volume of 3.5 ml when it was prepared at 11.30 h. If you wished to inject 23 MBq from this dose into a patient at 16.30 h, what volume would you administer? (Half-life of Tc-99m = 6 h.)

4 A radiopharmaceutical has a physical half life of 6 h and a biological half-life of 15 h. How long will it take for the activity in the patient to drop to 10% of that injected?

5 List the main characteristics of an ideal radiopharmaceutical.

6 What are the possible disadvantages of preparing a radiopharmaceutical a long time before it is administered to the patient?

7 Why is the ideal energy for gamma rays used in clinical radionuclide imaging in the range 100–200 keV?

8 In nuclear medicine, why are interactions of Tc-99m gamma rays in the patient primarily by the Compton effect whilst those in the sodium iodide crystal are mainly photoelectric processes?

9 The sodium iodide crystal in a gamma camera is 8 mm thick. Calculate the fraction of the gamma rays it will absorb at

(a) 140 keV,
(b) 500 keV.

Assume the gamma rays are incident normally on the crystal and that the linear absorption coefficient of sodium iodide is 0.4 mm^{-1} at 140 keV and 0.016 mm^{-1} at 500 keV.

10 Why is it necessary to use a collimator for imaging gamma rays but not for X-rays produced by a diagnostic set?

11 What is the difference between a collimator used to image low energy radionuclides and one used to image high energy radionuclides?

12 Why is pulse height analysis used to discriminate against scatter in nuclear medicine but not in radiology?

13 Compare and contrast the methods used to reduce the effect of scattered radiation on image quality in radiology with those used in clinical radionuclide imaging.

14 How is the spatial resolution of a gamma camera measured and what is the clinical relevance of the measurement?

15 What factors affect

(a) the sharpness,

(b) the contrast of a clinical radionuclide image?

16 List the steps you would take in setting up a gamma camera to produce a brain image.

17 Explain, with a block diagram of the equipment, how a dynamic study is performed with a gamma camera.

18 A radiopharmaceutical labelled with Tc-99m and a gamma camera system were used for a renogram. Curves were plotted of the counts over each kidney as a function of time and although the shapes of both curves were normal, the maximum count recorded over the right kidney was higher than over the left. Suggest reasons.

9

Assessment and Enhancement of Image Quality

9.1 Introduction

In earlier chapters the principles and practice of X-ray production were considered and the ways in which X-rays are attenuated in different body tissues were discussed. Differences in attenuation create 'contrast' on an X-ray film and differences in 'contrast' provide information about the object. However, when an X-ray image, or indeed any other form of diagnostic image, is assessed subjectively, it must be appreciated that the use made of the information is dependent on the observer, in particular the performance of his or her visual response system. Therefore, it is important to consider those aspects of the visual response system that may influence the final diagnostic outcome of an investigation. Ways in which the visual process may be facilitated by image enhancement or even, in some circumstances, by-passed completely so as to provide a more objective assessment of information content in the image will also be considered. Finally, some methods of assessing image quality will be discussed.

9.2 Factors affecting image quality

A large number of factors may control or influence image quality. They may be subdivided into three general categories.

Image parameters

1 The signal to be detected—the factors to consider here will be the size of the abnormality, the shape of the abnormality and the inherent contrast between the suspected abnormality and non-suspect areas.

2 The number of possible signals—for example the number and angular frequency of sampling in computed tomography.

3 The nature and performance characteristics of the image system—spatial resolution (this is important when working with an image intensifier and in nuclear medicine), sensitivity, linearity, noise (both the amplitude and character of any unwanted signal such as scatter), speed, especially in relation to patient motion, and its effect on the dose to the patient.

4 Non-signal structure—interference with the wanted information may arise from grid lines, anatomy and overlapping structures.

Observation parameters

1 The display system—features that can affect the image appearance include the brightness scale, gain, offset, non-linearity (if any) and the magnification or minification.

2 Viewing conditions—viewing distance and ambient room brightness.

3 Detection requirements.

4 Number of observations.

Psychological parameters

1 *A priori* information given to the observer.

2 Feed back (if any) given to the observer.

3 Observer experience from other given parameters—this may be divided into clinical and non-clinical factors and includes for example familiarity with the signal and with the display, especially with the types of noise artefacts to be expected.

Some of these points have been considered in earlier chapters and others will be considered later in this chapter. However, it is important to realize that final interpretation of a diagnostic image, especially when it is subjective, depends on far more than a simple consideration of the way in which the radiation interacts with the body.

9.3 Analogue and digital images

These terms would not have formed part of a radiologist's vocabulary 15–20 years ago but nowadays it is essential to appreciate the difference between them. In an **analogue** image, there is a direct spatial relationship between the X-ray photon that interacts with the recording medium and the response of that medium. The result is a continuously varying function describing the image. Developed photographic film is an example of an analogue image, providing in radiology a detailed, permanent record of the distribution of X-ray photons transmitted by the body but only in a form suitable for visual inspection. It is not easy to extract quantitative data from an analogue image.

To manipulate and extract numerical information, the distribution pattern of photon interactions must be collected and stored in a computer. In principle the x and y co-ordinates of every X-ray interaction with the recording medium could be registered and stored. This is sometimes known as 'list mode' data collection and is used for a few specialized studies in nuclear medicine. However, apart from practical problems associated with the recording medium, in view of the large number of photon interactions in a conventional radiograph there would be a massive data storage problem not capable of being solved by computers used in hospitals. Therefore for the purpose of data collection and storage the image space is subdivided into a number of compartments, which are usually but not necessarily square and of equal size. In a **digitized** image the X-ray interaction is assigned to the appropriate compartment but the position of the interaction is located no more precisely than this. Thus the image has a discontinuous aspect that is absent from an analogue image.

The number and size of compartments is variable. For example in the extreme case of a 2 × 2 matrix illustrated in Fig. 9.1a, all interactions

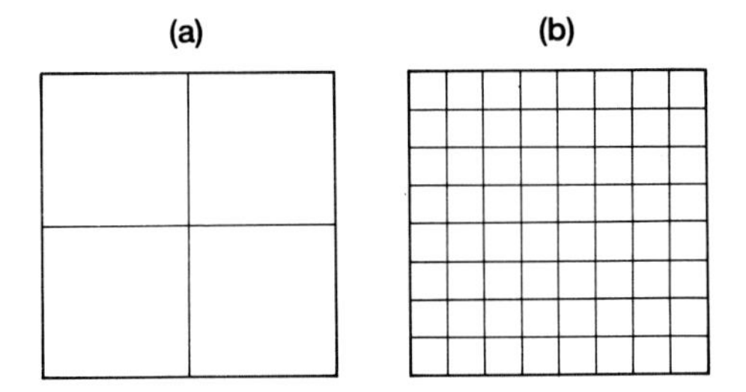

Fig. 9.1. Examples of coarse pixellation. (a) 2 × 2 matrix. (b) 8 × 8 matrix.

would be assigned to just one of four areas. Figure 9.1b illustrates an 8 × 8 matrix.

Matrices used for diagnostic imaging range from 64 × 64 to 1024 × 1024 and the choice of a suitable matrix size depends on a number of factors:

1 Resolution cannot be better than the size of an individual matrix element or pixel. For a 128 × 128 matrix and a 40 cm × 40 cm field of view, the pixel size is about 3 mm. Note that the pixel size is governed by the field of view. For the head, where the field of view might be only 20 cm × 20 cm, a 128 × 128 matrix would give a 1.5 mm pixel size.

2 If pixellation is very coarse, a finer pixellation will improve resolution. Consider the extreme example of a 2 × 2 matrix in Fig. 9.1a.

3 There is nothing to be gained by decreasing pixel size below the resolving capability of the imaging equipment. Thus, whereas a 512 × 512 matrix (0.4 mm pixel size) might be fully justified for a high resolution image intensifier screen used in cine angiography of the skull, it has no place in nuclear medicine where the system resolution of a gamma camera is no better than 5 mm.

4 As the pixel size becomes smaller, the size of the signal gets smaller and the ratio of signal to noise gets smaller. When counting photons, the signal size N is subject to Poisson statistical fluctuations (noise) of $N^{\frac{1}{2}}$ (see section 9.6). So the signal/noise ratio is proportional to $N^{\frac{1}{2}}$ and decreases as N decreases. In magnetic resonance imaging (see Chapter 13), electrical and other forms of noise cannot be reduced below a certain level and thus the size of the signal, for example from protons, is a limiting factor in determining the smallest useful pixel size.

5 Finer pixellation places a greater burden on the computer in terms of data storage and manipulation. A 256 × 256 matrix contains over 65,000 pixels and each one must be stored and examined individually.

9.4 Operation of the visual system

This is a complex subject and it would be inappropriate to attempt a detailed treatment here. However, four aspects are of particular relevance to the assessment and interpretation of diagnostic images and should be considered.

9.4.1 Response to different light intensities

It is a well established physiological phenomenon that the eye responds logarithmically to changes in light intensity. This fact has already been mentioned in section 4.4 and is one of the reasons for defining contrast in terms of log (intensity).

9.4.2 Rod and cone vision

When light intensities are low, the eye transfers from cone vision to rod vision. The latter is much more sensitive but this increased sensitivity brings a number of disadvantages for radiology.

First, the capability of the eye to detect contrast differences is very dependent on light intensity (Fig. 9.2). Whereas, for example, a contrast difference of 2% would probably be detectable at a light intensity or brightness of 100 millilambert (typical film viewing conditions), a contrast difference of 20% might be necessary at 10^{-3} millilambert (the light intensity that might be emitted from a fluorescent screen).

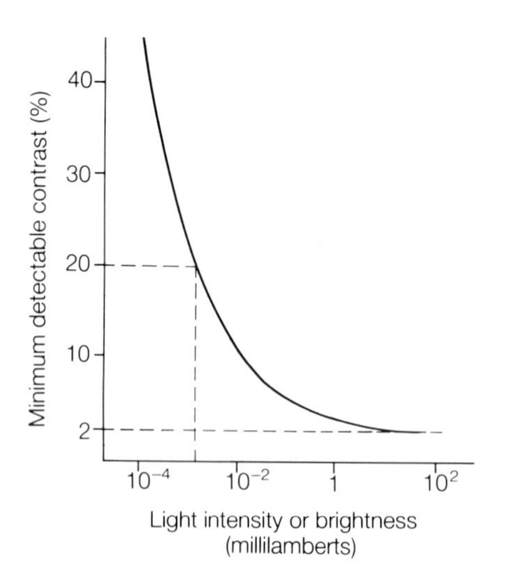

Fig. 9.2. Variation of minimum detectable contrast with light intensity, or brightness, for the human visual system.

Second, there is a loss of resolving ability, or visual acuity, at low light intensities. With cone vision the minimum detectable separation between two objects is better than 1 mm. Since rods respond as bundles of fibres rather than individually, visual acuity is worse and strongly dependent on light intensity. At 10^{-4} millilambert, which is a typical brightness for a traditional fluoroscopic screen, the minimum detectable separation is probably no better than 3 mm.

Loss of contrast perception and visual acuity are the two main reasons why fluoroscopic screens have been abandoned in favour of image intensifiers. Other disadvantages of rod vision are the need for dark adaptation, which may take up to 30 min, and a loss of colour sensitivity, although the latter is not a real problem in radiology.

9.4.3 Interrelationship of object size, contrast and perception

Even when using cone vision, it is not possible to decide whether an object of a given size will be discernible against the background unless the contrast is specified. The exact relationship between minimum perceptible contrast difference and object size will depend on a number of factors including the signal/noise ratio and the precise viewing conditions. A typical curve is shown in Fig. 9.3 and illustrates the general principle that the higher the object contrast, the smaller the detectable object size.

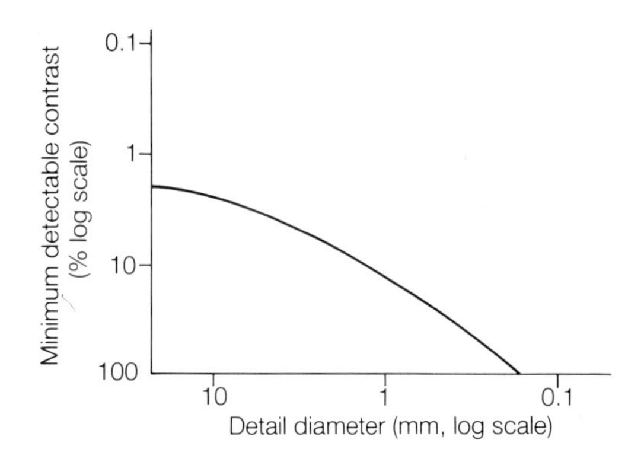

Fig. 9.3. Typical relationship between minimum perceptible contrast difference and object size for an image intensifier TV camera screening unit.

Some typical figures for a modern intravenous angiographic system will illustrate the relationship. Suppose an incident dose rate in air of 10 μGy per image is used. At a mean energy of 60 keV this will correspond to about 3×10^5 photons mm^{-2}. If the efficiency of the image intensifier for detecting these photons is 30%, about 10^5 counts mm^{-2} are used per image. Assuming a typical signal/noise ratio of 5, then objects as small as 0.2 mm can be detected if the contrast is 10% but the size increases to 2.5 mm if the contrast is only 1%.

9.4.4 Response of the eye to different spatial frequencies

When digitized images are displayed, the matrix of pixels will be imposed on the image and it is important to ensure that the matrix is not visually intrusive, thereby distracting the observer.

To analyse this problem, it is useful to introduce the concept of spatial frequency. Suppose a 128 × 128 matrix covers a square of side 19 cm. Each matrix element will measure 1.5 mm across, or alternatively expressed, the frequency of elements in space is 1/0.15 or about 7 cm^{-1}. Thus the spatial frequency is 7 cm^{-1}, 700 m^{-1} or more usually expressed in radiology as 0.7 mm^{-1}.

Clearly the effect of these pixels on the eye will depend on viewing distance. The angle subtended at the eye by pixels having a spatial frequency of 0.7 mm^{-1} will be different depending on whether viewed from 1 m or 50 cm. Campbell (1980) has shown that contrast sensitivity of the eye–brain system is very dependent on spatial frequency and demonstrates a well-defined maximum. This is illustrated in Fig. 9.4 where a viewing distance of 1 m has been assumed. Exactly the same curve would apply to spatial frequencies of twice these values if they were viewed from 50 cm.

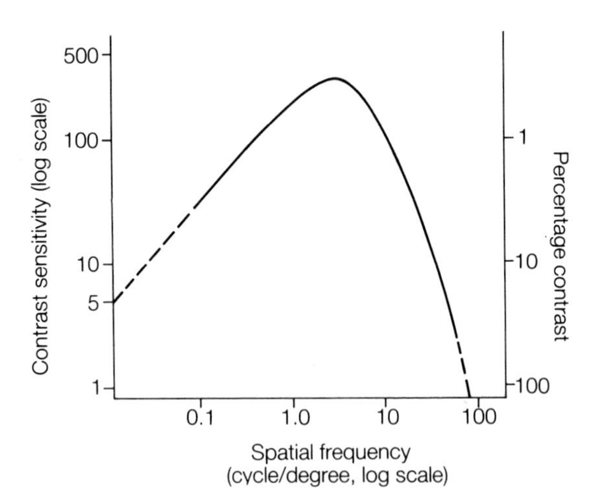

Fig. 9.4. Contrast sensitivity of the human visual system defined as the reciprocal of threshold contrast and measured with sinusoidal gratings, plotted against spatial frequency (after Campbell 1980).

Thus it may be seen that the regular pattern of matrix lines is unlikely to be intrusive when using a 512 × 512 matrix in digital radiology but it may be necessary to choose both image size and viewing distance carefully to avoid or minimize this effect when looking at 128 × 128 or 64 × 64 images sometimes used in nuclear medicine.

9.4.5 Limitations of a subjective definition of contrast

In Chapter 4 a definition of contrast based on visual or subjective response was given. However, this definition alone is not sufficient to determine whether or not a given boundary between two structures will be visually detectable because the eye-brain will be influenced by the type of boundary as well as by the size of the boundary step. There are two reasons for this. First, the perceived contrast will depend on the sharpness of the boundary. Consider for example Fig. 9.5. Although the intensity change across the boundary is the same in both cases, the boundary illustrated in Fig. 9.5a would appear more contrasty on X-ray film because it is sharper. Second, if the boundary is part of the image of a small object in a rather uniform background, contrast perception will depend on the size of the object.

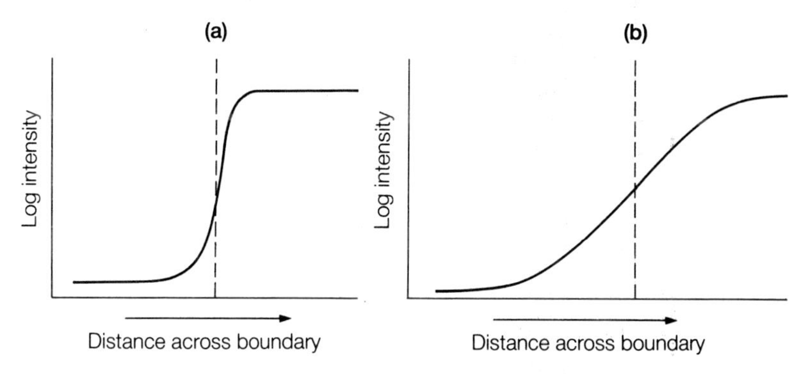

Fig. 9.5. Curves showing the difference in transmitted intensity across (a) a sharp boundary, (b) a diffuse boundary.

9.5 Alternative definition of contrast—signal/noise ratio

The previous definition of contrast is quite appropriate for X-ray film where it is only necessary to densitometer the film. However, it has limitations

1 because film is a non-linear device,

2 because it cannot readily be extended to other imaging systems.

An alternative definition is in terms of signal/noise ratio. Imagine for example that an object consisted of a number of parallel strips of lightly attenuating material. The image would appear as in Fig. 9.6, with regular fluctuations in the pattern on a uniform background.

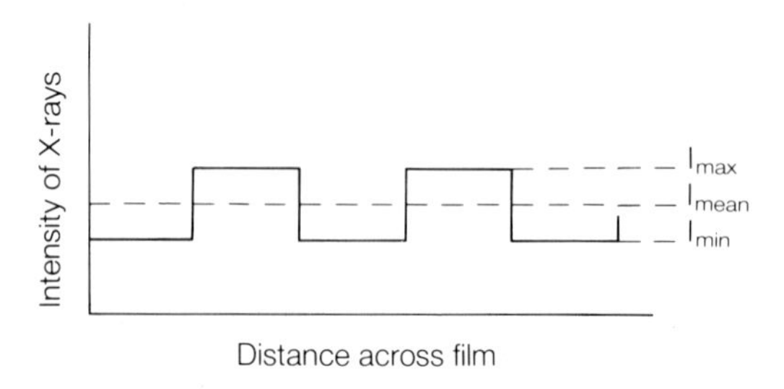

Fig. 9.6. Appearance of a perfect image of an object comprising a number of parallel strips of lightly attenuating material.

The signal may be represented by $I_{max} - I_{min}$ and the noise is the uniform background I_{min}. Hence the signal/noise ratio is $(I_{max} - I_{min})/I_{min}$ or if contrast is small, to a good approximation $(I_{max} - I_{min})/I_{mean}$. Since $I_{mean} = \frac{1}{2}(I_{max} + I_{min})$, contrast is sometimes given as $(I_{max} - I_{min})/(I_{max} + I_{min})$. This will give values that differ by a factor of two from those obtained using I_{mean}.

This definition can be easily extended to digital systems simply by replacing 'intensity' by 'number of photon interactions in the recording medium' or any other variable that is a measure of signal. It is now easier to take account of statistical fluctuations in the signal.

9.6 Quantum noise

When a radiographic screen–film system is exposed to X-rays with a uniform intensity distribution, the macroscopic density distribution of the developed film fluctuates around the average density. This fluctuation is called radiographic noise or radiographic mottle and the ability to detect a signal will depend on the amount of noise in the image. There are many possible sources of noise, for example noise associated with the imaging device itself, film graininess and screen structure mottle, and noise is generally difficult to analyse quantitatively. However, the effect of one type of noise, namely quantum noise, can be predicted reliably and quantitatively. Quantum noise is also referred to variously as quantum mottle or photon noise.

The interaction of a flux of photons with a detector is a random process and the number of photons detected is governed by the laws of Poisson statistics. Thus if a uniform intensity of photons were incident on an array of identical detectors as shown in Fig. 9.7a, and the mean number of photons recorded per detector were N, most detectors would record either more than N or less than N as shown in Fig. 9.7b. The width of the distribution is proportional to $N^{\frac{1}{2}}$ and thus, as shown in Table 9.1,

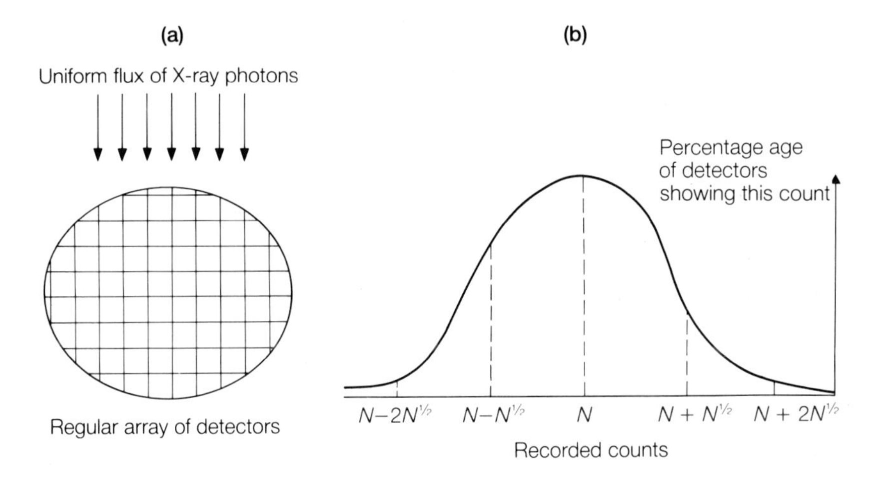

Fig. 9.7. Demonstration of the statistical variation of detected signal (a) uniform flux of X-ray photons onto a regular array of detectors (b) the spread in recorded counts per detector. If the mean count per detector is N, 66% of the readings lie between $N - N^{\frac{1}{2}}$ and $N + N^{\frac{1}{2}}$; 95% of the readings lie between $N - 2N^{\frac{1}{2}}$ and $N + 2N^{\frac{1}{2}}$.

Table 9.1. Variation in counts due to statistical fluctuations as a function of the number of counts collected (or the number of counts per pixel), N

N	$N^{\frac{1}{2}}$	$(N^{\frac{1}{2}}/N) \times 100$ (%)
10	3	30
100	10	10
1000	30	3
10000	100	1
100000	300	0.3
1000000	1000	0.1

the variation in counts due to statistical fluctations is dependent on the value of N and, expressed as a percentage, always increases as N decreases. Since the number of photons detected depends on the patient dose, quantum noise always increases as the patient dose is decreased. Whether or not a given dose reduction and the consequent increase in quantum noise will significantly affect the examination can only be determined by more detailed assessment of the problem. For example a radiographic image might contain 3×10^5 photons mm^{-2} for an incident skin dose of 10 mGy. Thus $N^{\frac{1}{2}} \simeq 6 \times 10^2$ and $N^{\frac{1}{2}}/N \simeq 0.2\%$. Changes in transmission of 1% or more will be little affected by statistical fluctuations. However, a nuclear medicine image comprising 10^5 counts *in toto* may be digitized into a 64×64 matrix (approx. 4000 pixels) giving a mean of 25 counts per pixel. Now $N^{\frac{1}{2}}/N = 20\%$!

Quantum noise is always a major source of image degradation in nuclear medicine and there are a number of situations in radiology where it must also be considered as a possible cause for loss of image quality or diagnostic information. These include the following:

1 The use of very fast intensifying screens may reduce the number of photons detectable to the level where image quality is affected (see section 4.10.2).

2 If an image intensifier is used in conjunction with a television camera, the signal may be amplified significantly by electronic means. However, the amount of information in the image will be determined at an earlier stage in the system, probably by the number of X-ray photon interactions with the input phosphor to the image intensifier. The smallest number of quanta at any stage in the imaging process determines the quantum noise and this stage is sometimes termed the quantum sink.

3 Enlargement of a radiograph decreases the photon density in the image and hence increases the noise.

4 In digital radiology (see section 9.7) the smallest detectable contrast over a small, say 1 mm^2, area will be determined by quantum noise.

5 In computed tomography (see Chapter 10) the precision with which a CT number can be calculated will be affected. For example, the error on 100,000 counts is 0.3% (see Table 9.1) and the CT number for uniformly attenuating material must vary accordingly.

Note that quantum mottle confuses the interpretation of low contrast images. In section 9.8.1 imager performance will be assessed in terms of modulation transfer functions which provide information on the resolution of small objects with sharp borders and high contrast. Other sources of noise may be the ultimate limiting factor in these circumstances.

9.7 Digital radiology

A good illustration of the use of signal/noise ratio is in digital radiology where the primary objective is to present the information in a projected X-ray image in a numerical form. For this purpose the detector system must

1 record X-ray quanta with a high efficiency, and

2 be capable of providing spatial information about the distribution of X-rays.

There are several ways in which this can be achieved. Assuming a point source of X-rays the possibilities are:

1 A small detector that scans across the area occupied by the image. This is a very simple geometry but will require a long time to acquire all the data in a 256 × 256 matrix (Fig. 9.8a).

2 A linear array of detectors, for example photodiodes coupled to a gadolinium oxysulphide screen which only has to make a linear movement to cover the image space. Data collection is now speeded up but may still take 100 s or more (Fig. 9.8b).

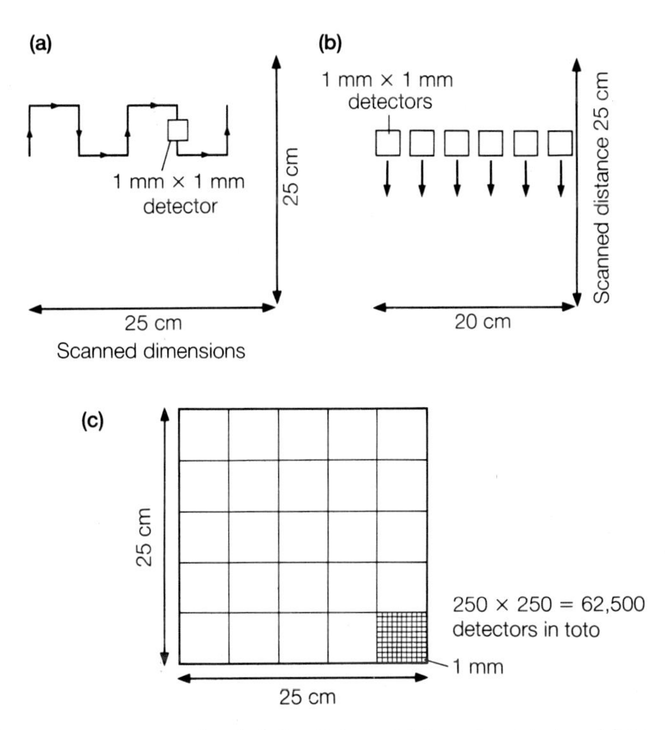

Fig. 9.8. Detector arrangements for digital radiology. (a) Scanning detector. (b) Linear array of detectors. (c) Static array of independent detectors: 250 × 250 = 62,500 detectors *in toto*.

3 A static array of detectors, each 1 mm × 1 mm (Fig. 9.8c).
4 An extended static primary imaging device such as an image intensifier screen, the light output of which is sampled by a rapid raster mechanism to provide the digitized image.

The fourth of these options is currently the most frequently used in practice. As shown schematically in Fig. 9.9 the input face of the image intensifier is a CsI phosphor that must retain high resolution but must also have a high detective quantum efficiency. The output from the X-ray image intensifier is coupled optically to a TV video tube. The data display capability of the image TV receiving the signals should match that of the image intensifier. To match the intrinsic resolution of the CsI phosphor during intravenous angiography (approximately two line pairs per mm) requires about 100 TV raster lines per inch.

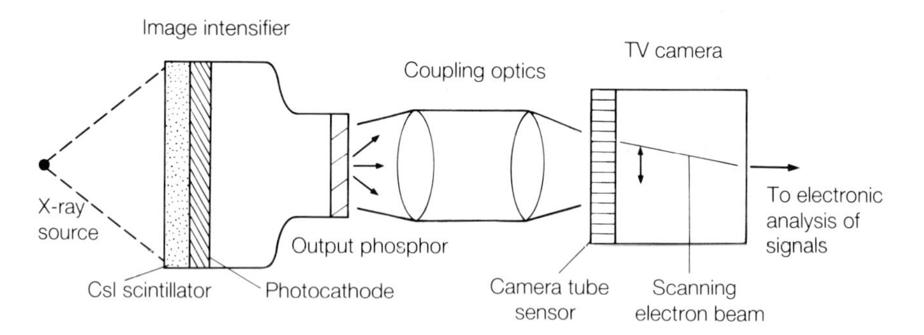

Fig. 9.9. An image intensifier and TV camera system arranged for digital radiology.

Once the data have been stored in numerical form, the capability for manipulation provides several advantages. For example, if a dose in air of 10 mGy is incident on the skull, approximately 10^6 photons mm^{-2} will be transmitted. The fluctuation in this signal due to Poisson statistics is $\pm 10^3$ which is only 0.1% of the signal. Thus changes in signal of this order produced in the skull should be detectable. So far there is no fundamental difference from a conventional radiograph. However, the eye can only distinguish about 20 grey levels and not the 1000 statistically distinguishable levels that are known to be present in the image. If the image data are present in numerical form, they can be manipulated to provide the requisite information.

Examples of data manipulation that may be useful are:
1 Background subtraction to enhance contrast (Fig. 9.10a).

$$\text{Contrast} = \log \frac{n_1 - n_0}{n_2 - n_0} > \log \frac{n_1}{n_2}$$

where n_0 is the background level of counts above zero.

2 Use of a window to examine features over a limited range of count densities (Fig. 9.10b).

3 Edge enhancement or use of gradients (Fig. 9.10c).

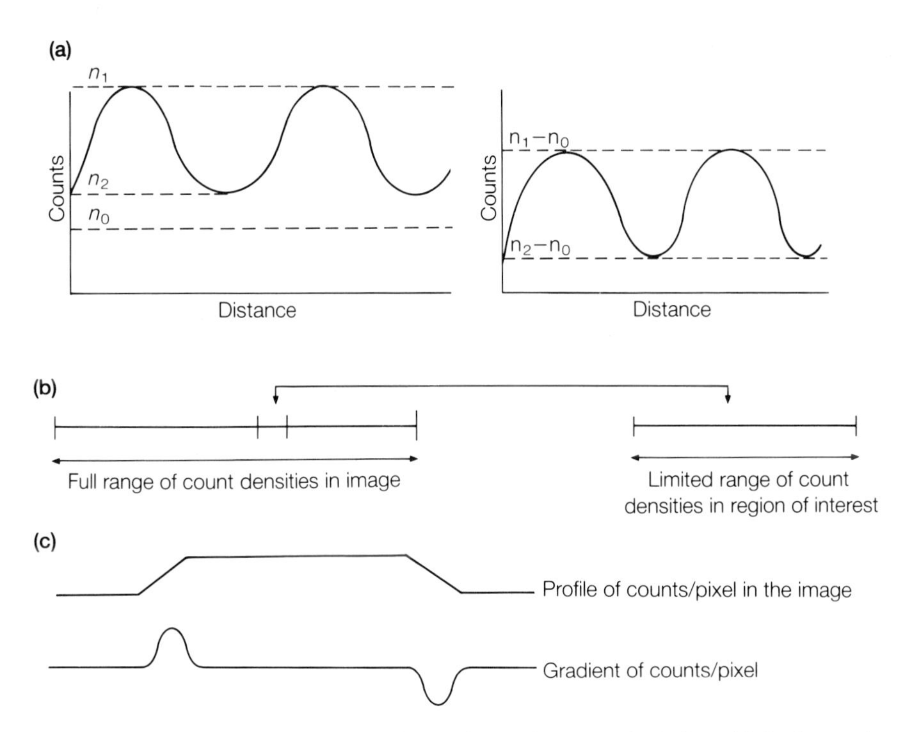

Fig. 9.10. Examples of potentially useful methods of data manipulation. (a) Background subtraction. (b) Use of a window. (c) Edge enhancement or use of gradients.

4 Data shifting—horizontally, vertically or by rotation—to overcome image mismatch in subtraction studies, etc.

Techniques that exploit the potential of digital radiology for high statistical accuracy without generating a wide dynamic range of data are particularly powerful. One way to do this is to use a subtraction technique on digital radiographs obtained under slightly different conditions. For example, X-ray beams of different energies can be used or images can be obtained before and after injection of contrast medium. For this purpose the absorption edge of iodine at 35 keV is particularly useful. Pullan (1981) has shown that for a dose in air of about 10 mGy, a differential concentration of 1.5 mg ml^{-1} iodine may be sufficient to visualize 1 mm diameter vessels. Note that when the required information is dependent on the dynamic behaviour of a contrast agent, all the image space must be sampled very quickly. For example data collection with a linear array would be useless.

It is important to realize that the requirement for a modern intravenous angiographic system to be capable of recording 60 frames per second of 512 $\times$ 256 images causes major problems for data collection, manipulation, storage and display. For example to achieve high resolution, small focal spot sizes (0.5 mm) are necessary. Coupled with repeat pulsed exposures of short duration, especially in cardiac work, this places new demands on X-ray tube rating (see section 2.5). For optimum contrast, scatter must be greatly reduced or eliminated completely. Several rather complex procedures, beyond the scope of this book, for achieving this have been suggested. One concept, whereby grids placed both in front of and behind the patient and moved in synchrony might remove scatter more effectively than a single grid is shown in Fig. 9.11.

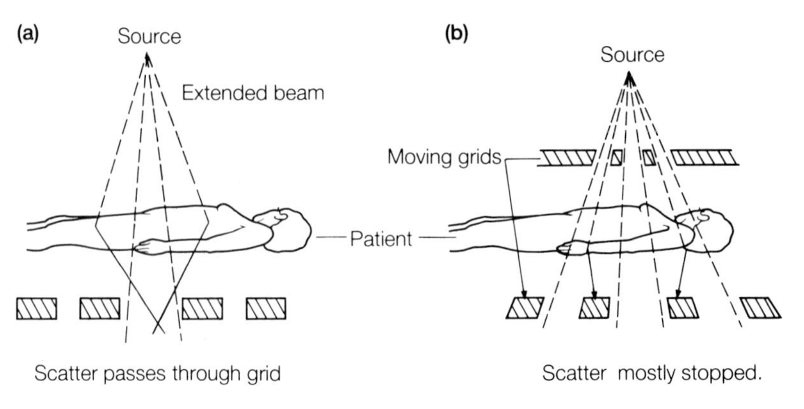

Fig. 9.11. (a) Single grid. (b) Use of grids both in front of the patient and between patient and detector and moved in synchrony to remove more scattered radiation than a single grid.

The main components of a digital radiographic system are shown in Fig. 9.12. Note that the real-time memory is very important because it can limit both image matrix size and frame frequency handling in real time. Also note that alternatives to film are now becoming available, e.g. hard discs, floppy disc or magnetic tape.

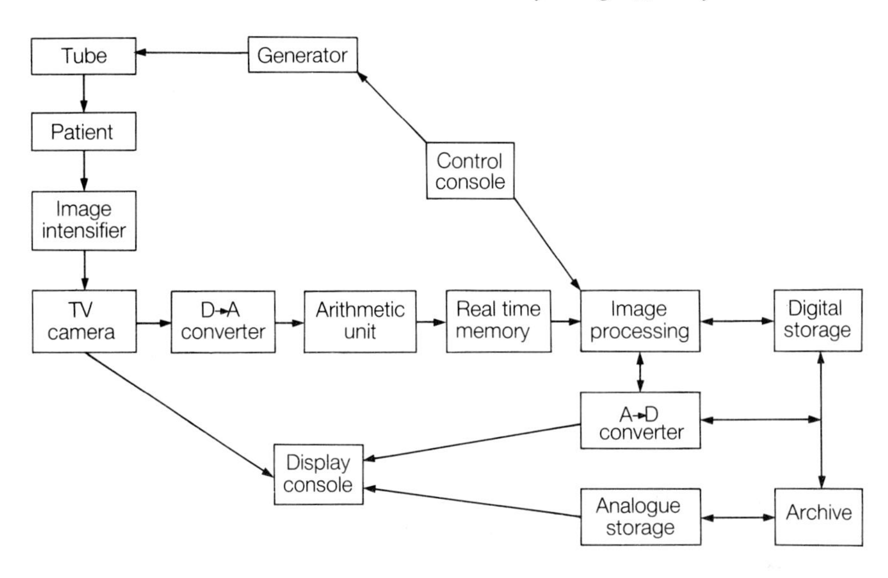

Fig. 9.12. Block diagram showing the main components of a digital radiographic system.

9.8 Assessment of image quality

9.8.1 Modulation transfer function

It is important that the quality of an image should be assessed in relation to the imaging capability of the device that produced it. For example it is well known that an X-ray set is capable of producing better anatomical images of the skeleton than a gamma camera—although in some instances they may not be as useful diagnostically.

One way to assess performance is in terms of the 'resolving power' of the image device—i.e. the closest separation of a pair of linear objects at which the images do not merge. The futility of pixellating digitized data to finer elements than the resolution capability of the system has already been mentioned. Unfortunately, diagnostic imagers are complex devices and many factors contribute to the overall resolution capability. For example, in forming a conventional X-ray image these will include the focal spot size, interaction with the patient, type of film, type of intensifying screen and other sources of unsharpness. Some of these interactions are not readily expressed in terms of resolving power and even if they were there would be no easy way to combine resolving powers.

A practical approach to this problem is to introduce the concept of modulation transfer function (MTF). This is based on the ideas of Fourier analysis, for detailed consideration of which the reader is referred elsewhere, e.g. Gonzalez & Wintz (1977). Only a very simplified treatment will be given here.

Starting from the object, at any stage in the imaging process all the available information can be expressed in terms of a spectrum of spatial frequencies. The idea of spatial frequency can be understood by considering two ways of describing a simple object consisting of a set of equally-spaced parallel lines. The usual convention would be to say the lines were equally spaced 0.2 mm apart. Alternatively, one could say that the lines occur with a frequency in space (spatial frequency) of five per mm.

Fourier analysis provides a mathematical method for relating the description of an object (or image) in real space to its description in frequency space. Two objects and their corresponding spectra of spatial frequencies are shown in Fig. 9.13. In general, the finer the detail, i.e. sharper edges in real space, the greater the intensity of high spatial frequencies in the spatial frequency spectrum (SFS). Thus, fine detail, or high resolution, is associated with high spatial frequencies.

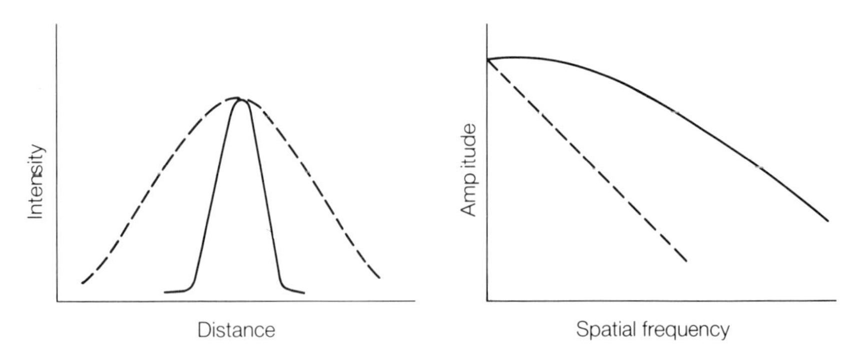

Fig. 9.13. Two objects and their corresponding spectra of spatial frequencies. *Solid line* = sharp object. *Dotted line* = diffuse object.

For exact images of these objects to be reproduced, it would be necessary for the imaging device to transmit every spatial frequency in each object with 100% efficiency. However each component of the imaging device has a modulation transfer function (MTF)* which modifies the spatial frequency spectrum of the information transmitted by the object.

* Rigorous mathematical derivation of MTF assumes a sinusoidally varying object with spatial frequency expressed in cycles mm^{-1}. In the graphs which follow it has been assumed that this sinusoidal wave can be approximated to a square wave and spatial frequencies are given in line pairs mm^{-1} for ease of interpretation by the reader.

Each component of the imager can be considered in turn so

$$\begin{matrix}\text{SFS out of} \\ \text{component M}\end{matrix} = \begin{matrix}\text{SFS into} \\ \text{component N}\end{matrix} \rightarrow \begin{matrix}\text{Component} \\ \text{N}\end{matrix} \rightarrow \begin{matrix}\text{SFS out of} \\ \text{component N}\end{matrix}$$

where $(SFS_{out})_N = (SFS_{in})_N \times MTF_N$.

Hence $(SFS)_{image} = (SFS)_{object} \times (MTF)_A \times (MTF)_B \times (MTF)_C \ldots$ where A, B, C are the different components of the imager.

No imager has an MTF of 1 at all spatial frequencies. In general, the MTF decreases with increasing spatial frequency, the higher the spatial frequency at which this occurs the better the device.

The advantage of representing imaging performance in this way is that the MTF of the system at any spatial frequency, ν, is simply the product of the MTFs of all the components at spatial frequency ν. This is conceptually easy to understand and mathematically easy to implement. It is much more difficult to work in real space.

Some examples of the way in which the MTF concept may be used will now be given. First, it provides a simple pictorial representation of the overall imaging capability of a device. Figure 9.14 shows MTFs for five imaging devices. It is clear that non-screen film transfers higher spatial frequencies, and thus is inherently capable of higher resolution, than screen film. Of course this is because of the unsharpness associated with the use of screens. Note that the MTF takes no account of the dose that may have to be given to the patient.

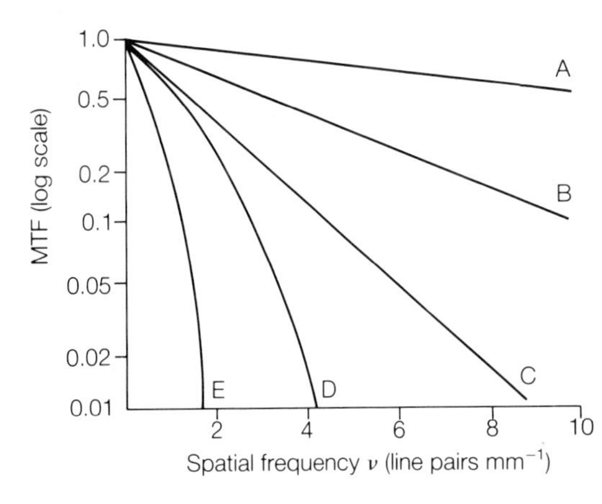

Fig. 9.14. Some typical MTFs for different imaging systems. A = non-screen film. B = film used with high definition intensifying screen. C = film with medium speed screens. D = a 150 mm CsI:Na image intensifier. E = the same intensifier with television display. (Adapted from Hay 1982.)

A similar family of curves would be obtained if MTFs were measured for different film–screen combinations. The spatial frequency at which the MTF fell to 0.1 might vary from 10 line pairs per mm for a film–screen combination of speed 1.0, to 2.5 line pairs per mm for a film–screen combination of speed 4.0. This confirms that slow film–screen combinations are capable of higher resolution than fast film–screens. The major source of this difference is in the choice of screen but recent work has shown small differences in MTF using the same intensifying screen and different films. The MTF is higher when light cross-over from one film to the other can be reduced.

Since the MTF is a continuous function, an imaging device does not have a 'resolution limit' i.e. a spatial frequency above which resolution is not possible, but curves such as those shown in Fig. 9.14 allow an estimate to be made of the spatial frequency at which a substantial amount of information in the object will be lost.

Second, by examining the MTFs for each component of the system, it is possible to determine the weak link in the chain, i.e. the part of the system where the greatest loss of high spatial frequencies occurs.

Figure 9.15 shows MTFs for some of the factors that will degrade image quality when using an image intensifier TV system. Since MTFs are multiplied, the overall MTF is determined by the poorest component—the vidicon camera in this example. Note that movement unsharpness will also degrade a high resolution image substantially.

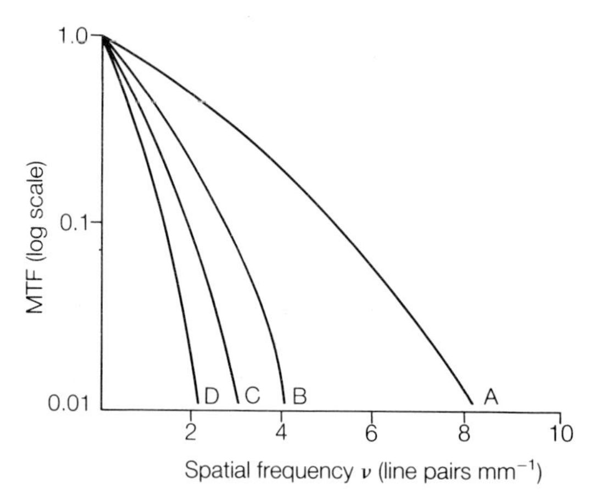

Fig. 9.15. Typical MTFs for some factors that may degrade image quality in an image intensifier TV system. A = 1 mm focal spot with 1 m focus film distance and small object–film distance. B = image intensifier. C = movement unsharpness of 0.1 mm. D = conventional vidicon camera with 800 scan lines.

As a third example, the MTF may be used to analyse the effect of varying the imaging conditions on image quality. Figure 9.16 illustrates the effect of magnification and focal spot size in magnification radiography. Curves B and C show that for a fixed focal spot size, image quality deteriorates with magnification and curves C and D show that for fixed magnification image quality deteriorates with focal spot size. Note that if it were possible to work with $M = 1$, then the focal spot would not affect image quality and an MTF of 1 at all spatial frequencies would be possible (curve A). All these changes could of course have been predicted in a qualitative manner from the discussions in Chapters 5 and 6 of magnification radiography. The point about MTF is that it provides a quantitative measure of these effects, and one that can be extended by simple multiplication to incorporate other factors such as the effect of magnification on the screen MTF, and the effect of movement unsharpness (see for example Curry *et al.* 1984). Hence it is the starting point for a logical analysis of image quality and the interactive nature of the factors that control it.

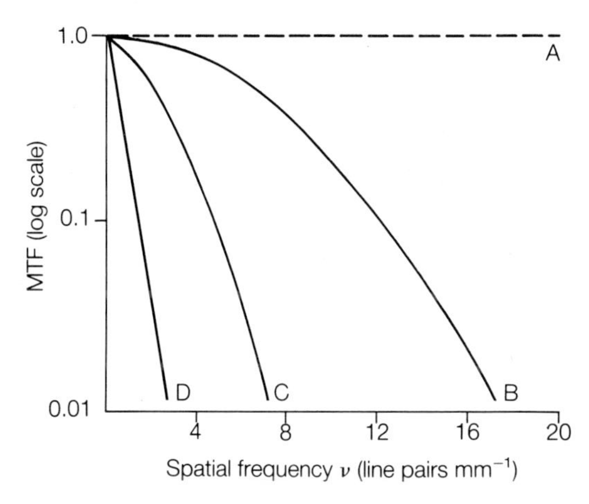

Fig. 9.16. MTF curves for magnification radiography under different imaging conditions. A = magnification of 1. B = magnification of 1.2 with a 0.3 mm focal spot size. C = magnification of 2.0 with a 0.3 mm focal spot size. D = magnification of 2.0 with a 1.0 mm focal spot size.

Finally, consider an unusual MTF (Fig. 9.17). Unlike the majority of systems, which transmit low spatial frequencies well, xeroradiography is not a good technique for visualizing large, low contrast structures and has a poor MTF at low spatial frequencies. However, the ability of xeroradiography to enhance edges means that it does transmit well the high spatial

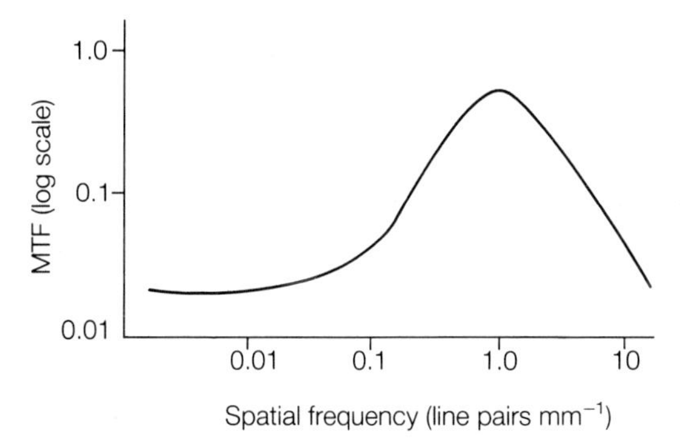

Fig. 9.17. MTF for a typical xeroradiographic system.

frequencies that carry information about the edges. Thus the MTF is better for higher spatial frequencies. Note that the fall in MTF in the region of 10 line pairs per mm is not related to the xeroradiographic process but is caused by other problems (e.g. focal spot size and patient motion).

9.8.2 Physical/physiological assessment

Although the MTF provides a useful method of assessing physical performance, its meaning in terms of the interpretation of images is obscure. For this purpose, as already noted in the discussion on contrast, it is necessary to involve the observer and in this context image quality may be defined as a measure of the effectiveness with which the image can be used for its intended purpose.

The simplest perceptual task is to detect a detail of interest, i.e. the signal, in the presence of noise. Noise includes all those features that are irrelevant to the task of perception—e.g. quantum mottle, anatomical noise and visual noise (inconsistencies in observer response). Three techniques that have been used to discriminate between signal and noise will be mentioned briefly (for further details see Sharp 1984, Sharp *et al.* 1985).

It should be noted in passing that the visual threshold at which objects can be (a) detected, (b) recognized and (c) identified are not the same and this is a serious limitation when attempting to extrapolate from studies on simple test objects to complex diagnostic images.

Method of constant stimulus

Consider the simple task of detecting a small region of increased attenuation in a background that is uniform overall but shows local fluctuations due to the presence of noise. Because of these fluctuations, the ease with which an abnormality can be seen will vary from one image to another even though its contrast remains unchanged.

Figure 9.18 shows the probability of generating a given visual stimulus for a set of images of this type in which the object contains very little contrast (C_1). If the observer adopts some visual threshold T, only those images which fall to the right of T will be reported as containing the abnormality (about 25% in this example).

If a set of images is now prepared in which the object has somewhat greater contrast (C_2), the distribution curve will shift to the right (shown dotted) and a much greater proportion of images will be scored positive.

If this experiment is repeated several times using sets of images, each of which contains a different contrast, visual response may be plotted against

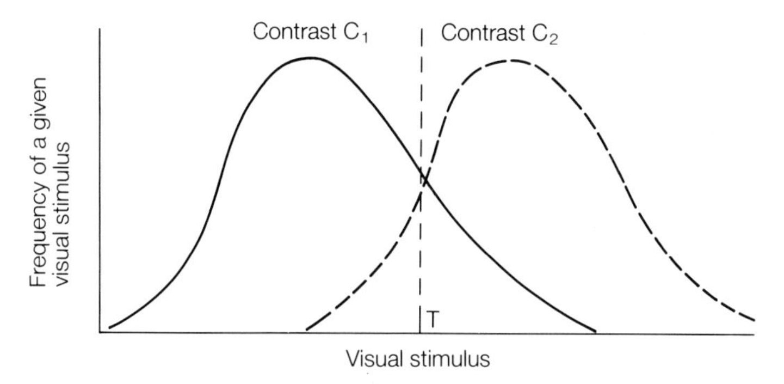

Fig. 9.18. Curves showing how, for a fixed visual stimulus threshold T, the proportion of stimuli detected will increase if contrast is increased from C_1 to C_2.

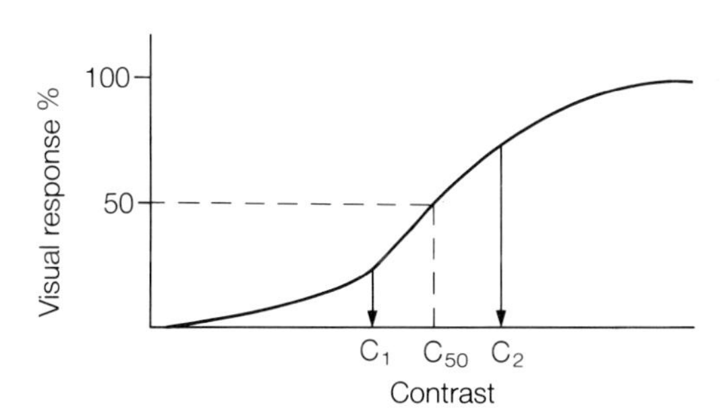

Fig. 9.19. Visual response (percent positive identification) plotted as a function of contrast on the basis of observations similar to those described in Fig. 9.18.

contrast as shown in Fig. 9.19. The contrast resulting in 50% visual response is usually taken as the value at which the signal is detectable.

Experiments of this type may be used to demonstrate, for example, that in relation to visual perception, object size and contrast are inter-related.

Signal detection theory

To apply the method of constant stimulus, contrast must be varied in a controlled manner. The method cannot therefore be applied to real images.

In signal detection theory, all images are considered to belong to one of two categories, those which contain a signal plus noise and those which contain noise only. It is further assumed that in any perception study some images of each type will be misclassified. This situation is represented in terms of the probability distribution used in the previous section in Fig. 9.20.

There are four responses: true positive, true negative, false positive and false negative. These four responses are subject to the following constraints:

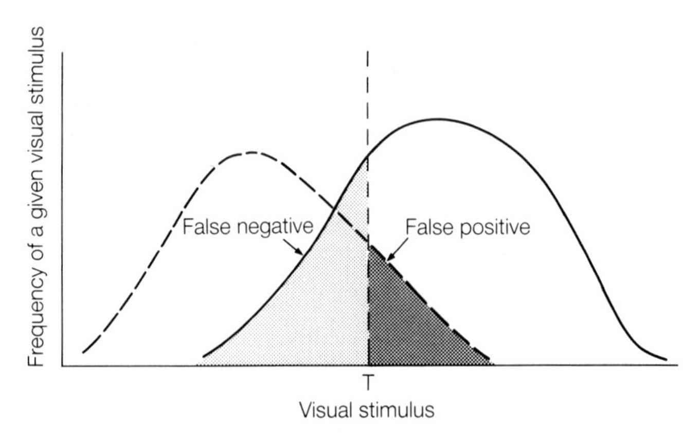

Fig. 9.20. Curves illustrating the concept of false negative and false positive based on the spread of visual stimuli for an object of fixed contrast. *Dotted line* = probability distribution of true negatives. *Solid line* = probability distribution of true positives.

True positive + False negative (all 'signal + noise' images) = constant
True negative + False positive (all 'noise only' images) = constant

The greater the separation of the distributions, the more readily is the signal detected.

Information about the complete distribution can be obtained if the observer can be persuaded to adopt different visual thresholds indicated by A, B, C, D and E in Fig. 9.21. Since all images falling to the right of the visual threshold are reported positive, A is a very lax criterion. Certainly all the real positives (*solid curve*) are reported positive but a high propor-

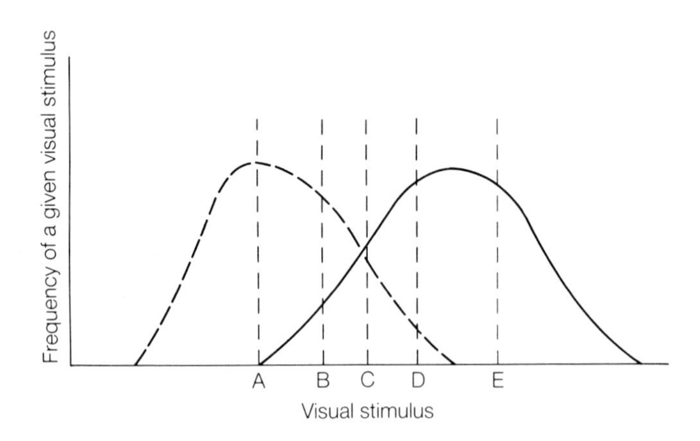

Fig. 9.21. Use of different visual thresholds with distributions of visual stimuli.

tion of the images containing no signal are also reported positive. They are therefore false positives. Conversely, E is a very strict criterion. None of the 'noise only' images is reported positive but a high proportion of the real positives are reported negative (false negatives).

If each working point is plotted on a curve of false positives against true positives (as shown in Fig. 9.22) the result is known as a receiver operating characteristic (ROC curve). The dotted curve represents a better discrimination between true positive images and true negative images because for any given level of true positive identifications (say 90%) there are fewer false positives on the dotted curve than on the solid curve.

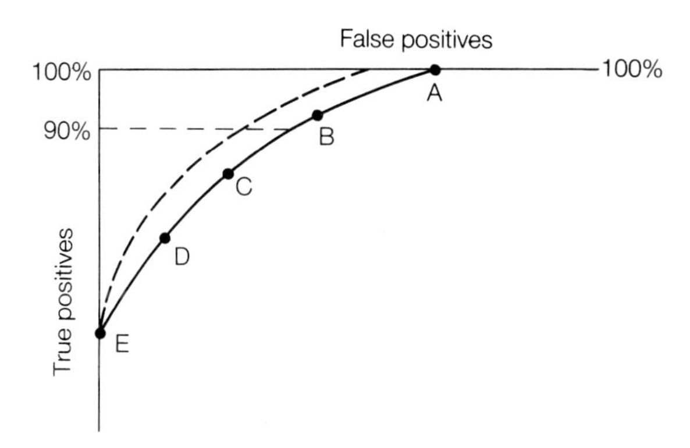

Fig. 9.22. A typical ROC curve, and a better ROC curve (shown *dotted*).

Ranking

In this method the observer is presented with a set of images in which some factor thought to influence quality has been varied. The observer is asked to arrange the images in order of preference. This approach relies on the fact that an experienced viewer is frequently able to say if a particular image is of good quality but unable to define the criteria on which this judgement is based.

When the ranking order produced by several observers is compared, it is possible to decide if observers agree on what constitutes a good image. Furthermore, if the image set has been produced by varying the imaging conditions in a controlled manner, for example by steadily increasing the amount of scattering medium, it is possible to decide how much scattering medium is required to cause a detectable change in image quality.

The simplest ranking experiment is to compare just two images. If one of these has been taken without scattering material, the other with scatter-

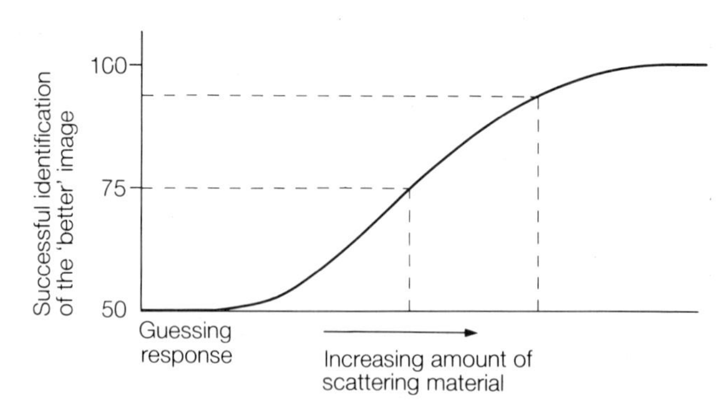

Fig. 9.23. Curve showing how the ability to detect the undegraded image might be expected to increase steadily as the amount of degradation was increased.

ing material, then the percentage of occasions on which the 'better' image is identified can be plotted against the amount of scattering material. The 75% level corresponds to correct identification 3/4 times, the 90% level to correct identification 9/10 times. It might be reasonable to conclude that when the better image was correctly identified 3/4 times, detectable deterioration in image quality had occurred (Fig. 9.23).

The ranking approach is particularly useful for quality control and related studies. All equipment will deteriorate slowly and progressively with time and although it may be possible to assess this deterioration in terms of some physical index such as resolution capability with a test object, it is not clear what this means in terms of image quality. If deterioration with time (Fig. 9.24a) can be simulated in some controlled manner (Fig. 9.24b), then a ranking method applied to the images thus produced may indicate when remedial action should be taken.

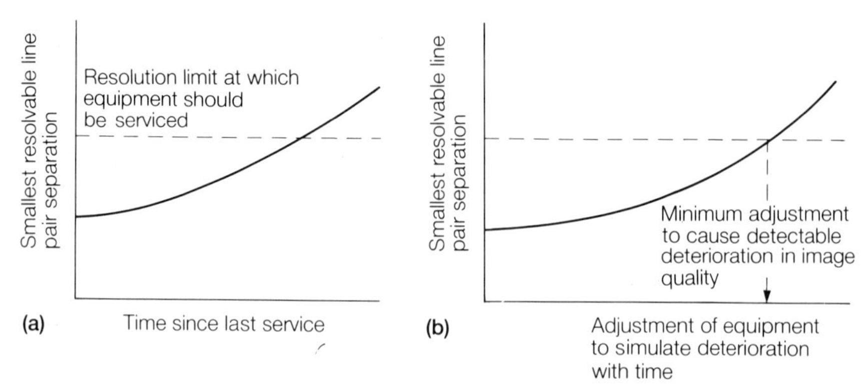

Fig. 9.24. Curves showing how a logical policy towards the frequency of imager servicing might be based on measurements of deterioration of imager performance.

A similar approach may be used to assess the benefit of new technology, for example the effect of different film screen combinations on the quality of tomographic images (Cohen *et al.* 1976) or in mammography (Sickles *et al.* 1977).

9.9 Design of clinical trials

By concentrating on the physics of the imaging process, it is easy to overlook the fact that *in vivo* imaging is not an end in itself but only a means to an end. Ultimately the service will be judged by the quality of the diagnostic information it produces and critical objective evaluation of each type of examination is essential for three reasons. First it may be used to assess diagnostic reliability and this is important when the results of two examinations conflict. Second, objective information allows areas of weakness to be analysed and provides a basis for comparison between different imaging centres. Finally, medical ethics is not independent of economics and increasingly in the future the provision of sophisticated and expensive diagnostic facilities will have to be justified in economic terms.

It is not difficult to understand why few clinical trials have been done in the past when the constraints are listed:

1 The evaluation must be prospective with images assessed within the reference frame of the normal routine work of the unit—not in some academic ivory tower.

2 A typical cross-section of normal images and abnormal images from different disease categories must be sampled since prevalence affects the predictive value of a positive test.

3 A sufficient number of cases must enter the trial to ensure adequate statistics.

4 Equivalent technologies must be compared. It is meaningless to compare the results obtained using a 1975 ultrasound scanner with those obtained using a 1986 CT whole body scanner or vice versa.

5 Evaluations must be designed so that the skill and experience of the reporting teams do not influence the final result.

6 Finally, and generally the most difficult to achieve, there must be adequate independent evidence on each case as to whether it should be classified as a true normal or a true abnormal.

If these constraints can be met, methods are readily available for representing the results. For example if images are simply classified as normal or abnormal, there are only four possible outcomes to an investigation, which can be expressed as a 2×2 decision matrix (Table 9.2).

Table 9.2. A 2×2 decision matrix for image classification

	Abnormal images	Normal images
True abnormal	a	b
True normal	c	d
Totals	$a+c$	$b+d$

$$\text{Overall diagnostic accuracy} = \frac{\text{no. of correct investigations}}{\text{total investigations}} = \frac{a+d}{a+b+c+d}$$

Other indices sometimes quoted are

$$\text{Sensitivity} = \frac{\text{abnormals detected}}{\text{total abnormals}} = \frac{a}{a+b}$$

$$\text{Specificity} = \frac{\text{normals detected}}{\text{total normals}} = \frac{d}{c+d}$$

and

$$\text{Predictive value of a positive test} = \frac{\text{abnormals correctly identified}}{\text{total abnormal reports}} = \frac{a}{a+c}$$

It is well known that prevalence has an important effect on the predictive value of a positive test and to accommodate possible variations in prevalence of the disease, Bayes Theorem may be used to calculate the posterior probability of a particular condition, given the test results and assuming different a priori probabilities.

If the prior probability of disease, or prevalence, is $P(D_+)$ then it may be shown that the posterior probability of disease when the test is positive (T_+) is given by

$$P(D_+/T_+) = \frac{\left(\dfrac{a}{a+b}\right) P(D_+)}{\left(\dfrac{a}{a+b}\right) P(D_+) + \left(\dfrac{c}{c+d}\right) P(D_-)}$$

Similarly the posterior probability of disease when the test is negative (T_-) is given by

$$P(D_+/T_-) = \frac{\left(\frac{b}{a+b}\right) P(D_+)}{\left(\frac{b}{a+b}\right) P(D_+) + \left(\frac{d}{c+d}\right) P(D_-)}$$

For a full account of Bayes Theorem see Shea (1978).

As an alternative to simple classification of the images as normal or abnormal, the data can be prepared for ROC analysis if a confidence rating is assigned to each positive response (Table 9.3).

Table 9.3. Examples of the confidence ratings used to produce an ROC curve

Rating	Description
5	Abnormality definitely present
4	Abnormality almost certainly present
3	Abnormality possibly present
2	Abnormality probably not present
1	Abnormality not present

These ratings reflect a progressively less stringent criterion of abnormality and correspond to points E to A respectively in Fig. 9.21. Thus by comparing the ROC curves constructed from such rating data, the power of two diagnostic procedures may be compared.

Much work has been done in recent years on the design of trials for the critical clinical evaluation of diagnostic tests and it is now possible to assess better the contribution of each examination within the larger framework of patient health care.

9.10 Conclusions

The emphasis in this chapter has been on the idea that there is far more information in a radiographic image than it is possible to extract by subjective means. Furthermore, many factors contribute to the quality of the final image and for physiological reasons, the eye can easily be misled over what it thinks it sees. Thus there is a strong case for introducing numerical or digital methods into diagnostic imaging. Such methods allow greater manipulation of the data, more objective control of the image quality and greatly facilitate attempts to evaluate both imager performance and the overall diagnostic value of the information that has been obtained.

References and further reading

Campbell F. W. (1980) The physics of visual perception. *Philos. Trans. R. Soc. Lond.* **290**, 5–9.

Chesters M. S. (1982) Perception and evaluation of images. In *Scientific Basis of Medical Imaging*, ed. P. N. T. Wells, pp. 237–280. Churchill Livingstone, Edinburgh.

Cohen G., Barnes J. O. & Peria P. M. (1976) The effects of filmscreen combination on tomographic image quality. *Radiology,* **129**, 515.

Curry T. S., Downey J. E. and Murry R. C. (1984) *Christensen's Introduction to the Physics of Diagnostic Radiology*, 3rd edn. Lea & Febiger, Philadelphia.

Gonzalez R. C. & Wintz P. (1977) *Digital Image Processing*. Addison Wesley, Massachusetts.

Harrison, R. M. & Isherwood I., eds (1984) *Digital Radiology Physical and Clinical Aspects*. (IPSM 1) The Hospital Physicists' Association, London.

Hay, G. A., ed. (1976) *Medical Images: Formation, Perception and Measurement*. The Institute of Physics, and Wiley, Chichester.

Hay G. A. (1982) Traditional X-ray imaging. In *Scientific Basis of Medical Imaging,* ed. P. N. T. Wells, pp. 1–53. Churchill Livingstone, Edinburgh.

Pullan B. R. (1981) Digital radiology. In *Physical Aspects of Medical Imaging*, eds B. M. Moores, R. P. Parker & B. R. Pullan, pp. 275–288. The Hospital Physicists' Association, and Wiley, Chichester.

Shea G. (1978) An analysis of the Bayes procedure for diagnosing multistage disease. *Comput. Biomed. Res.* **11**, 65–75.

Sharp P. F. (1984) Physical limitations to the quality of X and gamma ray images and the presentation of photon limited images. In *Technical Advances in Biomedical Physics*, eds. P. P. Dendy, D. W. Ernst & A. Şengün, pp. 219–258. Martinus Nijhoff, The Hague.

Sharp P. F., Dendy P. P. & Keyes W. I. (1985) *Radionuclide Imaging Techniques*. Academic Press, London.

Sickles E. A., Genant H. K. and Doi K. (1977) Comparison of laboratory and clinical evaluations of mammographic screen-film systems. In *Applications of Optical Instruments in Medicine* Vol. 4, eds J. E. Gray & W. R. Hendee, pp. 30. SPIE, Boston.

Exercises

1 What factors affect
(a) the sharpness,
(b) the contrast of clinical radionuclide images?

2 List the factors affecting the sharpness of a radiograph. Draw diagrams to illustrate these effects.

3 What do you understand by the 'quality' of a radiograph? What factors affect the quality?

4 Explain the terms subjective and objective definition, latitude and contrast when used in radiology.
5 Explain what is meant by the term 'perception' in the context of diagnostic imaging. Explain how perception studies may be used to assess the quality of diagnostic images.
6 Why does the MTF of an intensifying screen improve if the magnification of the system is increased?
7 Show that enlargement of an image such that the photons are spread over a larger area increases quantum noise by $(N/m^2)^{\frac{1}{2}}$ where N is the original number of photons mm^{-2} and m is the magnification. Is quantum noise increased by magnification radiography?
8 Explain how quantum mottle may limit the smallest detectable contrast over a small, 1 mm^2, area in a digitized radiograph.

10

Tomographic Imaging

10.1 Introduction

Three fundamental limitations that apply equally to imaging with both X-rays and gamma rays from radionuclides can be identified.

Superimposition
The final image is a two-dimensional representation of an inhomogeneous three dimensional object with many planes superimposed. In radiology the data relate to the distribution of attenuation coefficients in the different planes through which the X-ray beam passes. In nuclear medicine the data relate primarily to the distribution of radioactivity, although the final result

is also affected by attenuation of the gamma rays as they emerge from the patient.

The confusion of overlapping planes results in a marked loss of contrast making detection of subtle anomalies difficult or impossible. A simplified example of the consequence of superposition taken from nuclear medicine is shown in Fig. 10.1.

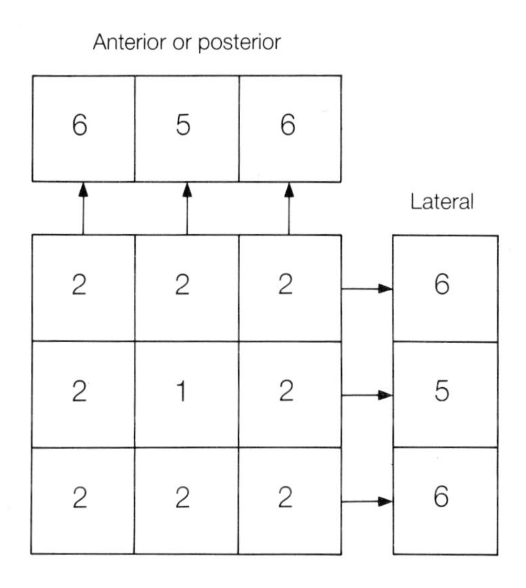

Fig. 10.1. Illustrating how superposition of signals from different planes reduces contrast. If the numbers in nine squares represent radioactive concentrations in the nine compartments, the theoretical contrast is 2:1. If, however, the activity is viewed along any of the projection lines shown, the contrast is only 6:5.

Geometrical effects

The viewer can be confused about the shapes and relative positions of various structures displayed in a conventional X-ray picture and care must be taken before drawing conclusions about the spatial distribution of objects even when an orthogonal pair of radiographs is available.

Attenuation effects

When the intensity of X-rays striking a film is described by the equation $I = I_0 \, e^{-\mu x}$, it must be appreciated that beam attenuation, I/I_0, depends on both μ and x. This can cause ambiguity since an observed difference in attenuation can be due to changes in thickness alone (x), in composition alone (μ) or a combination of these factors.

Two methods have been used to overcome the effect of superposition of information in different planes. The first is longitudinal or blurring tomography. This relies on the principle that if the object is viewed from different angles, such that the plane of interest is always in the same orientation relative to the detector, or film, the information in this plane will superimpose when the images are superimposed, but information in other planes will be blurred.

Longitudinal tomography with X-rays is considered in greater detail in section 10.2. Longitudinal tomography using gamma ray emissions from radionuclides is also possible using either a slant hole collimator or a multiple pinhole collimator, both of which can be used to take views from different angles. However, longitudinal tomography is not widely used in nuclear medicine.

In longitudinal tomography, the angle between the normal to the plane of interest and the axis of the detector, ϕ, can be anything from 1° to 20°. Since the object space is not completely sampled, the in-focus plane is not completely separated from its neighbouring planes, but the tomographic effect increases with increasing ϕ. The second method, transverse tomography, can be considered as a limiting form of longitudinal tomography with $\phi = 90°$. The object is now considered as a series of thin slices and each slice is examined from many different angles. Isolation of the plane of interest from other planes is now complete and is determined by the width of the interrogating X-ray beam, collimator design and the amount of scatter. Transverse tomography provides an answer to geometrical and attenuation problems as well as that of superposition and the majority of this chapter is devoted to consideration of the many physical factors that contribute to the production of high quality transverse tomographic images.

10.2 Longitudinal tomography

10.2.1 The linear tomograph

This is the simplest form of tomography and is a good way to introduce the principles As shown in Fig. 10.2, the X-ray tube is constrained to move along the line SS′ so that the beam is always directed at the region of interest P in the subject. As the X-ray source moves in the direction S → S′, shadows of an object P will move in the direction F′ → F. Hence if a film placed in the plane FF′ is arranged to move parallel to that plane at

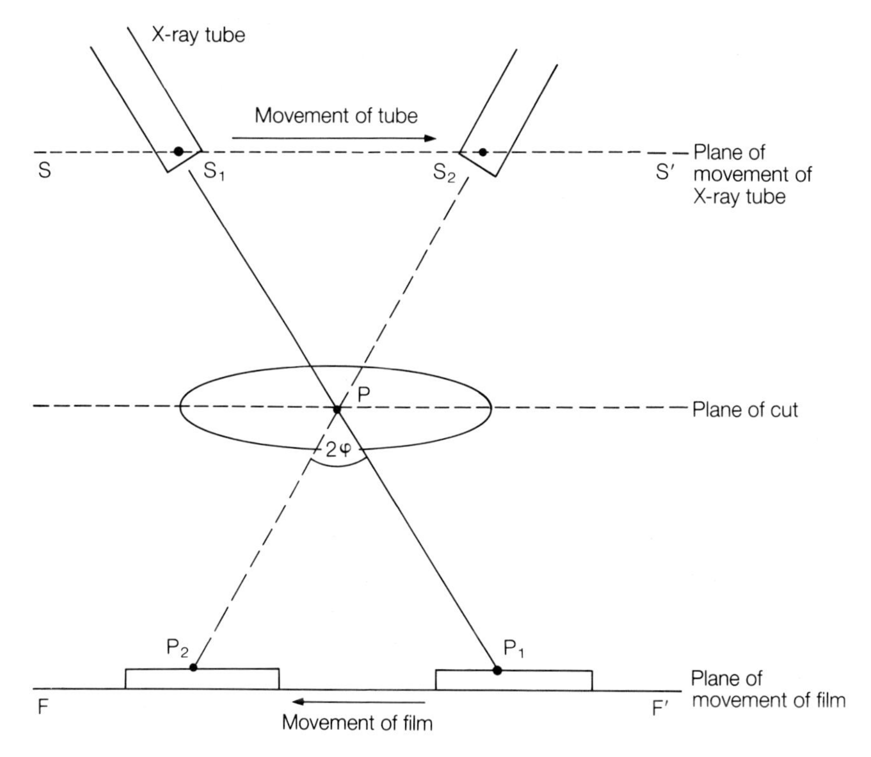

Fig. 10.2. Diagram illustrating the movement of the X-ray tube and the film in linear longitudinal tomography.

such a rate that when S_1 has moved to S_2, the point P_1 on the film has moved to P_2, all the images of the object at P will be superimposed.

Consider however the behaviour of a point in a different plane (Fig. 10.3). As S_1 moves to S_2, the film will move a distance P_1P_2 as before, so the point F_1, which coincides with X_1, will move to F_2. However, the shadow of X will have moved to X_2. Hence the shadows cast by X for all source positions between S_1 and S_2 will be blurred out on the film between F_2 and X_2 in a linear fashion. A similar argument will apply to points below P. Thus only the image of the point P, about which the tube focus and film have pivoted, will remain sharp.

Consideration by similar triangles of other points in the same horizontal plane as P, for example Q, shows that they also remain sharply imaged during movement. The horizontal plane through P is known as the *plane of cut*.

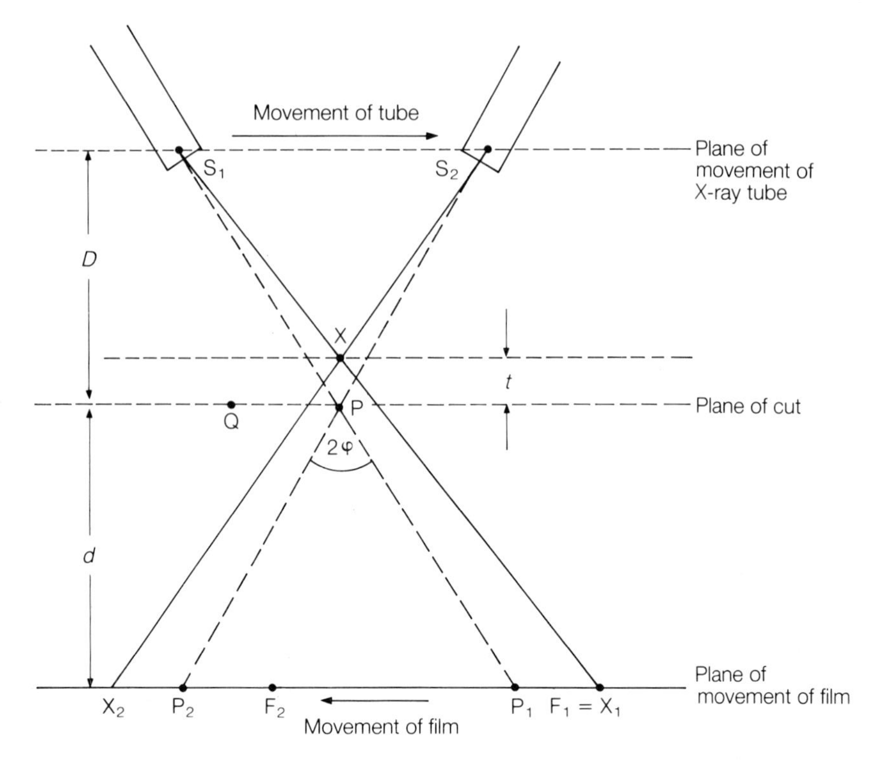

Fig. 10.3. Demonstration of the blurring effect that will occur in longitudinal tomography for points that are not in the plane of cut.

10.2.2 Thickness of plane of cut

Although the plane of cut is not completely separated from adjacent planes, there is a minimum amount of blurring that the eye can discern and this effectively determines the 'thickness' of the plane that is in focus.

It is instructive to calculate the thickness of the plane in terms of the minimum amount of blurring, B say, and the geometry of the system. Referring again to Fig. 10.3, suppose the point X suffers the minimum amount of blurring that is just detectable. From the previous discussion:

Film blurring $= X_2F_2$

By geometry this is equal to $X_2X_1 - X_1F_2 = X_2X_1 - P_2P_1$

This must be set equal to B

Now $P_1P_2 = S \cdot \dfrac{d}{D}$

$$\text{and } X_1X_2 = S \cdot \frac{(d+t)}{(D-t)}$$

$$\text{Hence } B = S\left[\frac{(d+t}{(D-t)} - \frac{d}{D}\right] \simeq S\frac{(D+d)}{D}t \quad (\text{since } t \ll \text{D})$$

$$\text{Since } \frac{S}{D} \simeq 2\phi$$

$$B = 2t\left(\frac{D+d}{D}\right)\phi$$

The same argument can be applied to points below P so the thickness of cut,

$$2t = B \cdot \frac{1}{\phi} \cdot \frac{D}{(D+d)} \qquad \text{(equation 10.1)}$$

Thus the thickness of cut increases if the value assumed for minimum detectable blurring, B, or the focus–plane of cut distance D increases. However, it decreases if the angle of swing 2ϕ or the focus–film distance $(D + d)$ increases.

Assuming typical values:

Angle of swing $2\phi = 10° = 10 \times 2\pi/360$ radians
Focus–film distance $(D + d) = 90$ cm
Focus–plane of cut $D = 75$ cm
Taking $B = 0.7$ mm
Thickness of cut, $2t = 6.6$ mm.

Note the following additional points:

1 The plane of cut may be changed by altering the level of the pivot.
2 Since the tube focus and film move in parallel planes, the magnification of any structure that remains unblurred throughout the movement remains constant. This is important because the relatively large distance between the plane of cut and the film means that the image is magnified (in this worked example by $90/75 = 1.2$).
3 Care is required to ensure that the whole exposure takes place whilst tube and film are moving.
4 The tilt of the tube head must change during motion—this minimizes reduction in exposure rate at the ends of the swings due to obliquity factors but there is still an inverse square law effect.

10.2.3 Miscellaneous aspects of longitudinal tomography

There is insufficient space to discuss in detail other aspects of longitudinal tomography but they will be summarized briefly.

Alternative linear movements

When ϕ is large it may be difficult to compensate adequately for inverse square law effects using the movement discussed in section 10.2.1. If both the tube and film move along the arcs of circles that are centered on a point in the plane of cut, the focus–film distance remains fixed. This movement allows easier engineering design for the X-ray tube but is more difficult to achieve for the film.

Non linear movements

An alternative way to overcome inverse square law problems is to use non-linear movement. For example, if both tube and film execute a circular motion (Fig. 10.4), the desired blurring will be achieved in a circular fashion but the source–film distance will not change. Such a movement also eliminates another major problem of linear tomography, namely that

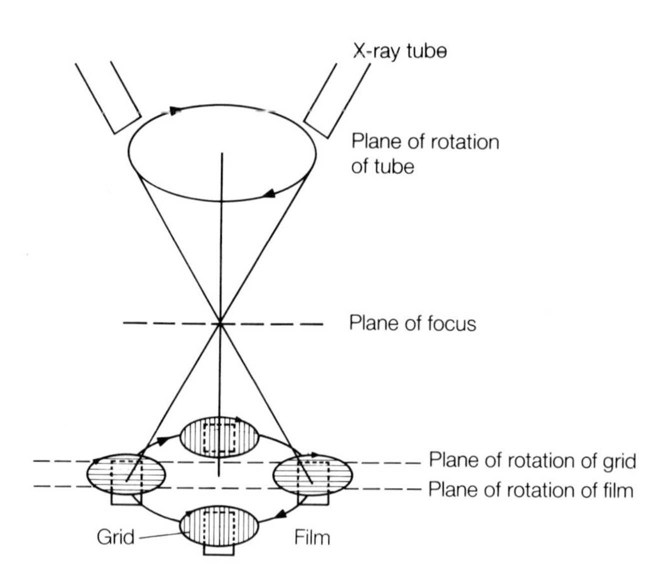

Fig. 10.4. Cyclic movement in longitudinal tomography ensures a fixed focus–film distance. Note that the grid rotates with the X-ray tube so that it maintains the same orientation relative to the target.

structures lying parallel to the line of motion of the tube are not well blurred because the blurred image of one part of the object is simply superimposed on another part of the object. Although any planar movement is theoretically possible, circular and elliptical movements illustrate adequately the realistic alternatives.

Choice of angle of swing

As shown in equation 10.1, the angle of swing will determine slice thickness. Whereas a thin slice may be desirable, contrast between two structures will be proportional to $(\mu_2 - \mu_1)t$. Thus large angle tomography giving thin slices may be used when there are large differences in atomic number and/or density between structures of interest, but if $(\mu_2 - \mu_1)$ is small, a somewhat larger value of t may be essential to give detectable contrast.

Zonography

This is the term given to very small angle tomography ($2\phi \sim 1°–5°$). The slice is now very thick and structures within it appear almost as on a normal radiograph. The technique may be useful if very thick structures are being examined or if the difference $(\mu_2 - \mu_1)$ is very small.

Use of multicassettes

Tomography will normally be performed after taking plane films and if several repeat exposures are necessary, either to find the correct plane of cut or because information from several planes is required, the dose to the patient can be quite high. Repeat exposures can be largely eliminated by using a multicassette box, the principle of which is shown in Fig. 10.5. Although shadows of Y will not be in register in the plane F, they will be in register in the plane G which is some distance below F. Thus if a second film is placed at G_1G_2 and moved at the same speed as that at F_1F_2, two planes of cut can be imaged with the same exposure. By similar triangles, the separation of the planes of cut PY is related to the separation of film planes FG by

$$\frac{\mathrm{PY}}{\mathrm{FG}} = \frac{D}{(D + \mathrm{d})}$$

Note that since the X-ray beam will be attenuated on passing through the cassette, faster intensifying screens must be used with the underlying films to give adequate blackening.

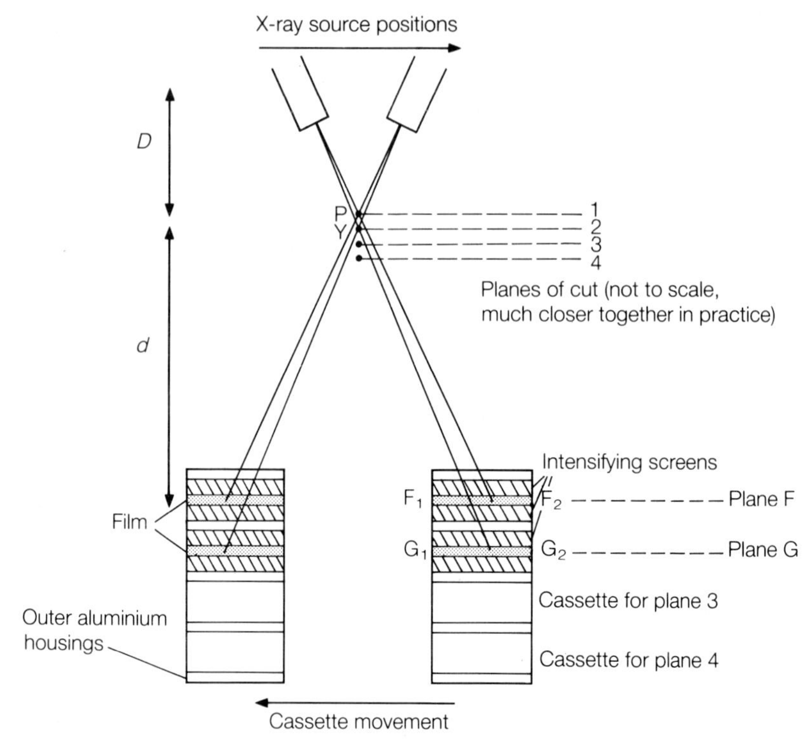

Fig. 10.5. Design of a multicassette box to image several tomographic planes simultaneously.

10.3 Computed axial transmission tomography (CT)

10.3.1 Principles

As discussed in Chapter 9, there is an inverse relationship between the contrast of an object and its diameter at threshold perceptibility. Thus the smaller the object the larger the contrast required for its perception, and loss of contrast resulting from superposition of different planes has an adverse effect on detectability. Hence, although the average film radiograph is capable of displaying objects with dimensions as small as 0.25 mm, when due allowance is made for film resolution, geometric movement blur and scattered radiation, a contrast difference of about 10% is required to achieve this.

The objective of CT scanning is to take a large number of one-dimensional views of a two-dimensional transverse axial slice from many different directions and reconstruct the structure within the slice. Digital techniques are normally used so the object plane can be considered as a slice of variable width subdivided into a matrix of attenuating elements with linear attenuation coefficients $\mu(x, y)$ (Fig. 10.6). A typical matrix size will be 256 × 256 or 512 × 512 corresponding to a pixel side of 1.0 or 0.5 mm. Slice thickness z may vary from 1 mm to 15 mm.

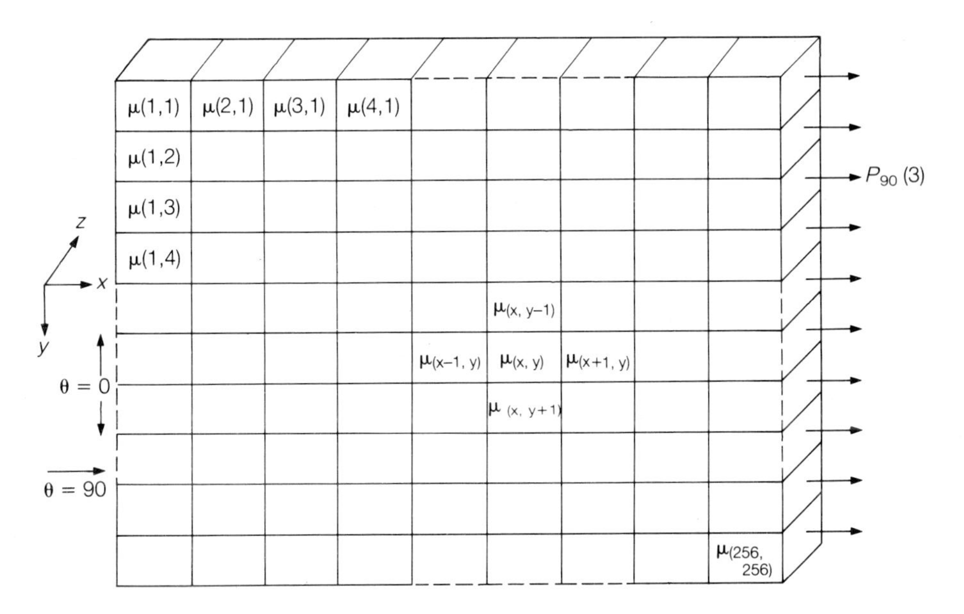

Fig. 10.6. A slice through the patient considered as an array of attenuation coefficients. If θ = 0 is chosen arbitrarily to be along the y axis, the arrows show the P_{90} projections.

When data are collected at one particular angle, θ, the intensity of the transmitted beam I_θ will be related to the incident intensity I_0 by

$$P_\theta = \ln\left(\frac{I_0}{I_\theta}\right)$$

where P_θ is the projection of all the attenuation coefficients along the line at angle θ.

For example for θ = 90°, *one* of the values of P_{90}, say $P_{90}(3)$ to indicate that it is the projection through the third pixel in the y-direction, is given by

$$P_{90}(3) = \mu(1,3) + \mu(2,3) + \mu(3,3) \ldots + \mu(256,3)$$

Note that, strictly, each μ value should be multiplied by x the dimension of the pixel in the direction of X-ray travel. All values of x are the same.

The problem now is to obtain sufficient values of P to be able to solve the equation for the $(256)^2 = 65{,}000$ values of $\mu(x,y)$. The importance of computer technology to this development now becomes apparent, since correlating and analysing all this information is beyond the capability of the human brain but is ideal for a computer, especially since it is a highly repetitive numerical exercise.

10.3.2 Data collection

The process of data collection can best be understood in terms of a simplified system. Referring to Fig. 10.7 which shows a single source of X-rays and a single detector, all the projections $P_{90}(y)$, i.e. from $P_{90}(1)$ to $P_{90}(256)$, can be obtained by traversing both the X-ray source and detector in unison across the section. Both source and detector now rotate through a small angle $\delta\theta$ and the linear traverse is repeated. The whole process of rotate and traverse is repeated many times such that the total rotation is at least 180°. If $\delta\theta = 1°$ and there is 360° rotation, 360 projections will be formed.

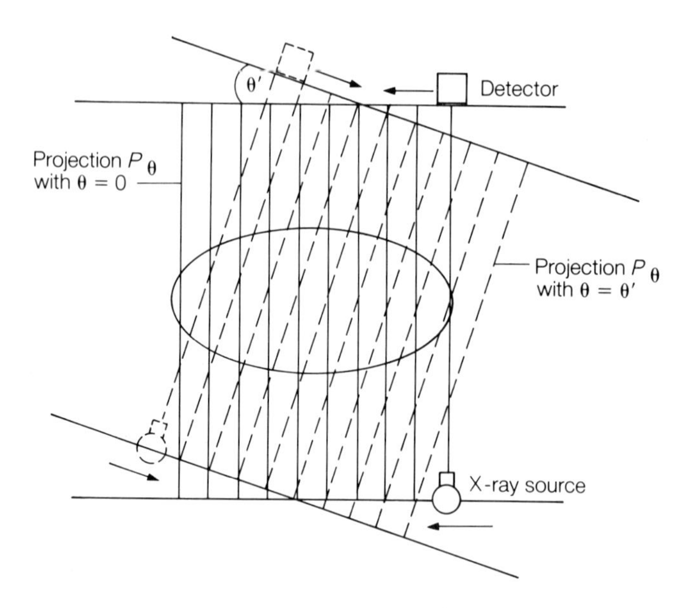

Fig. 10.7. Illustration of the collection of all the projections using a single source and detector and a translate–rotate movement.

Although this procedure is easy to understand and was the basis of the original or first generation systems, it is slow to execute, requires many moving parts and requires a scanning time of at least 10–20 s. Thus a major technological effort has been to collect data faster and hence reduce scan time, thereby reducing patient motion artefacts. Scan time can be reduced by:

1 using several pencil beams and an array of detectors;

2 using a fan-shaped beam, wide enough to cover the whole body section, and an array of detectors;

3 arranging for source and detectors to rotate continuously, maintaining a fixed geometrical relationship between them;

4 utilizing a complete ring of detectors and arranging for the source only to rotate continuously (Fig. 10.8). Modern equipment may contain as many as 1200 detectors and scan times of 1–2 s can be achieved.

CT places a number of new demands on the design and construction of X-ray tubes. Data reconstruction procedures assume an exact geometrical relationship between the X-ray source and the detectors and a very stable X-ray output. Voltage fluctuations should not exceed about 0.01%. High resolution requires a small focal spot size, typically of the order of 0.6 mm and the tube must be arranged with its long axis perpendicular to the fan beam to avoid heel effect asymmetry.

Use of rotating anode tubes with a high heat loading (typically about 70 kW) is essential and even so not all desirable features can be optimized. If operated continuously, tubes are limited in output so they are usually operated in pulse mode with say 300 2–3 ms pulses during a 5 s scan. Advantages of pulsing the beam are

1 higher X-ray outputs in a short time,

2 use of smaller spots giving better collimation,

3 facility to recalibrate the electronics between pulses giving a more uniform output. Nevertheless target cooling may limit the frequency of repeat scans.

Choice of X-ray energy is a compromise between high detection efficiency and good image contrast on the one hand and low patient dose on the other. Calculation of a unique matrix of linear attenuation coefficients assumes a monoenergetic beam. Heavy filtration is required to approximate this condition—for example 0.5 mm Cu might be used with a 120 kVp beam to give a mean energy of about 70 keV. When corrections are made for differences in mass attenuation coefficient and density, 0.5 mm Cu is approximately equivalent to 8 mm Al. Note however that the characteristic radiation that would be emitted from copper ($Z = 29$, K shell

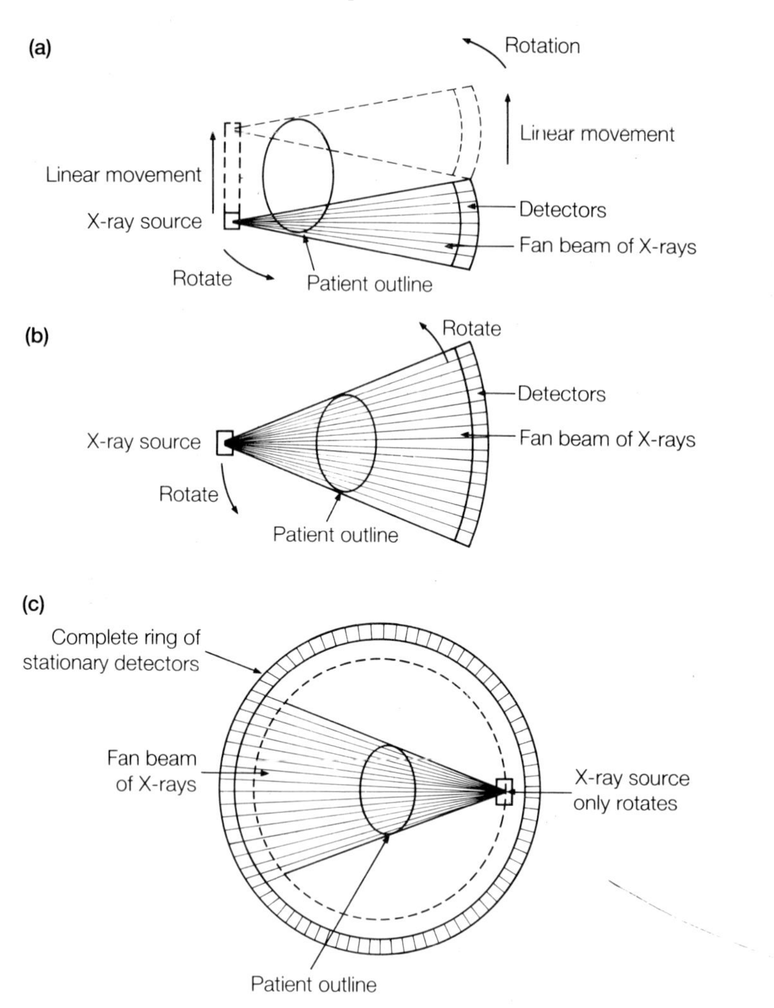

Fig. 10.8. Design of different generations of CT imagers. (a) Translate–rotate. (b) Rotate–rotate. (c) Source only rotates.

energy = 9.0 keV) as a result of photoelectric interactions would be sufficiently energetic to reach the patient and increase the skin dose. Thus the copper filter is backed by 1 mm Al to remove this component.

Modern scanners are tending to use less filtration, perhaps 3 mm Al. The beam is now more heterochromatic but the tube output is higher.

Since Compton interactions predominate, especially at the higher energies, good collimation is required to minimize scattered radiation. Normally only a few per cent of scattered radiation is detected.

The requirements of radiation detectors for CT are: a high detection efficiency, high dynamic range, fast response, linearity, stability, reliability and, because many detectors have to be accommodated in a limited volume, small physical size and low cost.

This has been an area of intensive commercial development. Originally thallium doped sodium iodide NaI (Tl) crystals and photomultiplier tubes (PMTs) were used. One problem of the NaI (TI) crystal is afterglow—the emission of light after exposure to X-rays has ceased. This effect is particularly bad when the beam has passed through the edge of the patient and suffered little attenuation because the flux of X-ray photons incident on the detector is high. Thallium doped caesium iodide or bismuth germanate gives less afterglow than NaI (Tl) and PMTs have now been replaced by high purity temperature stabilized silicon photodiodes. These detectors have a wide dynamic range, close detector spacing, high efficiency and no high voltage power supply is required.

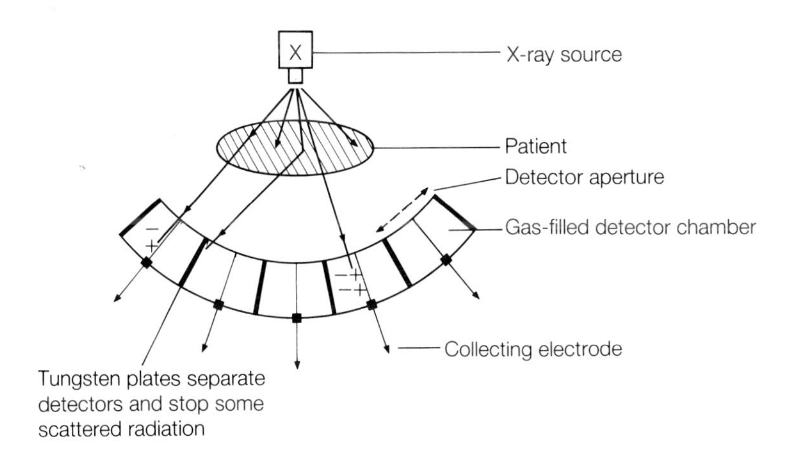

Fig. 10.9. Design of an array of gas-filled detectors for CT.

In spite of the inherently low sensitivity of gas filled ionization chambers, with suitable design (Fig. 10.9) they provide a realistic alternative for CT scanning. Use of high atomic number gases (Xe/Kr) at high pressure gives a sensitivity of about 50% and the detectors are closely spaced. There is little variation in sensitivity and since for an ion chamber the current produced is independent of the voltage across the ion chamber, they are very stable.

Variations in detector sensitivity reproduce as ring artefacts in the reconstructed image. Thus the detectors must be recalibrated frequently and in the latest generation of scanners the leading edge and trailing edge of the rotating fan beam are used for continuous recalibration.

10.3.3 Data reconstruction

The problem of reconstructing two-dimensional sections of an object from a set of one-dimensional projections is not unique to radiology and has been solved more or less independently in a number of different branches of science. As stated in the introduction, in diagnostic radiology the input is a large number of projections, or values of the transmitted radiation intensity as a function of the incident intensity, and the solution is a two-dimensional map of X-ray linear attenuation coefficients.

Mathematical methods for solving problems are known as algorithms and a wide variety of mathematical algorithms has been proposed for solving this problem. Many are variants on the same theme so only two, which are both fundamental and are quite different in concept, will be presented. The first is filtered back projection sometimes called convolution and back projection, and the second is an iterative method.

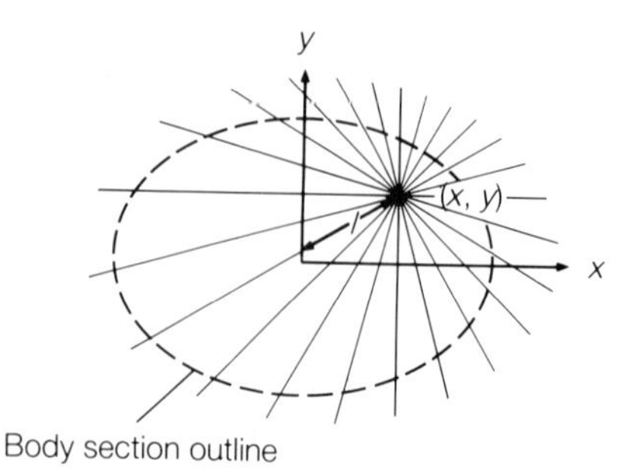

Fig. 10.10. The projections that will contribute to the calculation of linear attenuation coefficient at some arbitrary point (x,y).

Filtered back projection

Figure 10.10 shows several projections passing through the point (x,y) which is at some distance l from the centre of the slice. The philosophy of back projection is that the attenuation coefficient $\mu(x,y)$ associated with this element contributes to all these projections, whereas other elements contribute to very few. Hence if the attenuation in any one direction is assumed to be the result of equal attenuation in all pixels along that profile, then the total of all the projections should be due to the element that is

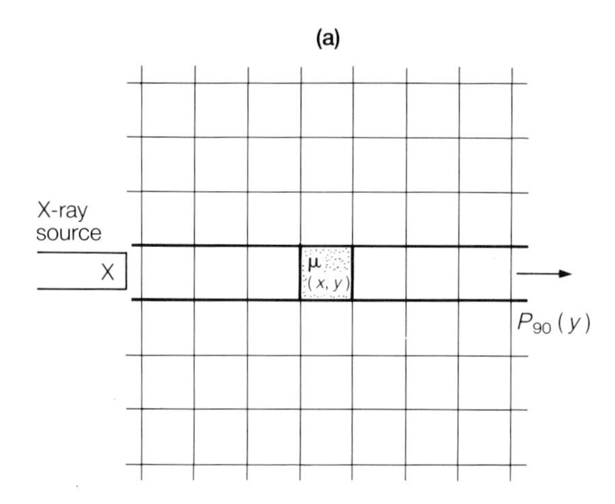

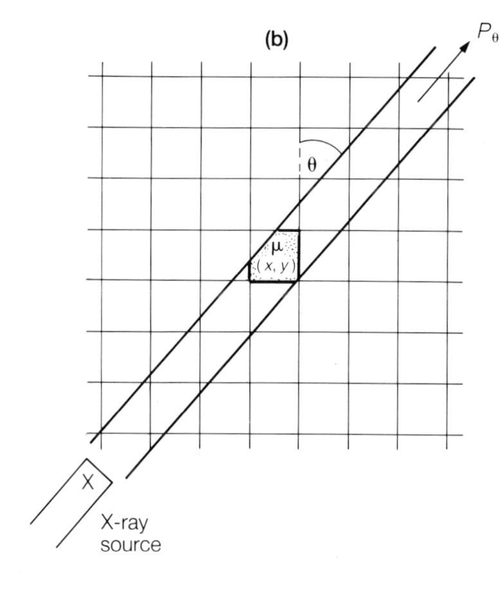

Fig. 10.11. Illustration of a partial volume effect. Assuming, for convenience, that the beam width is equal to the pixel width, then for the P_{90} projection (a) it can be arranged for the beam to match the pixel exactly. However this is not possible for other projections. Only a fraction of the pixel will contribute to the projection P_θ (b) and allowance must be made for this during mathematical reconstruction.

common to all of them. Hence one might expect an equation of the form

$$\mu(x,y) = \sum_{\substack{\text{over all} \\ \text{projections}}} P_\theta(l)$$

where $P_\theta(l)$ is the projection at angle θ through the point (x,y) and $l = (x \cos\theta + y \sin\theta)$, to give an estimate of the required value of $\mu(x,y)$.

Note that because the beam has finite width, the pixel (x,y) may contribute fully to some projections (Fig. 10.11a) but only partially to others (Fig. 10.11b). Due allowance must be made for this in the mathematical algorithm. Similar corrections must be applied if a fan beam geometry is being used.

If this procedure is applied to a uniform object that has a single element of higher linear attenuation coefficient at the centre of the slice, it can be shown to be incomplete. As shown in Fig. 10.12a, for such an object each projection will be a top-hat function, with constant value except over one pixel width. These functions will back project into a series of strips as shown in Fig. 10.12b. Thus simple back projection creates an image corresponding to the object but it also creates a spurious image. For the special case of an infinite number of profiles, the spurious image density is inversely proportional to radial distance, r, from the point under consideration.

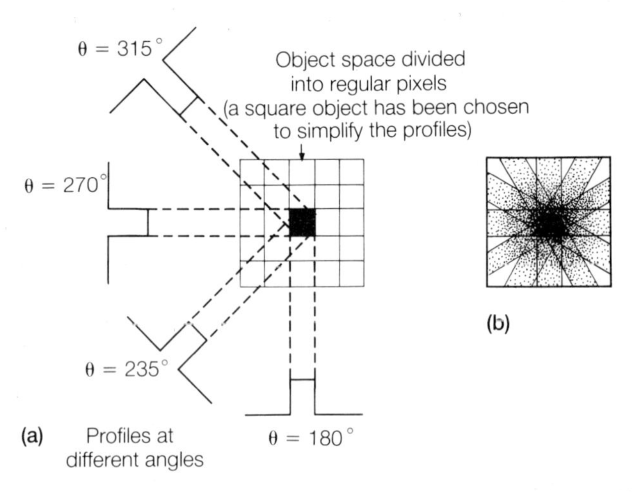

Fig. 10.12. (a) The projections for a single pixel of higher linear attenuation coefficient at the centre of the field of view. (b) Back-projection of these projections shows maximum attenuation at the centre of the field of view but also a star-shaped artifact.

Further rigorous mathematical treatment is beyond the scope of this book and the reader is referred to one of the numerous texts on the subject. Suffice to say that the back-projected image in fact represents a blurring of the true image with a known function which tends to 1/r in the limiting case of an infinite number of profiles. Furthermore, correction for this effect can best be understood by transforming the data into spatial frequencies (see section 9.4.4). In terms of the discussion given there, the blurred image can be thought of as the result of poor transmission of high spatial frequencies and enhancement of low spatial frequencies. Any process that tends to reverse this effect helps to sharpen the image and it may be shown that in frequency space correction is achieved by multiplying the data by a function that increases linearly with spatial frequency.

An upper limit to the spatial frequency is imposed by the size of the detector aperture, the effective spot size of the X-ray tube and the frequency of sampling. If this upper limit is ν_m, the value of ν_m imposes a constraint on the system resolution since for an object of diameter D reconstructed from N profiles, the cut-off frequency ν_m should be of the order of $N/(\pi D)$ and ensures spatial resolution of $1/2\nu_m$. For example if $N = 360$ and $D = 40$ cm, the limit placed on spatial resolution by finite sampling is about 2 mm.

Finally, the correction function cannot rise linearly to ν_m and then stop suddenly or multiple low intensity images (aliasing) will result. Hence it must reduce smoothly to zero at the cut-off frequency. These stages in the image reconstruction process are summarized in Fig. 10.13. The exact shape of the filter function has a marked effect on image quality in radiology.

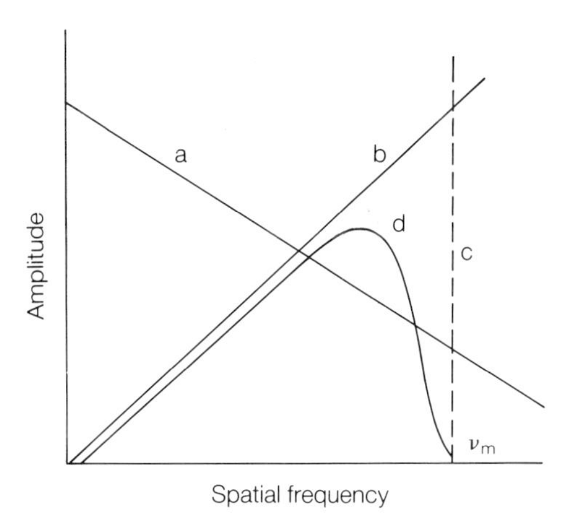

Fig. 10.13. (a) The effect of simple back projection on the amplitudes of different spatial frequencies. (b) The ideal correction function. (c) Upper limit on spatial frequency imposed by physical limitations of the system. (d) A practical filter function in frequency space.

Iterative methods

The starting point for such a method is to guess a distribution of values for $\mu(x,y)$. This is quite arbitrary, for example the μ for each pixel could be set equal to the average for the whole slice. One projection to be expected from the assumed values of μ is now compared with the actual value and the difference between observed and expected projections is found. Values of μ along this projection are then adjusted so that the discrepancy is attributed equally to all pixels. This process is repeated for all projections and when all μ values have been adjusted in this way, a revised set of assumed projections is calculated and the whole process is repeated.

Clearly this sequence can be repeated indefinitely but there is a high cost in computing time and a procedure that requires large numbers of iterations is probably unsatisfactory. Also care is necessary to ensure that the iterative process converges to a unique solution. If μ values are changed too much, the revised image may be less similar to the object than the original image. A further disadvantage of iterative methods is that all data collection must be completed before reconstruction can begin.

No one reconstruction method holds absolute supremacy over all others and it is essential to assess the efficiency of a particular algorithm for a particular application. With respect to X-ray transmission tomography, minor modifications, for example to the correction filter (curve d in Fig. 10.13) may improve diagnostic accuracy for certain examinations but such considerations are beyond the scope of this book.

10.3.4 Data presentation and storage

For each pixel in the reconstructed image, a CT number is calculated which relates the linear attenuation coefficient for that pixel $\mu(x,y)$ to the linear attenuation coefficient for water μ_w according to the equation:

$$\text{CT number} = \mathrm{K} \cdot \frac{\mu(x,y) - \mu_w}{\mu_w}$$

where K is a constant equal to 500 on the EMI scale and 1000 on the Hounsfield scale. Some typical values for CT numbers are given in Table 10.1.

Table 10.1. Typical CT values for different biological tissues

Tissue	Range (Hounsfield units)
Air	−1000
Lung	−200 to −500
Fat	−50 to −200
Water	0
Muscle	+25 to +40
Bone	+60 to +1000

An interactive display is normally used so that full use can be made of the wide range of CT numbers. Since only about 10 distinguishable grey levels can be shown on the screen between 'black' and 'white', adjustment of both the mean CT number and the range of CT numbers covered by the grey levels must be possible quickly and easily. The range of CT numbers covered is known as the window width. A wide window is used when comparing structures widely different in μ value, but a narrow window must be used when variations in μ are small. Consider for example a 5% change in μ, which represents a range of about 25 CT numbers. On a scale from +500 to −500 this may show as the same shade of grey. If a narrow window, ranging from say, −10 to +50 is selected, 25 represents almost half the range and several shades of grey will be displayed. Data manipulation in the manner discussed for digitized nuclear medicine images in section 8.3.5 is possible and by setting the window width equal to 1 and changing the threshold, the precise CT number of a pixel may be obtained. Selecting a region of interest and finding the average CT number in it may sometimes help to decide the exact nature of the imaged tissues.

Data storage and retrieval present a formidable problem in CT. Clearly the final images must be stored and these can consist of several slices with 512 pixels across a diameter ($\pi D^2/4$ = about 200,000 pixels), together with information on more than 1000 grey levels. If each attenuation value is stored with a 12 bit accuracy, each image requires about 300 kbytes of storage (1 byte = 8 bits). Furthermore, it may be necessary to store the original projections if several reconstruction algorithms are available and the radiologist is unsure which one will give the best image.

Access to the data must be rapid so that reconstructions can be completed whilst the patient is still in the department and so that the radiologist may quickly recall earlier studies. Finally, the long-term storage medium must be reliable, physically compact and not too expensive.

For short-term storage, floppy discs are frequently used. They allow ready transfer between the scanner and remote viewing console, but have limited storage capacity—as little as six images for body scans. The viewing console itself has limited storage.

For longer-term storage, magnetic tape will hold about 250 images, but retrieving pictures is then rather time-consuming. Thus there are advantages to using large storage discs—one with a capacity of say 200 Mbytes can accommodate a complete day's work for even a busy department. This permits any image to be accessed quickly and the data can be archived onto magnetic tape at the end of the day. Video discs, with a capacity of the order of 100 Mbytes, may eventually replace magnetic tape for long-term storage via a suitable computer interface.

10.3.5 Quantum mottle, contrast detection limit and dynamic range

If a uniform water phantom were imaged, then even assuming perfect imager performance, not all pixels would show the same CT number. This is because of the statistical nature of the X-ray detection process which gives rise to quantum mottle as discussed in section 9.6. CT in fact operates close to the limit set by quantum mottle. A 0.5% change in linear attenuation coefficient, corresponding to a change in CT number of about 5, can only be detected if the statistical fluctuation on the number of counts collected ($n^{\frac{1}{2}}$) is less than 0.5%. Hence

$$\frac{n^{\frac{1}{2}}}{n} < \frac{5}{1000}$$

which gives a value for n of about 4×10^4 photons. This is close to the collection figure for a single detector in one profile.

Note that attenuation in the patient could be a factor of 1000 or more so the number of photons incident on a detector without attenuation would be in excess of 4×10^7. No detector has a dynamic range of this magnitude and hence shaped wedge filters must be used around the patient to reduce the photon flux to a measurable value.

10.3.6 System performance

For a high contrast object the system resolution is determined by the focal spot size, the width of detector aperture, the separation of measurement points or data sampling frequency, and the display pixel size. The focal spot can affect resolution as in any other form of X-ray imaging although in practice with the 0.6 mm focal spots now used in CT this is not a problem. However, the width of the detector aperture is important and if resolution is considered in terms of the closest line pairs that can be resolved, simple ray optics shows that the line pair separation must be greater than the detector aperture. The effect of sampling frequency on spatial resolution was discussed in section 10.3.3. The pixel size on the image display must match the resolving capability of the remainder of the system. For example a 5 mm × 5 mm matrix is useless for displaying 1 mm images. Equally there would be little point in using a 1 mm × 1 mm display matrix if the resolving capability of the rest of the system were no better than 5 mm.

As for conventional imaging, resolution becomes a function of contrast and dose at low contrast. When a CT imager is used in the normal 'fast scan' mode, a typical relationship between resolution and contrast will be as shown in Fig. 10.14.

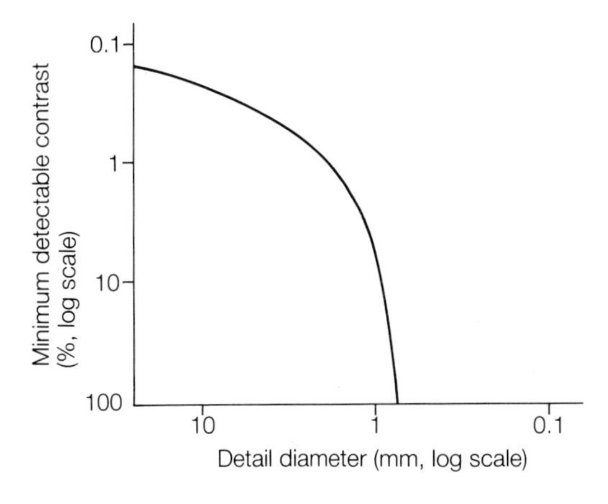

Fig. 10.14. Relationship between minimum detectable contrast and detail size for a CT scanner.

Any attempt to reduce the dose will increase the standard deviation in the attenuation coefficient due to statistical noise in the detected photons. When measurements are photon limited, statistical noise increases if the patient attenuation increases. However it decreases if the slice thickness increases or the pixel width increases because more photons contribute to a given attenuation value.

Most CT examinations can be performed with skin doses in the region of 10–50 mGy but the dose distribution within the body depends on a number of factors including:

1 kVp—this affects the dose as in conventional imaging;

2 the scanning motion—if the radiation source moves through only 180° instead of 360°, a movement that will give sufficient profiles for reconstruction, the dose distribution is non-uniform;

3 the beam profile—a large focal spot for example will irradiate a larger volume than that being interrogated;

4 the number of scans in the study—adjacent slices are partially exposed to stray radiation during each scan and the cumulative dose can be substantial;

5 the required picture quality—higher doses reduce quantum noise. Doses as high as 350 mGy have been measured during low noise multiple slice imaging (McCullough & Payne 1978).

Exact comparison of CT doses with conventional radiology is difficult because the pattern of energy deposition is different. The somatic risks (see Chapter 11) are probably comparable when CT is used in the normal 'fast scan' mode but as shown above the doses associated with the 'higher accuracy' mode may be substantially higher and this technique should be used sparingly.

10.3.7 Artifacts

A number of artifacts may be present in CT images and these will be summarized briefly.

Mechanical misalignment and patient movement
These will generate streak artifacts. Patient movement is more serious than in conventional imaging because the superimposed profiles have not passed through the same part of the body.

Detector non-uniformities
These cause ring artifacts. In X-ray CT, variations in X-ray output or detector response of as little as 1 in 5000 can cause problems. Non-uniformities are a particular problem when a gamma camera is used for tomography in nuclear medicine (see section 10.4), with much better uniformity being required for this purpose than for conventional gamma camera imaging.

Beam hardening
Even though the X-ray beam is heavily filtered, it is still not truly monochromatic. A certain amount of beam hardening will occur on passing through the patient and this will affect the measured linear attenuation coefficients.

Aliasing
This is the phenomenon whereby high frequency noise generated at sharp, high contrast boundaries appears as low frequency detail in the image. In filtered back projection the extent to which aliasing occurs depends on how the filter function falls to zero at high spatial frequencies. The effect is less marked when iterative techniques are used but cannot be eliminated entirely.

10.3.8 Quality control

Since a CT scanner is a complex and expensive piece of equipment, frequent calibrations and adjustments are necessary. Hence a regular programme of quality control should be followed.

Precision and accuracy can be measured daily by scanning a water bath with rotations in each direction. A simple computer programme will print out the mean value and standard deviation of the CT numbers in each pixel. Precision requires that the mean CT number should be close to the assumed value of zero for water. As regards detection of small contrast differences, precision is less important than accuracy. The readings are accurate if the recorded standard deviation is close to that expected from statistical variations on the number of counts collected. A higher standard deviation could be evidence for a failing X-ray tube or noisy detectors.

At weekly intervals the unprocessed data should be examined to check detector and integrator uniformity, noise and speed corrections and spatial uniformity. At monthly intervals the contrast scale should be checked using composite phantom blocks of various plastics with known linear attenuation coefficients and alignment may be checked using an aluminium pin in a water bath. Finally, beam profile should be checked and dose measurements should be made after routine preventative maintenance, say at 3-monthly intervals.

10.4 Single photon emission computed tomography (SPECT)

Although many of the principles of SPECT are similar to those of CT, the objective is somewhat different, namely to recover from a series of projections a map of the concentration of radionuclide which is varying continuously throughout the volume of interest.

If $C(x,y)$ is the number of counts per unit time recorded in a normal gamma camera image at an arbitrary point (x,y,z), this is related to the concentration of radionuclide [activity per unit volume $A(x',y',z)$] at some arbitrary point (x',y',z) in the same slice by the equation.

$$C(x,y) = \int_{\substack{\text{all the volume} \\ \text{occupied by activity}}} A(x',y',z)\, S(x-x',y-y',z)\, e^{-\mu t} \mathrm{d}x'\, \mathrm{d}y'\, \mathrm{d}z$$

where $S(x - x',y - y',z)$ represents the response of the detector (at x,y,z) to a point source of activity (at x',y',z), μ is the linear attenuation coefficient of the medium and t is the thickness of attenuating medium traversed by the gamma rays. The recovery of the function $A(x',y',z)$ for the whole slice from the available data $C(x,y)$ represents a complete solution to the problem.

Use of a single value of μ is of course an approximation. Ideally it should be replaced by a matrix of values for the linear attenuation coefficient in different parts of the slice, so one could say that the end point of CT is the starting point for SPECT!

Three fundamental limitations on emission tomography can be mentioned. The first is collection efficiency. Gamma rays are emitted in all directions but only those which enter the detector are used. Thus collection efficiency is severely limited unless the patient can be surrounded by

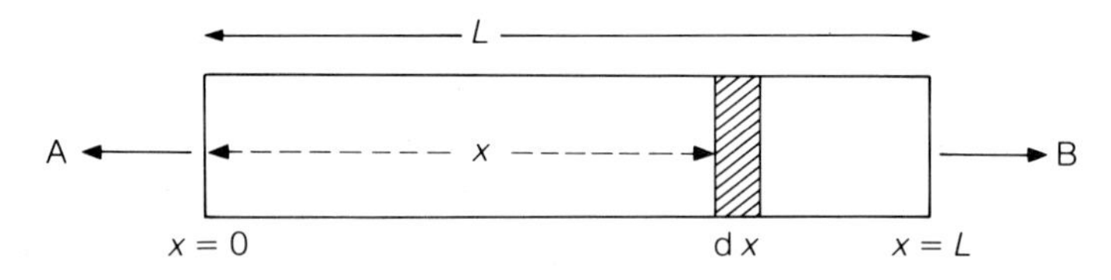

Fig. 10.15. Correction factor to be applied for gamma ray attenuation in the patient when the source of radioactivity is uniformly distributed. If the total activity is I, then the activity per unit length is I/L and the activity in the strip dx is $I\mathrm{d}x/L$. The signal recorded at A is

$$\frac{I}{L}\int_0^L e^{-\mu x}\,\mathrm{d}x$$

where μ is the linear attenuation coefficient of the medium. The signal recorded at B is

$$\frac{I}{L}\int_0^L e^{-\mu(L-x)}\,\mathrm{d}x$$

Both expressions work out to

$$\frac{I}{L}\cdot\frac{(1-e^{-\mu L})}{\mu}$$

and since, in the absence of attenuation, the signal recorded at A and B should be I, the total activity in the strip, the required correction factor is

$$\frac{\mu L}{(1-e^{-\mu L})}$$

detectors. The second is attenuation of gamma rays in the patient. Allowances can be made and corrections simplified by adding counts registered in opposite detectors. As shown in Fig. 10.15, for a uniformly distributed source a correction factor $\mu L/(1 - e^{-\mu L})$ where L is the patient thickness can be applied. However, experimental work indicates that the correct value of μ is neither that for narrow beam attenuation, nor that for broad beam attenuation, but somewhere between the two. The third problem is common to all nuclear medicine studies, namely that the time of collection is only a small fraction of the time for which gamma rays are emitted. Hence the images are seriously photon limited.

The required projections are normally collected by rotating a gamma camera round the patient, although a few machines surround the patient with banks of NaI (Tl) crystals to improve detection efficiency and operate on the translate–rotate principle. Both systems will typically collect data at 6° intervals for 15–20 min. The scanner has greater sensitivity for a single slice, but since several slices can be reconstructed from the gamma camera projections, camera sensitivity is higher for multisection images.

Tomography places more stringent demands on the design and performance of gamma cameras than conventional imaging. For example, multiple views must be obtained at precisely known angles and the centre of rotation of the camera must not move, for example under its own weight, during data collection. The face of the camera must remain accurately parallel to the long axis of the patient and the mechanical and electronic axes of the camera must be accurately aligned. Camera non-uniformities are more serious than in conventional imaging since they frequently reconstruct as 'ring' artifacts. If views are corrected with a non-uniformity correction matrix collected at a fixed angle, care must be taken to ensure that the pattern of non-uniformity does not change with camera angle. Such changes could occur, for example, as a result of changes in PM tube gain due to stray magnetic fields.

For all these reasons, especially very poor counting statistics, resolution is inferior in SPECT to that in conventional gamma camera imaging and much inferior to CT. Resolution is rarely better than 10 mm so 6° sampling ($N = 30$) is quite adequate. For example with objects 20 cm in diameter

$$\frac{N}{\pi D} \simeq 0.05 \text{ mm}^{-1} \qquad \text{so } \frac{1}{2\nu_m} \simeq 10 \text{ mm}$$

Further, matrices finer than 64 × 64 (about 6 mm) are not really necessary for reconstruction. Slices are usually two or three pixels (12–18 mm) thick to improve statistics.

10.5 Positron tomography

Since radionuclide imaging provides functional or physiological information, it would be highly desirable to image elements such as carbon, oxygen and nitrogen which have a high abundance in the body. The only radioisotopes of these elements that are suitable for imaging are short half-life positron emitters (carbon-11 with a half-life of 20.5 min, nitrogen-13 with a half-life of 9.9 min and oxygen-15 with a half-life of 2.0 min).

For positron emitters, the origin of the detected radiation is the gamma rays released when the positron comes to rest and annihilates with an electron.

$$\beta^+ + \beta^- \rightarrow 2\gamma$$

The gamma rays have a well-defined energy of 0.51 MeV (the energy equivalence of the rest mass of each particle according to $E = mc^2$) and are released simultaneously in nearly opposite directions. Thus coincident detection of these two 'annihilation photons' in a pair of opposed detectors establishes the line on which the positron came to rest.

For tomographic imaging, positron emitters have some advantages. For example in tomographic projections it is only the line of origin, not the point of origin that is important. Second, since coincidence detection climinates stray and scattered radiation, conventional lead collimators are not necessary, leading to increased geometric efficiency and sensitivity. Finally, the total path travelled by the two gamma rays in the attenuating medium is always equal to L the thickness of the patient in that projection, irrespective of the point of origin of the gamma rays (Fig. 10.16). So for coincident detection the attenuation correction is $e^{\mu L}$ for all coincident events arising along that projection.

A number of interesting studies have been reported using positron emitters, for example C-11 deoxyglucose has been used extensively for studies of cerebral glucose metabolism. However, there are a number of

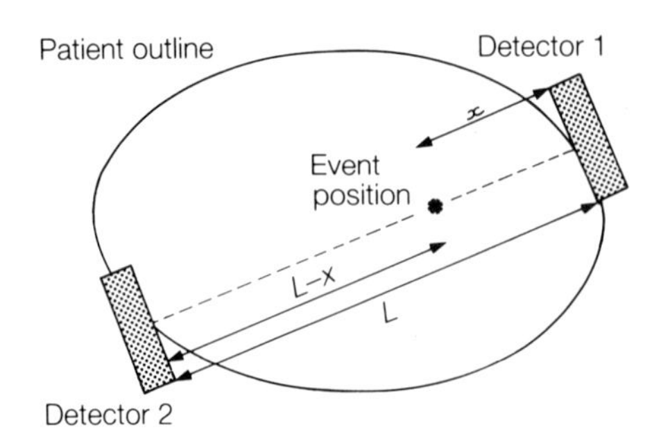

Fig. 10.16. Demonstration that the total path travelled by two coincident gamma rays is equal to the length of the patient in that projection and that a fixed attenuation correction can be applied. If one photon travels a distance x, the other one travels $L - x$. If $I_1 = I_0e^{-\mu x}$, then $I_2 = I_0e^{-\mu(L-x)}$, and $I_1I_2 = I_0e^{-\mu L}$, which is independent of x.

technological problems. For example, to achieve adequate sensitivity a ring of detectors is required. In one commercial model there are 66 sodium iodide detectors in a hexagonal arrangement around a single transverse section plane. Because the gamma ray energy is high, detector sensitivity is low. Sensitivity is further reduced because random coincidences give a high background unless some collimation is used and additional electronic circuits are also used to monitor random events. Although the short physical half-lives might suggest that high activities could be used, the emitted positrons have a high kinetic energy which is deposited locally before annihilation takes place, thereby contributing a significant dose to the patient. Finally, because of the short half-lives, the positron emitters considered here can only be used close to the cyclotron where they are produced. Thus in the foreseeable future the technique is likely to remain restricted to rather specialized applications in research centres.

10.6 Conclusions

A radiographic image is a two-dimensional display of a three-dimensional structure and in a conventional image the required detail is always partially obscured by the superposition of information from underlying and overlying planes. The overall result is a marked loss of contrast.

Tomographic imaging provides a method for eliminating, either partially or totally, contributions from adjacent planes. Longitudinal tomography essentially relies on the blurring of structures in planes above and below the region of interest. It is a well-established technique and the main consideration is choice of the thickness of plane of cut. If the focus–plane of cut distance and focus–film distance are fixed, the thickness is determined by the angle of swing, decreasing with increasing angle. Large angle tomography may be used when there are large differences in atomic number and/or density between the structures of interest, but if differences in attenuation are small, somewhat larger values of slice thickness may be desirable.

In CT imaging a large number of views are taken of a transverse slice of the patient from different angles. Mathematical methods, using a computer to handle the large amount of data, are then used to map the linear attenuation coefficients $\mu(x,y)$ for an array of small elements in the slice of interest. Values of $\mu(x,y)$ are expressed as CT numbers, and a grey scale which can be adjusted to cover different ranges of CT numbers is used to display the data for visual interpretation.

Several generations of CT scanner have been introduced, each designed primarily to give a faster scan time, which is now of the order of 1–2 s per slice. Contrast, resolution and dose are inter-related. At high contrast, resolution of the order of 1 mm or better can be achieved for a thin slice. At low contrast typical approximate figures are that a 0.4% contrast might permit a resolution of about 4 mm for a skin dose at the entry surface of the patient of about 50 mGy.

It is difficult to predict future developments in CT. Scan times can be reduced further by using multiple X-ray sources as well as a complete ring of detectors but this will increase cost. Image quality can only be improved by increasing the dose to the patient irrespective of whether one X-ray source is used for a long time or several sources are used for a short time (setting aside the question of movement artifacts). Consideration has been given to using an image intensifier as the input to the computer system. If the patient were then rotated in front of the image intensifier or the image intensifier rotated round the patient, data collection would be analogous to the use of a gamma camera in SPECT. Each video line scan on the image intensifier is in effect a profile of high spatial resolution.

Sagittal and coronal section images have already been obtained that can be examined from any angle. The eventual role of CT may be determined not so much by its cost-effectiveness in the diagnostic process but in terms of its impact on patient morbidity and mortality.

Resolution with SPECT is markedly inferior to that with CT for a number of reasons. For example radionuclides produce a low photon flux for a long time whereas X-ray generators produce a high photon flux for a short time. Furthermore, emission from radionuclides is isotropic. Many of the photons are wasted and use of collimators means that only about 1 in 2000 of the photons are detected. An X-ray beam is highly collimated and hence for equal patient dose very many more photons can be collected in the CT image.

The time of examination is much longer for SPECT, so movement of the patient is a greater problem and there is even time for movement of activity within the patient. Scattering of the uncollimated beam of gamma photons from the radionuclide means that the appropriate attenuation coefficient is difficult to determine. Indeed, the end point for CT, calculation of a matrix of μ values, is the starting point for calculating attenuation corrections that are a necessary part of high precision SPECT. Hence complex iterative reconstruction techniques are required.

For all these reasons the limiting resolution of SPECT is only about 10 mm and its role in nuclear medicine is still unclear.

There are a number of theoretical points in favour of positron emission tomography. For example the biologically important radionuclides C-11, N-13 and O-15 are all positron emitters and the fact that two gamma rays are emitted simultaneously makes coincidence detection possible. However the requirement for on-site cyclotron production of most physiologically useful radionuclides and the high cost of the imaging equipment limits this technique to major research centres at present.

References and further reading

Boyd D. P. & Parker D. L. (1983) Basic principles of computed tomography. In *Computed Tomography of the Body*, eds A. A. Moss, G. Gamsu & H. Genant, pp. 1–22. Saunders, Philadelphia.

Claussen C. & Lochner B. (1985) *Dynamic Computed Tomography* (*Basic Principles and Clinical Applications*). Springer, Berlin.

Davison M. (1982) X-ray computed tomography. In *Scientific Basis of Medical Imaging*, ed. P. M. T. Wells, pp. 54–92. Churchill Livingstone, Edinburgh.

Ell P. J., Khan O., Jarritt P. H. & Cullum I. D. (1982) *Radionuclide Section Scanning—an Atlas of Clinical Practice*. Chapman & Hall, London.

Fullerton G. D. & Zagzebski J. A., eds (1980) *Medical Physics of CT and Ultrasound: Tissue Imaging and Characterisation*. (Medical Physics Monograph 6) American Association of Physicists in Medicine.

Gordon R., Herman G. T. & Johnson S. A. (1975) *Image Reconstruction from Projections. Sci. Am.* **233**, 56–68.

McCullough E. C. & Payne J. T. (1978) Patient dosage in computed tomography. *Radiology*, **129** 457–463.

Mansfield B. A. (1975) Body section radiography. In *Principles of Diagnostic X-ray Apparatus*, ed. D. R. Hill, pp. 269–289. Macmillan, London.

Pullan B. R. (1979) The scientific basis of computerised tomography. In *Recent Advances in Radiology and Medical Imaging*, Vol. 6, eds T. Lodge & R. E. Steiner. Churchill Livingstone, Edinburgh.

Williams E. D., ed. (1985) *An Introduction to Emission Computed Tomography*. (Report 44) Institute of Physical Sciences in Medicine, London.

Exercises

1 Explain why the technique of tomography can eliminate shadows cast by overlying structures. Suggest reasons why the dose to parts of the patient might be appreciably higher than in many other radiographic examinations.

2 Explain briefly, with the aid of a diagram, why an X-ray tomographic cut is in focus.

3 List the factors that determine the thickness of cut of a longitudinal X-ray tomograph and explain how the thickness will change as each factor is varied.

4 Describe and explain the appearance of objects that are in the plane of cut of a longitudinal tomogram but are not parallel to it.

5 Compare and contrast the production of a tomogram by a 'linear' movement X-ray set and a gamma camera.

6 Explain the meaning of the terms pixel and CT number and discuss the factors that will cause a variation in CT numbers between pixels when a uniform water phantom is imaged.

7 Explain why the use of a fan beam geometry in CT without collimators in front of the detector would produce an underestimate of the μ values for each pixel.

8 Discuss the factors that would make a radiation detector ideal for CT imaging and indicate briefly the extent to which actual detectors match this ideal.

9 Describe the production of a tomogram using an X-ray CT scanner and explain how the production of a tomogram using a radionuclide and a gamma camera differs.

10 Figures 9.3 and 10.14 show the relationship between contrast and resolution for an image intensifier and CT scanner respectively. Explain the differences.

11 Describe and explain how the CT number for a tissue might be expected to change with kVp.

11

Radiobiology and Radiological Protection

11.1 Introduction

This chapter deals with a problem that is central to the theme of the book—namcly that ionizing radiation, even at very low doses, is potentially capable of causing serious and lasting biological damage. If this were not so, steps that are taken to reduce patient doses, for example the use of intensifying screens, would be unnecessary and generally undesirable. Furthermore, the amount of physics a radiologist would need to know would be greatly reduced and this book might not be necessary!

As shown in Table 11.1, medical exposure is the highest contributor, other than natural radiation, to the average national radiological body burden and the majority of this can be attributed to diagnostic radiology. Since this radiation can cause deleterious effects, it is essential for the radiologist to know what these effects are and to be aware of the risk when a radiological examination is undertaken.

Table 11.1. Average annual per caput effective dose equivalents to the UK population. (Reproduced with permission from Hughes & Roberts 1984. Copyright NRPB)

Source of radiation	Dose equivalent* (μSv)	Percentage (%)
Natural		
Cosmic	300	14
Terrestrial gamma	400	19
Internal irradiation	370	17
Radon	700	32
Thoron	100	5
Artificial		
Medical	250	11.5
Miscellaneous	11	0.5
Fall out	10	0.5
Occupational	9	0.4
Waste disposal	2	0.1

* For medical exposures the dose equivalent will be almost identical numerically to the dose in mGy (see section 11.4).

11.2 Radiation sensitivity of biological materials

11.2.1 Evidence for high radiosensitivity

Extrapolation from experiments with animals suggests that the dose of acute whole body radiation required to kill about 50% of a human population within the first 30 days (LD 50/30) would be about 4.5 Gy to the bone marrow. This would correspond to a much higher skin dose at 100 keV owing to self-absorption in the body.

Several aspects of this statement need to be discussed, but first consider its meaning in terms of absorbed energy.

$$\text{Energy absorbed} = \text{mass} \times \text{specific heat} \times \text{temperature rise}$$

4.5 Gy corresponds to an absorbed energy of 4.5 J in 1 kg. Thus for water, of specific heat capacity 4.2×10^3 J kg^{-1} K^{-1},

$$\text{the temperature rise} = \frac{\text{energy absorbed}}{\text{mass} \times \text{specific heat}}$$

$$= \frac{4.5}{1 \times 4.2 \times 10^3}\ \text{K} = 10^{-3}\ \text{K}$$

Hence the temperature rise is only about one thousandth of one degree and would be virtually undetectable. Thus in terms of energy deposited in the body, ionizing radiation is by far the most potent agent known to man, being some 10^4 times more potent that the next most harmful (ultraviolet light) and about 10^9 times more harmful than cyanide. It is against this background of extreme sensitivity of cells and tissues to ionizing radiation that its use in diagnosis must be assessed.

11.2.2 Cells particularly at risk

There is a long established 'law' in radiobiology, first proposed by Bergonié & Tribondeau (1906) which, in modern parlance, states that 'the more rapidly a cell is dividing, the greater its radiosensitivity'. From this it follows that the lower the degree of morphological and functional differentiation, the higher the radiosensitivity.

The law applies well to rapidly dividing cells such as spermatogonia, haemopoetic stem cells, intestinal crypt cells and lymphoma cells, which are all very radiosensitive. Differentiated cells are generally relatively radioresistant but three important exceptions should be noted. Small lymphocytes, primary oocytes, especially just before release, and neuroblasts are all radiosensitive.

At the lowest dose at which radiation-induced death is likely to occur, the primary effect will be severe depletion of the bone marrow stem cells. Hence dose to bone marrow is the most meaningful LD 50/30 to quote.

11.2.3 Time course of radiation-induced death and molecular explanation

After a potentially lethal dose of about 4.5 Gy, a typical time sequence might be

1 0–48 h: loss of appetite, nausea, extreme sweating and fatigue;
2 48 h to 2–3 weeks: latent period of apparent well-being;
3 2–3 weeks to 4–5 weeks: the manifest illness stage, which may include fever, loss of hair, extreme susceptibility to infection, haemorrhage and a number of other symptoms;
4 5 weeks onwards: the situation will have resolved itself one way or the other.

This time scale reflects the changes taking place at the molecular and cellular level.

After exposure to ionizing radiation, physical processes of absorption of photons of energy h*f*, ionization and excitation will be complete within about 10^{-15} s. During the next millisecond, a number of chemical and biomolecular changes will occur. Direct dissociation of an excited macromolecule, RH* where R is the organic radical and H is hydrogen, into free radicals according to the equation

$$RH + hf \rightarrow RH^* \rightarrow R^o + H^o$$

may occur, but since living tissue is 70–90% water by weight, primary interactions with water molecules are more likely.

A typical sequence resulting in the release of a free electron is

$$H_2O + hf \rightarrow H_2O^* \rightarrow H_2O^+ + e^-$$

The free electron is then captured by another water molecule

$$e^- + H_2O \rightarrow H_2O^-$$

Further reactions occur rapidly such as

$$H_2O^+ \rightarrow H^+ + OH^o$$

and

$$H_2O^- \rightarrow H^o + OH^-$$

The **radicals** R^o, H^o and OH^o must be distinguished from the **ions** R^-, H^+ and OH^-. The latter take part only in normal chemical reactions whereas the former are highly reactive and within 10^{-5} s will take part in further reactions such as

$$H^{o} + H_2O \rightarrow H_2 + OH^{o}$$

$$OH^{o} + RH \rightarrow R^{o} + H_2O$$

If there is oxygen present, the longer lived hydroperoxy radical may be formed as shown by the equation

$$H^{o} + O_2 \rightarrow HO_2^{o}$$

and finally toxic products such as hydrogen peroxide may be formed

$$OH^{o} + OH^{o} \rightarrow H_2O_2$$

Hence during the first few hours the body is flooded by toxic products and a general feeling of malaise results.

The steps leading from these initial physico-chemical changes to the observation of cell death (see section 11.3) are still poorly understood. Nevertheless according to the law of Bergonié & Tribondeau, the differentiated cells will have resisted the radiation well and will continue to fulfil their specialized functions, so a period of relative well-being should ensue. However, as these cells die, their replacement from the stem cell pool will have been severely depleted or will have stopped completely. The patient becomes manifestly ill when there is a marked loss of a wide range of differentiated, mature cells, particularly in the circulating blood. Ultimately, the cause of death will usually be failure to control infection or failure to prevent internal haemorrhage.

11.2.4 Other mechanisms of radiation-induced death

Above 4.5 Gy, damage to the gastrointestinal stem cells in the crypts of Lieberkuhn becomes increasingly important and this effect dominates in the dose range 10–100 Gy with loss of body water and body salts into the gut being the major cause of death. The time scale is now much shorter and death will occur in about 3–4 days.

At even higher doses, death may be caused by disturbance of the central nervous system (100–150 Gy), loss of lung function (150 Gy) and ultimately simultaneous disruption of the total body chemistry (above 200 Gy).

Survival time plotted as a function of dose might be as shown in Fig. 11.1, but note that doses shown are only orders of magnitude and in any given situation death will probably be due to a combination of causes.

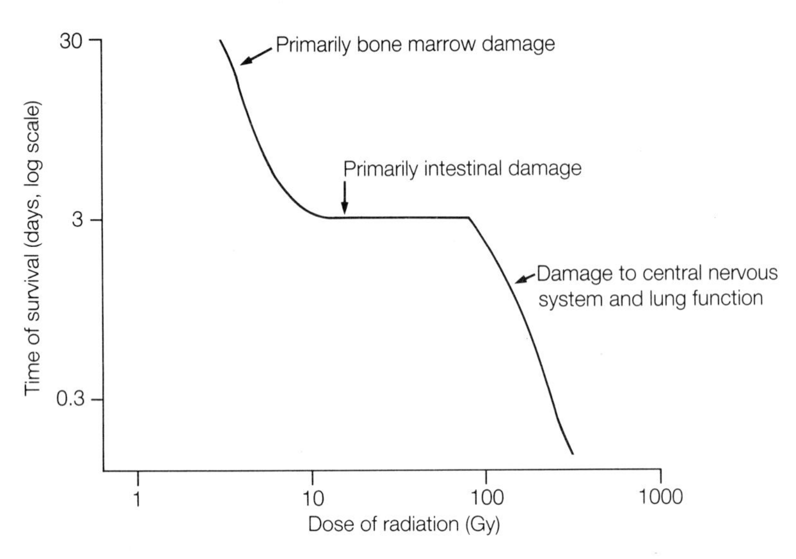

Fig. 11.1. Approximate representation of survival time plotted as a function of dose for acute exposure to whole body radiation.

11.3 Evidence on radiobiological damage from cell survival curve work

About 30 years ago, the first reports appeared of successful attempts to use a clonogenic assay of cell survival following irradiation. Basically the technique is as follows. A small number of single cells is placed in a Petri dish with medium and incubated for 10–14 days. The cells settle on the base of the dish and, if they are capable of cell division (reproductive integrity) they develop into submacroscopic colonies which may be counted. Not all cells are capable of growing into colonies, even if unirradiated. Suppose that 100 cells are seeded and 90 colonies grow. If now a second sample of the same cells is irradiated before incubation, from 1000 cells only 180 colonies might develop. The expected colonies from 1000 cells would be 900 and therefore 180/900 or 20% of the irradiated cells have survived. By repeating this experiment at different doses, a survival curve may be obtained and for mammalian cells exposed to X-rays it would resemble Fig. 11.2.

For further details of the extensive literature on survival curves, including the evidence that qualitatively similar curves are obtained *in vivo*, the reader is referred to more specialized texts (e.g. Hall 1978, Coggle 1983). However, two aspects which are of direct relevance to this chapter will be discussed.

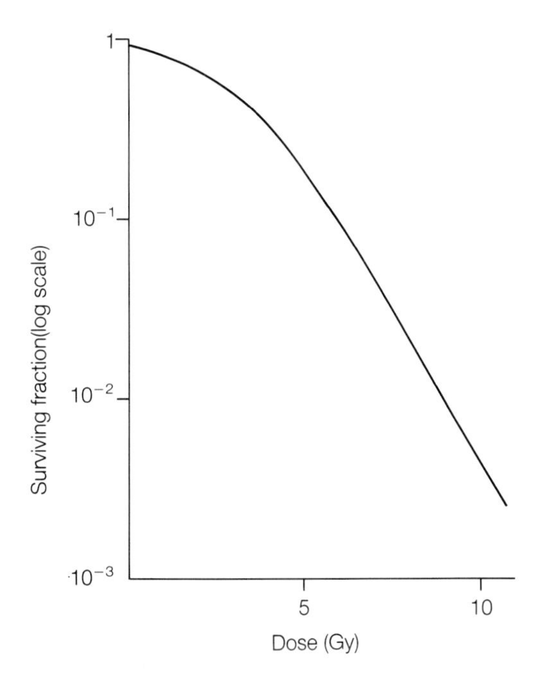

Fig. 11.2. A typical clonogenic survival curve for mammalian cells irradiated *in vitro* with X-rays.

11.3.1 Cellular recovery and dose rate effects

After a dose of 1–2 Gy, some cells are killed and others damaged. It may be demonstrated convincingly that some damaged cells will recover by performing a 'split dose' experiment. In Fig. 11.3 the dotted curve is reproduced from Fig. 11.2. The solid curve would be obtained if a dose of, say, 4 Gy were given but a time interval of 10 h elapsed before any further irradiation. The 'shoulder' to the curve has reappeared and the total dose required to achieve a given surviving fraction is higher for the split dose than for the single dose. The recovery effect is detectable by 2 h and reaches a maximum by 24 h.

One consequence of recovery is that when radiation exposure is protracted, the effect may be dose rate dependent (see Fig. 11.4). A simple explanation is that recovery is occurring during radiation exposure. Conversely, dose rate effects are evidence of recovery. They indicate that cell killing is occurring as a result of a sequence of events following interaction of more than one X-ray photon with the cell.

Cellular recovery is also observed *in vivo* but now a further recovery mechanism is also observed. This has a much longer time scale and is primarily a result of homeostatic mechanisms involving stem cells in the whole animal. One consequence is shown in Fig. 11.5. The LD 50/30 for

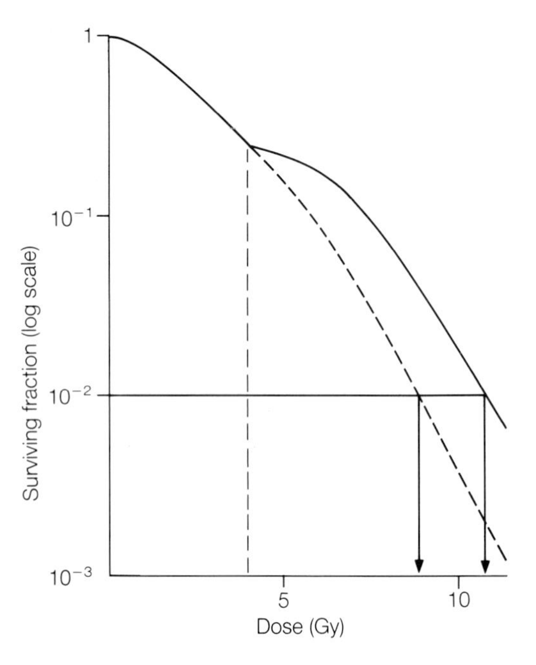

Fig. 11.3. Typical results for a 'split dose' experiment to demonstrate radiation recovery. If an initial dose of 4 Gy is delivered but there is a time delay of a few hours before any further irradiation, the survival curve follows the *solid line* rather than the dotted line. Note that the shoulder to the curve reappears and the total dose to produce a given surviving fraction is now higher.

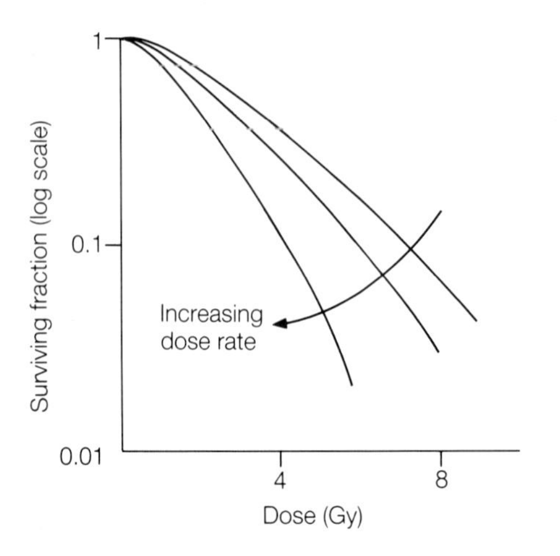

Fig. 11.4. Graphs demonstrating that because of recovery the killing effect of X-rays may be dose rate dependent.

acute exposure is about 5 Gy (5 Gy at 100 Gy per day corresponds to an exposure time of about 1 h), but this figure increases when the radiation is protracted, first because of cellular repair and subsequently because of homeostatic repair. The results suggest that animals could tolerate 0.1 Gy per day for a long time.

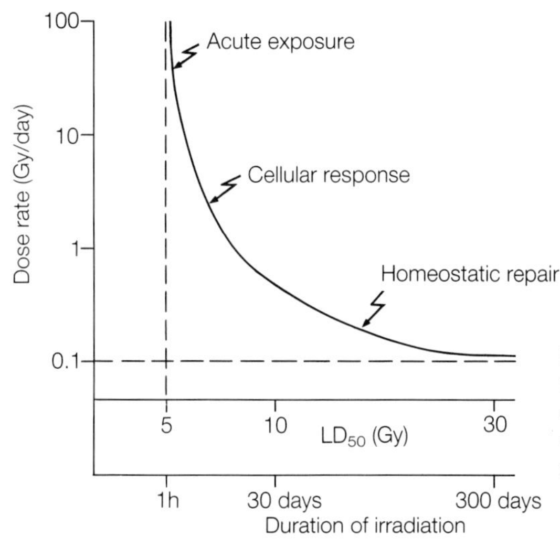

Fig. 11.5. Curve to demonstrate that the LD_{50} *in vivo* is very dependent on dose rate because of cellular and homeostatic recovery mechanisms.

11.3.2 Radiobiological effectiveness (RBE)

If the survival curve experiment is repeated with neutrons, the result shown in Fig. 11.6 will be obtained. A smaller dose of radiation is required to produce a given killing effect and the curve has a smaller shoulder, indicating less capacity to repair sub-lethal damage.

As shown on the curve, the RBE is defined as

$$\text{RBE} = \frac{\text{Dose of 200 kVp X-rays}}{\text{Dose of radiation under test required to cause the same biological end point}}$$

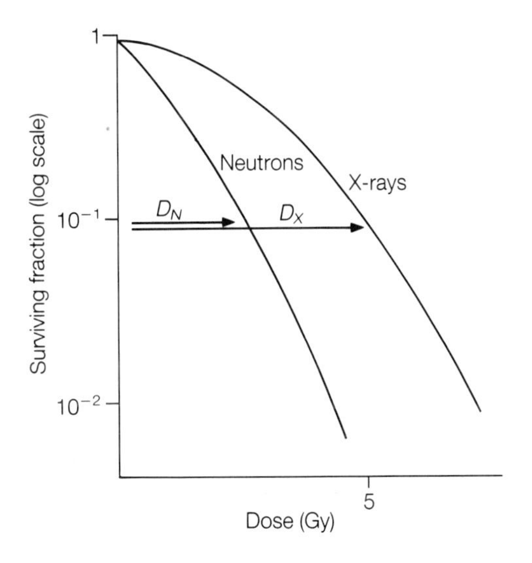

Fig. 11.6 Comparative survival curves for X-rays and neutrons. RBE $= D_X/D_N$ for a given biological end-point (10% survival in this example).

The RBE of neutrons when determined in this way is frequently between 2 and 3.

Except at very high LET values (see section 1.14), the RBE of a radiation increases steadily with LET. However, it is an incomplete answer simply to state that 'neutrons cause more damage than X-rays because they are a higher LET radiation and therefore produce a higher density of ionization'. For equal doses measured in grays, the number of ion pairs released by each type of radiation is the same. Hence over a relatively large volume, e.g. a cell, the ion density for each radiation is the same. Therefore, for reasons not yet fully understood and beyond the scope of this book, it is differences in the spatial distribution of ion pairs in the cell nucleus at the submicroscopic level, illustrated in Fig. 11.7, that cause the difference in biological effect. For further discussion see, for example, Alper (1979).

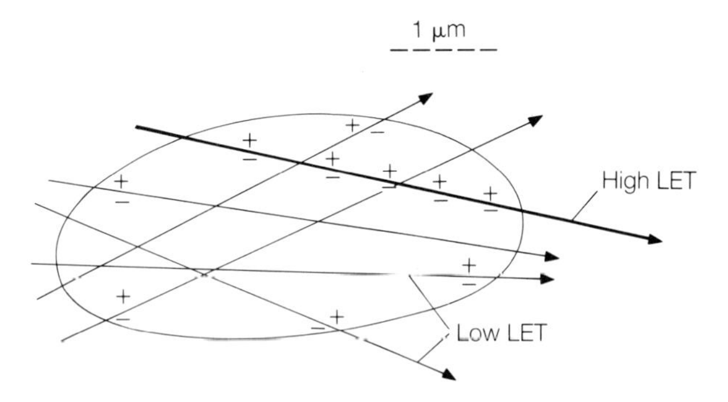

Fig. 11.7. Illustration of the difference in spatial distribution of ion pairs across a cell nucleus for low LET and high LET radiations. In each case five ion pairs are formed (same dose) but whereas these are likely to result from the same high LET particle, and thus be quite close together, they are more likely to result from five different low LET photons and hence be much more widely separated.

11.4 Quality factor and the sievert

RBE-type experiments demonstrate clearly that, when attempting to predict the possible harmful effects of radiation, the purely physical concept of dose as measured by the gray is inadequate. However, in numerical terms, RBE itself is a very difficult concept since it varies with dose, dose rate and fractionation, physiochemical conditions such as the presence or absence of oxygen, the biological end point chosen, the biological species and the time after irradiation at which measurements are made. Furthermore, since the shapes of the survival curves are different, inspection of Fig. 11.6 shows that, at the very low doses that are important in radiological protection, the RBE may be rather higher than the value of 2 or 3 quoted for higher doses.

Largely for these reasons, the International Commission on Radiation Units (ICRU) introduced a new term, 'Quality Factor'. This is a dimensionless, invariant quantity for a given type of radiation and is determined solely by the LET of the radiation. The numerical value is based on RBE but, as shown in Table 11.2, also includes a measure of conservatism to allow for many unknown factors.

Table 11.2. Values of quality factor as a function of linear energy transfer

LET (keV μm^{-1})	Radiation type	QF
3.5 or less	X-rays, γ rays, electrons	1
3.5–7.0	Low energy electrons (100 keV)	2
7.0–23	High energy protons (100 MeV)	5
23–53	Neutrons, protons	10
175 and above	Alpha particles	20

A new unit is now required and this is the Dose Equivalent (H) measured in sieverts (Sv). The sievert has replaced the older unit of the rem (1 Sv = 100 rem). Dose equivalent is related to dose by:

$$\text{Dose equivalent} = \text{dose} \times \text{quality factor} \times \text{other dose modifying factors}$$

$$H\,(\text{Sv}) = D(\text{Gy}) \times QF \times N$$

N might take into account absorbed dose rate, fractionation and non-uniformity of absorbed dose over the volume of the organ. ICRU has recommended a value of 1.0 for N at present. For a recent report on radiation quantities and units see International Commission on Radiation Units and Measurements (1980). (Note added in proof: An even more recent report from ICRU (1986) has recommended changes in the numerical values for quality factor but the principles remain essentially as described here.)

Note that the concepts of quality factor and dose equivalent should only be applied in the context of radiological protection.

11.5 Radiation effects on humans

11.5.1 Stochastic and non-stochastic effects

The biological effects of ionizing radiation on humans can be divided into two general categories. For some effects there appears to be a definite threshold dose below which no damage, in terms of measurable biological response, can be detected. Effects such as skin erythema, epilation and opacification of the lens of the eye and death resulting from acute exposure are in this category.

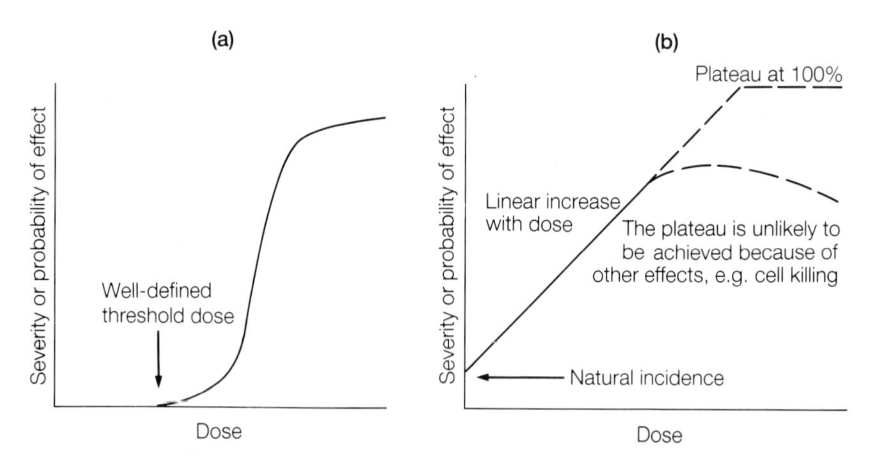

Fig. 11.8. Typical dose–response curves for (a) non-stochastic effects, (b) stochastic effects.

Such effects are now described as **non-stochastic** and are characterized by a dose–response curve of the type shown in Fig. 11.8a. No effect occurs below a threshold dose although the threshold varies from one individual to another, and the severity of the effect then increases nonlinearly with dose. The shape of the curve implies that recovery processes operate and the International Commission on Radiological Protection (1977) assumes that non-stochastic effects are non-additive. However, if the time interval between two exposures is short compared to cellular recovery times, there will be partial additivity and the severity of the effect depends on dose rate. Some typical thresholds for non-stochastic effects are shown in Table 11.3.

For a **stochastic** effect, which is defined as one that is governed by the laws of chance, the probability of an effect occurring increases linearly from the level of natural incidence at zero dose (Fig. 11.8b). This response

Table 11.3. Some approximate thresholds of radiation dose for non-stochastic effects

Effect	Threshold absorbed dose (Gy) for X-rays
Erythema	4–10
Epilation	5
Lens opacification	10–15
Skin damage (repairable)	15–40
(irreparable)	60

implies that no recovery processes are operating and hence all doses are strictly additive and the effect is independent of dose rate.

The two most important long-term effects of radiation, namely carcinogenesis and mutagenesis, are thought to be stochastic effects. Life shortening may also be a stochastic effect, but the evidence for any appreciable life shortening due to causes other than malignancy has been questioned and at the doses of interest to radiologists the effect is minimal.

Note that carcinogenesis and mutagenesis may be contrasted in that the former is **somatic**, that is to say the effect is observed in the irradiated individual, whereas the latter is **hereditary**, with the effect being detected in the descendants.

11.5.2 Carcinogenesis

There is ample evidence from a wide range of sources that ionizing radiation can cause malignant disease. For example, occupational exposure results in a greatly increased incidence of lung cancer among uranium miners, and in the period 1929–1949 American radiologists contracted nine times as many leukemias as other medical specialists. A frequently quoted example of industrial radiation-induced carcinogenesis is the 'radium dial painters'. They were mainly young women employed during and after the First World War to paint the dials on clocks and watches with luminous paint. It was their custom to draw the brush into a fine point by licking or 'tipping' it. In so doing, the workers ingested appreciable quantities of radium-226 which passed via the blood stream to the skeleton. Years later a number of tumours, especially relatively rare osteogenic sarcomas were reported.

A limited amount of evidence comes from approved medical procedures. For example, between 1939 and 1954, radiotherapy treatment was given to the whole of the spine for more than 14,000 patients suffering from ankylosing spondylitis. Statistically significant excesses of death due to malignant disease were subsequently observed, especially for leukemia and carcinoma of the colon. Other data comes from the use in radiology of thorotrast which contains the alpha emitter thorium-232, X-ray pelvimetry and radiation treatment for enlargement of the thymus gland. Unfortunately, radiation exposure in medical procedures is associated with a particular clinical condition. Therefore it is difficult to establish suitable controls for the purpose of quantifying the effect.

Finally, there is information gathered from survivors of the Japanese atomic bombs. This has been fully reported in a series of articles published over many years in *Radiation Research* (e.g. Wakabayashi *et al.* 1983).

The most salient features of all the data may be summarized as follows:

1 Whereas the incidence of excess leukemia appears to have reached a peak by about 7 years after exposure, there is evidence to suggest that the incidence of excess solid tumours is still rising at 30 years (Fig. 11.9).

2 Estimates based on at least two independent sets of data can be made of the cancer risk for most major organs. The agreement is quite good.

3 The risk varies quite markedly for different sites in the body. Results are summarized in Table 11.4: 125 deaths for 10^4 persons exposed to 1 Sv is usually expressed as a rounded mortality risk factor of 10^{-2} Sv^{-1}.

4 The Japanese work has confirmed that the carcinogenic effect, considered without regard to type, and measured in terms of relative risk, is highest amongst those under 10 years of age at the time of the bomb.

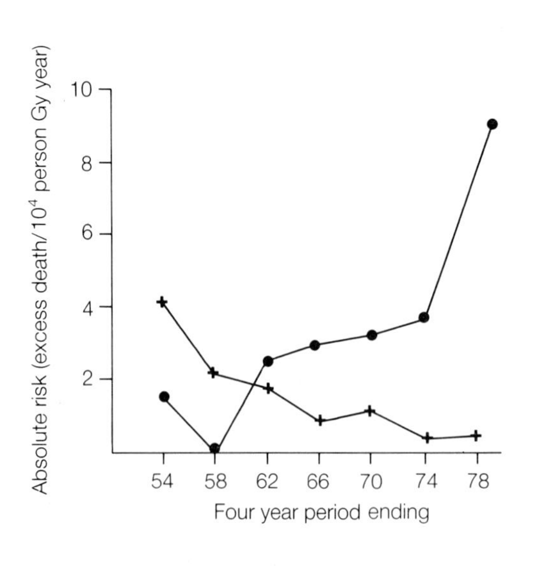

Fig. 11.9. Incidence of leukaemias and solid tumours as a function of time after irradiation (based on the Japan data) ● = leukaemia; ■ = all cancers except leukaemia.

Table 11.4. Excess deaths from cancer for a population of 10^4 persons exposed to an average dose equivalent of 1 Sv

Organ	Excess deaths
Breast	25
Bone marrow	20
Lung	20
Thyroid	5*
Total	125

* Although many more cases of thyroid cancer would occur, the majority would be curable.

11.5.3 Mutagenesis

The circumstantial evidence that ionizing radiation will cause mutations in humans is overwhelming. For example, the mutation frequency per locus per sievert has been calculated for drosophilia fruit flies and for mice, and radiation is known to impair the learning ability of mice and rats.

Furthermore, radiation is known to cause extensive and long-lasting chromosomal aberrations in cells circulating in the peripheral blood of humans. Some of the mechanisms by which these aberrations can arise from breaks and faulty rejoining are shown in Fig. 11.10. For a dose of 0.05 Sv whole body radiation, about one dicentric or ring chromosome would be scored for every 3000 mitotic cells examined. The normal incidence is negligible.

Notwithstanding, researchers have so far failed to demonstrate convincing statistical evidence of hereditary or genetic changes in humans as a result of radiation, even for offspring of the Japanese survivors. This failure is presumably caused by the statistical difficulty of showing a significant increase in the presence of a high and variable natural incidence of both physical and mental genetically related abnormalities. For severe disability the natural incidence is between 4 and 6%.

There are additional problems in assessing genetic risk. For example, only radiation exposure to the gonads is important and even this component can be discounted after the child-bearing age. Second, many radiation-induced mutations will be recessive, so their chance of 'appearing' may depend on the overall radiation status of the population. Finally,

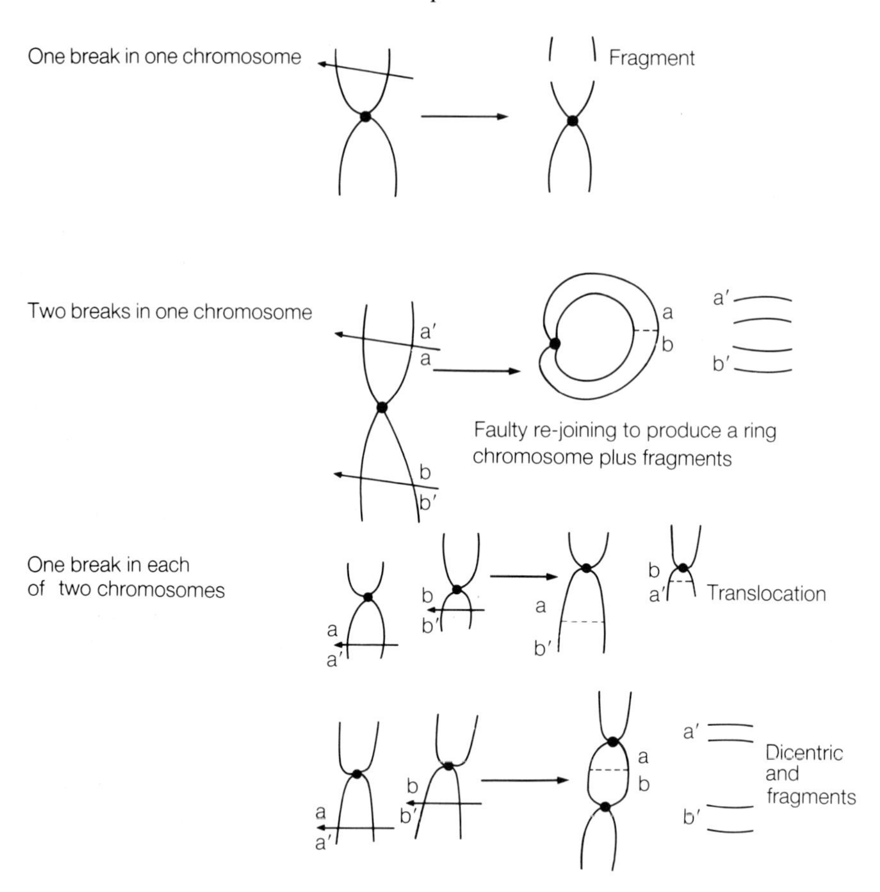

Fig. 11.10. Examples of some typical radiation-induced chromosome aberrations.

the risk to subsequent generations will depend on the stability of the mutation once formed.

The International Commission on Radiological Protection (1977) suggested that, following an average dose equivalent of 1 Sv to the population, about 100 cases of radiation-induced hereditary disease per 10^4 live births might occur in the first and second generations with 200 cases at equilibrium. However, a measure of the uncertainty comes from the more recent report of the National Academy of Sciences Committee on the Biological Effects of Ionizing Radiation (BEIR III 1980) who gave figures in the ranges 5–75 (first generation) and 60–1100 (at equilibrium).

11.6 Maximum permissible dose equivalent limits

Establishing dose equivalent limits has been the responsibility of the International Commission on Radiological Protection (ICRP) and the current general statement of their recommendations was published in 1977.

For non-stochastic effects it is fairly easy to establish maximum permissible levels that are sufficiently far below the threshold for the effect not to occur. The current limit is 500 mSv in a year for all tissues except the lens of the eye for which the limit is 150 mSv.

It is more difficult to set limits for stochastic effects because there is no level of radiation exposure that can be declared safe. Therefore figures have been derived by comparing the risk of mortality with that in other industries. On this basis, the maximum permissible level for occupationally exposed workers has been set at 50 mSv. Assuming a mortality risk factor of 10^{-2} Sv^{-1} (Table 11.4), an annual whole body exposure of 50 mSv carries a risk of 5×10^{-4} which is higher than for coal mining (2.5×10^{-4}) but lower than for deep sea fishing. In Radiology Departments, staff doses are normally less than one-tenth (5 mSv) of the annual exposure limit so the risk is comparable to that associated with other forms of employment in the Western World.

For members of the general public, the limit is 5 mSv in any one year. This is a planning limit and is used when designing, say an X-ray room adjacent to an area used by members of the public. The dose to a member of the public averaged over several years should not exceed 1 mSv per annum.

Maximum permissible doses are summarized in Table 11.5.

Table 11.5. Summary of maximum permissible dose equivalents

	Annual dose equivalent limit (mSv)		
Category	Whole body	Individual organs	Lens of eye
Persons at work aged 18 years or over	50	500	150
Trainees under 18 years*	15	150	50
Any other persons	5	50	300

* These figures also apply *pro rata* for the duration of a pregnancy once diagnosed.

11.6.1 Effective dose equivalent and weighting factors

As already stated, the annual permissible dose equivalent is 50 mSv when the whole body is irradiated and 500 mSv when a single organ is irradiated. (150 mSv for the eye lens). In practice most doses are neither whole body doses nor are they received solely by a single organ. To allow for this, the ICRP has established a system for apportioning the 50 mSv annual whole body limit whilst still retaining the 500 mSv individual organ limit.

This was done by identifying the five most sensitive tissues for the production of fatal cancers or leukemias at low doses of ionizing radiation together with the gonads for mutagenic effects and establishing the risk per sievert of radiation received for each of them. The risks to all other tissues were summed.

Each individual risk was then divided by the total risk to establish a 'weighting factor' W_T. Thus W_T is the proportion of the total risk attributable to tissue T when the whole body is uniformly irradiated. When radiation exposure is not uniform, the doses to individual tissues must first be found. The doses to six named tissues (H_T) are multiplied by the appropriate weighting factor. The five tissues giving the highest remaining doses are multiplied by a fixed weighting factor. All other issues are ignored. Weighting factors are shown in Table 11.6.

Table 11.6 Values of the weighting factor W_T recommended by the ICRP

Tissue	W_T
Gonads	0.25
Breast	0.15
Red bone marrow	0.12
Lung	0.12
Thyroid	0.03
Bone surfaces	0.03
Remainder*	0.3

* ICRP recommends a value of $W_T =$ 0.06 for each of the five remaining organs receiving the highest doses. Other tissues are then neglected.

This procedure gives several terms, (H_1W_1) for tissue 1, (H_2W_2) for tissue 2. The products are termed **effective dose equivalents** and when summed for a calendar year, the total must be less than 50 mSv, i.e.

$$\sum_T H_T W_T \leqslant 50 \text{ mSv}$$

Thus if a person received a dose of 10 mGy diagnostic X-rays ($QF = 1$) to the breast, lung, heart and a substantial part of the red bone marrow, but none to the rest of the body the effective dose equivalent would be

$$10 \times 0.15\ \text{(breast)} + 10 \times 0.12\ \text{(lung)} + 10 \times 0.12\ \text{(bone marrow)} + 10 \times 0.06\ \text{(heart)} = 4.5\ \text{mSv}$$

11.6.2 Special high risk situations

Although there are conflicting reports in the literature, the balance of evidence suggests that exposure *in utero* or during early childhood carries an enhanced radiation risk. If, for example, a dose of 2 Gy is delivered to mice at different stages in pregnancy, the type of result shown in Fig. 11.11 might be obtained. Initially, there will be a small number of rapidly

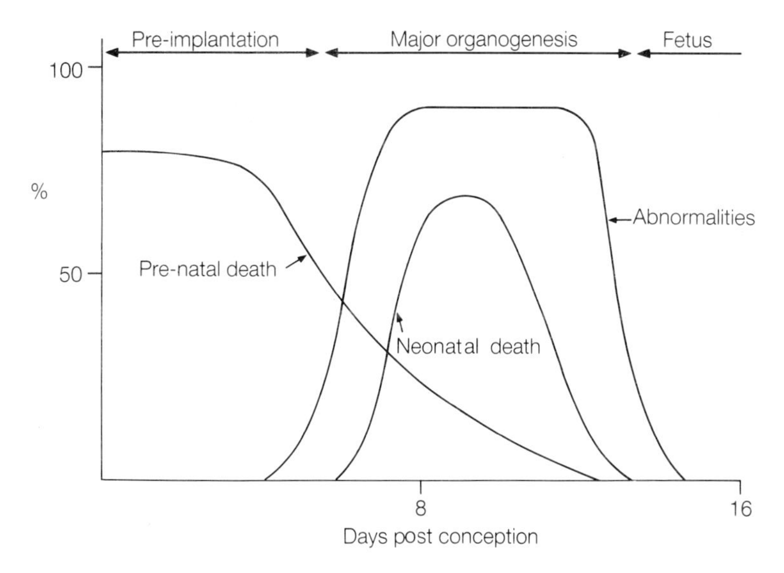

Fig. 11.11. Incidence of pre-natal death, abnormalities and neonatal death for mouse embryos irradiated with 2 Gy at different stages during pregnancy. (Redrawn from Russell & Russell 1954.)

dividing, highly radio-sensitive cells and damage is likely to result in pre-natal death. In the latter stages of pregnancy, there may be sufficient cells overall for the embryo to survive but only small numbers are performing any one specialized function so radiation damage causes abnormalities and neonatal death. Otake & Schull (1984) have recently re-evaluated the information on metal retardation following *in utero* exposure to A-bomb radiation and have found evidence of a similar pattern of effects. Extrapolation from animal experiments suggests that doses as low as 120 mGy at critical stages during pregnancy might cause abnormalities.

There is a body of literature, notably careful work by Stewart (see for example Bithell & Stewart 1975) to suggest that doses of 50 mGy or even less when delivered *in utero* in humans give a measurable increase in the incidence of leukemia. Cohen (1980) has published figures showing that the risk of leukemia is higher by a factor of about 1.6 and the risk of other malignancies is higher by a factor of about 2.9 for children irradiated under the age of 10 as compared to adults.

Thus the greatest care must be exercised when using X-rays on children or during pregnancy.

Until relatively recently, advice on the diagnostic exposure of women was directed towards confining less urgent examinations to the abdomen to the 10 days following the start of menstruation—the 10 day rule. However there is evidence to suggest that early pregnancy may not be the most sensitive period for radiation induced abnormal live births. For example:

1 Oocytes may be most sensitive just before release. Application of the 10 day rule makes irradiation at this time more likely.

2 Irradiation before organogenesis is less likely to produce abnormalities.

3 Stewart's work relates to the latter half of pregnancy.

The ICRP (1984) no longer recommends application of the 10 day rule and recent advice from the National Radiological Protection Board (March 1985) is that the importance of the 10 day rule should be reduced. Although this view may not be universally adopted, administrative arrangements within the X-ray department based on establishing whether or not the patient is pregnant are preferable to unthinking application of the 10 day rule.

11.7 Hazard from ingested radioactivity

11.7.1 Annual limit on intake

Since radiation to an organ may arise from ingested or injected radioactivity, intake of radionuclides must be such that the basic recommendations on dose limitation can be met.

ICRP recommends that the integrated dose equivalent to a tissue T

over an assumed working lifetime of 50 years ($H_{50,T}$), following intake of a radionuclide should be calculated.

For example assume

1 all exposure arises from internal sources,

2 the dose to a given tissue T over 50 years for unit intake (say 1 MBq) is $h_{50,T}$,

3 the intake is I (measured in MBq). Then the total committed dose equivalent is $I \cdot W_T \cdot h_{50,T}$ summed over all tissues. Hence the annual limit on intake (ALI) is that value of I which satisfies the equation

$$I \sum_T W_T \cdot h_{50,T} = 0.05 \text{ Sv}$$

For non-stochastic effects I is limited by

$$H_{50,T} \leqslant 0.5 \text{ Sv or } I \cdot h_{50,T} \leqslant 0.5 \text{ Sv}$$

except for the eye where the limit is 0.15 Sv.

11.7.2 Absorbed dose to an organ

To find the value of I, $h_{50,T}$ must be calculated. To do this, the factors to consider are:

1 the total energy radiated per unit time by the radionuclide, perhaps separated into that from charged particles and that from gamma rays;

2 absorption effects from gamma rays, both originating in that organ and elsewhere, and taking into consideration geometrical factors, the inverse square law and the mass absorption coefficient;

3 the distribution of radionuclide within the body;

4 the time-scale of exposure. If the radionuclide localizes quickly, this may be expressed in terms of the effective half-life, but note that after leaving an organ the activity may localize elsewhere—e.g. the kidney excretes via the bladder.

Methods of calculation and relevant data are contained in a series of publications from the Medical Internal Radiation Dose Committee (MIRD) of the Society of Nuclear Medicine (e.g. 1968, 1975). Exact calculation of the absorbed dose to an organ can be extremely difficult.

11.7.3 Derived limits for airborne and surface contamination

When the ALI is known, the permissible derived air concentration (DAC) can be calculated if simplifying assumptions are made. Assuming a working year of 2000 h (50 weeks at 40 h), a rate of breathing of 0.2 m^3 min^{-1} and that inhalation is the only route of intake

$$\text{DAC} = \frac{\text{ALI}}{2000 \times 60 \times 0.02} = \frac{\text{ALI}}{2.4 \times 10^3}\ \text{Bq m}^{-3}$$

For Tc-99m the ALI is 6×10^9 Bq, hence the maximum permissible DAC, rounded to the nearest whole number, is 2×10^6 Bq m^{-3}. For I-131 the ALI is 2×10^6 Bq and the DAC 7×10^2 Bq m^{-3} (ICRP 1979–81).

Note that the concept should be used with care since it only applies to the ICRP reference man working under conditions of light activity and makes a number of other assumptions about the metabolic breathing pattern.

The derived working limit for surface contamination that will ensure the maximum permissible DAC is not exceeded varies from one radionuclide to another. For radionuclides used in nuclear medicine, it may be as low as 370 Bq spread over an area of 10^{-2} m^2. This causes two problems:

1 The activity 'seen' by a small detector may be no more than 20–30 Bq. A Geiger–Müller counter is insufficiently sensitive to detect this level of activity and a purpose-built scintillation crystal monitor must be used.

2 It may be impossible to decontaminate to maximum permissible levels by washing and cleaning after even quite a small spill (say 1 MBq). If the radionuclide has a long half-life, contaminated equipment will then have to be removed from service.

11.8 Principles of radiological protection

11.8.1 Concept of dose limitation

The discussion has been concerned so far with establishing maximum permissible levels of radiation exposure. However, since carcinogenic effects are thought to be stochastic, it must be recognized that all exposure to radiation may carry a risk and all additional exposures will carry an additional risk.

Thus the ICRP has laid down three basic principles which should be applied in the following order:

1 No practice shall be adopted unless its introduction produces a positive net benefit.

2 All exposures shall be kept *as low as reasonably achievable* (ALARA), economic and social factors being taken into account.

3 The dose equivalent to individuals shall not exceed the limits recommended for the appropriate circumstances by the ICRP.

11.8.2 General radiation protection procedures

To ensure that the ALARA principle is implemented, a number of factors that will affect the total dose must be considered:
1 *Need*. It is essential to ensure that the examination is justified in terms of clinical necessity. Administrative arrangements must be designed to minimize the need for repeat examinations.
2 *Beam intensity or activity*. For a given time of exposure, the dose will be proportional to the X-ray beam intensity or to the activity of the radionuclide so dose rates should be as low as practicable.
3 *Beam attenuation*. After passing through the patient, an X-ray beam is further attenuated before reaching the film, for example by the couch. The dose to the patient can be reduced if this attenuation can be reduced. One way to do this is to use carbon fibre materials. At 70 kVp a carbon fibre table absorbs about 14% less radiation than a plastic table and a carbon fibre cassette may result in 10–30% less attenuation (ICRP 1982b).
4 *Time*. The total dose is the product of dose rate and time so investigations should be short if possible.
5 *Distance*. For a point source, dose rate in air will decrease according to the inverse square law. For example, if a vial of radioactivity is handled with tongs, the fingers will be 10 cm from the radiation source instead of 1 cm away and the dose rate is reduced by a factor of 100.
6 *Shielding*. At diagnostic X-ray energies, effective shielding materials are readily available and must be used where appropriate.

Three categories of person should be considered
1 patients,
2 staff,
3 general public,
but if all precautions given below for patients and staff are followed, the general public should be adequately protected too. Note that some of these measures are not primarily the responsibility of the radiologist.

General precautions for patients

1 Query unnecessary examinations or examinations not of direct relevance to the patient's condition whenever possible.
2 Confirm that the examination has not already been done at an outlying clinic or hospital from which the patient has been referred.

3 Examine the possibility of using alternative techniques (e.g. ultrasound).
4 Minimize the use of X-rays during pregnancy and shield the foetus. Note that ultrasound cannot be used to look at the foetus during the latter stages of pregnancy.
5 Provide adequate shielding of radiosensitive regions (e.g. gonads). Note that gonad shields can be difficult to use on children and without regular reminders tend to be set aside as 'too difficult to use'.
6 Collimate the beam to the smallest possible useful area. Minimizing scatter in this way not only reduces patient dose but also improves contrast.
7 Ensure that the correct beam filtration is used. The requisite total filtration is laid down in the Regulations and should be clearly displayed on the tube. It is the responsibility of the Radiological Protection Adviser to check this feature.
8 Use high kV techniques where possible.
9 Use carbon fibre table tops to reduce dose.
10 Take care over setting-up technique to reduce repeat films. This will generally be the responsibility of the radiographer, although the radiologist will have specified the views required.
11 Ensure that all views taken will contribute to the diagnosis.
12 Use image intensifiers not fluoroscopy screens.
13 Use small film format, e.g. 100 mm camera to reduce dose where possible. Note, however, that using a small format camera only reduces the dose if the same number of views is taken. If the reduction in dose per view is used as an excuse to increase the number of views taken, the overall dose reduction may be minimal.
14 Screen at dose rates that are below 50 μGy min^{-1} at the image intensifier if practicable. Some sets automatically adjust the kVp and mA to obtain the best picture.
15 Implement a strict quality control programme for film processing to minimize rejects.

General precautions for staff
1 Ensure that all staff, including nurses, have received adequate instruction. This is particularly important during screening procedures and in theatres.
2 Pay careful attention to room design. The shielding in the walls, doors and control panel screens will have been calculated by a physicist. Everyone should be behind a lead screen when radiographs are taken.

3 Check that illuminated warning signs or lights on all access doors to the X-ray room are working.
4 Check that a warning light on the tube head shows when a tube is 'on' and that only one tube per room may be energized at any one time. When two X-ray tubes are operated from behind the same control desk, care is required to ensure that the correct tube has been selected before an exposure is made. Instances have been recorded of inadvertent exposure of patients and radiographers in IVP rooms containing two units following the firing on the 'wrong' head.
5 Lead aprons must be provided, sufficient for each person in the room, and worn at all times during screening procedures. Aprons must always accompany mobile units. Lead gloves must also be provided. Both aprons and gloves must be checked regularly for cracks and a record of the tests must be kept even if it is a nil return.
6 Test equipment for performance, especially tube leakage which should give a dose rate in air of less than 1 mGy h^{-1} at 1 m, both before the unit is accepted from the manufacturer and periodically thereafter. These tests should normally be carried out by a physicist.
7 Provide adequate personnel monitoring.
8 Keep exposure records for staff.

11.8.3 Special precautions during fluoroscopy and screening

When using an image intensifier, the radiologist must take care to ensure that where possible the protective lead rubber skirt round the image intensifier is in position.

Radiologists must also ensure that their lead rubber aprons are fully fastened because there is a substantial amount of radiation scattered from the patient, especially when using overcouch tube screening units. These units are not recommended when staff have to remain close to the table.

When patients are manipulated or palpated, the hand should be shielded using a lead rubber glove. New interventional radiological techniques are producing many problems with regard to finger doses. Many of these procedures involve manipulations in which the fingers of the radiologist are very close to the main beam with the inevitable chance that they will stray into the main beam. Because of the delicate nature of the manipulations, normal lead rubber gloves cannot be worn. The screening times in these investigations can be very long and the dose rate in the main beam approaches 100 mGy min^{-1}.

If patients have to be held during an examination, this must be undertaken by a nurse or some other person not normally exposed to ionizing radiation. If children have to be held, one of their parents should be asked to assist—if this is the mother, care should be taken to ensure she is not pregnant. The person holding the patient must be given a lead rubber coat and gloves.

In paediatric clinics the baby must be mechanically restrained rather than held by the radiologist. For certain procedures this can be very difficult and a member of staff not normally involved in X-ray work should be asked to assist.

Do not use fluoroscopic methods when they can be avoided, e.g. when locating the position of a foreign body.

11.8.4 Special precautions in nuclear medicine

With unsealed radionuclides there are two hazards. One is the external hazard from gamma rays escaping from the source container, i.e. a bottle or syringe, and from the patient after injection. The other is the internal hazard following accidental ingestion of a radionuclide either by the person injecting or preparing the dose or by other persons following contamination of the area. Although small doses from the external hazard are most frequent, the internal hazard is potentially far more serious.

The external hazard can be greatly reduced by following relatively simple precautions:

1 Always keep bottles containing radionuclides in lead pots in a shielded area—never leave the vial unshielded.

2 Never handle bottles containing radionuclides but always use tongs.

3 If possible stay at least 0.5 m away from a patient who has received radioactivity, and check that nurses do not remain unduly close to the patient for unnecessarily long periods. The dose rate 25 cm away from a patient who has received 500 MBq of Tc-99m for, say, a bone scan is about 33 $\mu Sv\ h^{-1}$. At 1 m the dose rate is only 9 $\mu Sv\ h^{-1}$.

4 Whenever possible, use shielded syringes to give injections.

5 If a shield cannot be used, extra care should be taken when handling the syringe not to hold the end containing the radionuclide.

The internal hazard can be avoided by simple good house-keeping practices. Protective clothing, especially gloves, should always be worn. Syringes do back-fire and contaminated skin is difficult to clean. All manipulations of the radionuclide from bottle to syringe must be performed over a tray so that any drops can be contained. Syringes must always be vented into a swab, never squirted generally over the room. All contaminated and potentially contaminated materials must be disposed of in a container that has been clearly labelled as suitable for the purpose.

11.9 Personnel dosimetry

Two fundamentally different techniques are used for personnel dosimetry—the film badge and the thermoluminescent dosemeter (TLD). The physical principles of both these techniques of radiation measurement were discussed in Chapter 7 so only those aspects of their performance that are relevant to personnel dosimetry are discussed here. This will be done by considering the requirements of an ideal personnel dosemeter and the extent to which each method satisfies this ideal.

Range of response
The monitor must be sensitive to very small exposures since they will be the norm, but it must also be capable of recording accurately a high exposure should this arise in an accident. Hence a wide range is required.

Because of the shape of its characteristic curve, photographic film is useful over only a limited range of doses. However, the range can be extended if a fast film is backed by a much slower film. If the fast film is over-blackened, it is carefully removed and readings are obtained from the slower film.

There are no such problems with TLD which has a wide range of response from 0.1 mGy upwards.

Linearity of response
If the response is linear with dose, measurement at just two known doses will allow a calibration curve to be drawn. This is possible with TLDs although not recommended, but for film, because of the shape of the characteristic curve, calibration is necessary at a large number of doses so that the exact shape of the curve can be established. Furthermore, calibration is necessary each time a batch of film is developed because of variations in film blackening with development conditions.

Calibration against radiation standards
A measure in terms of a fundamental physical property—e.g. temperature

rise—would be desirable. However, neither film blackening nor thermoluminescence is in this category. Therefore calibration against a standard radiation source is necessary on each occasion.

Variation of sensitivity with radiation energy
As discussed in Chapter 7, because of the presence of silver and bromine in film, blackening per unit dose is much higher at low photon energies where the photoelectric effect dominates than at higher photon energies. Similar differences in sensitivity exist for electrons. Therefore the film badge holder contains a number of filters (Fig. 11.12a), which not only extend the range of radiation energies over which the blackening per unit dose is approximately constant (Fig. 11.12b), but also provide data which may be used to calculate the dose for low energy photons and electrons. The cadmium filter will capture neutrons with subsequent emission of gamma rays so additional blackening under the filter is evidence of neutrons.

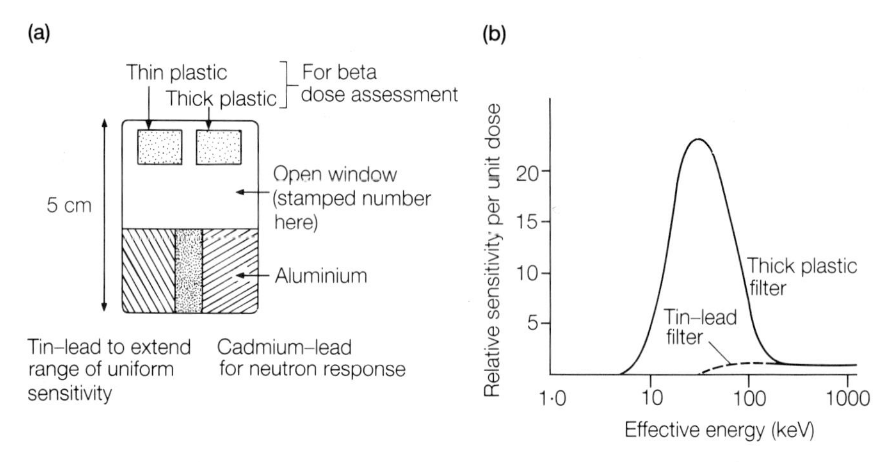

Fig. 11.12. (a) Details of a film badge holder. (b) Curves showing the variation of sensitivity (film blackening per unit dose) with radiation energy both without and with filter. Note that in the diagnostic range, correction for variation in sensitivity has to be made.

The sensitivity of a lithium fluoride TLD is independent of radiation energy down to about 100 keV, then increases slightly due to the small difference in atomic number between LiF and soft tissue. At even lower energies this effect is counter-balanced by self absorption in the lithium fluoride/ceramic chip and the overall change in sensitivity with energy is unlikely to exceed 20%.

Sensitivity to temperature and humidity
The fog level of film can increase markedly under conditions of elevated temperature or humidity and this may be misinterpreted as spurious radiation. TLDs show no such effects.

Uniformity of response within batches
Provided care is taken over storage and development, photographic films are nowadays uniformly sensitive within a batch. The sensitivity of a TLD can vary within a batch so individual calibration may be necessary. Also careful annealing is required after use or the TLD tends to 'remember' its previous radiation history.

Maximum time of use
This is primarily governed by the risk of latent image fading when radiation is accumulated in small amounts over a long period of time. For both systems the effect is negligible provided that the time-scale for calibration is comparable with the time-scale for use.

Compactness
As shown in Fig. 11.12 a film badge holder is quite small. TLDs can be extremely small (Fig. 11.13) and are especially useful for monitoring exposure to the hands and fingers.

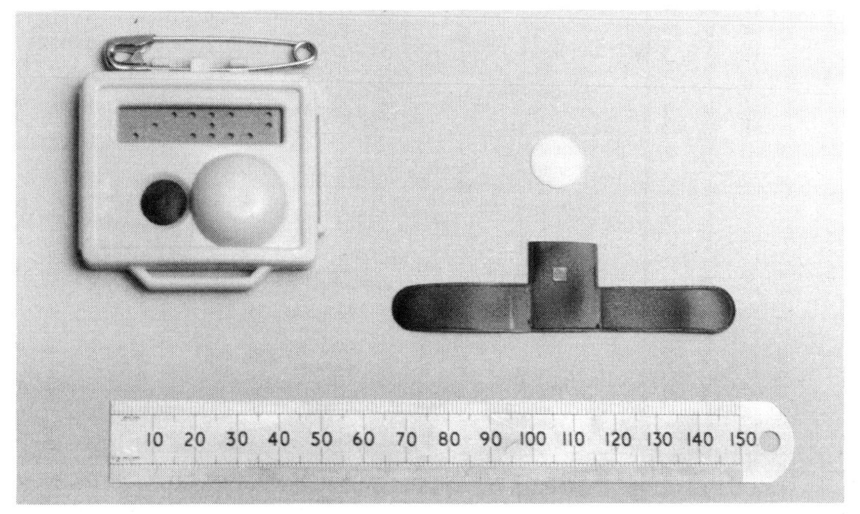

Fig. 11.13. An example of a TLD monitor. On the left is a badge holder for whole body monitoring. On the upper right is the LiF ceramic disc that would be useful for measuring extremity doses. A very small LiF chip is shown on the lower right.

Permanent visual record
When investigating the possible cause of a high reading (e.g. is it real or has it arisen because the dosemeter but not the individual has been exposed?), access to a permanent visual record can be useful. This is readily available with photographic film but with TLDs all the raw data are lost at read-out.

Indication of type of radiation
Similarly, the pattern of film blackening can give useful information on the type of radiation as discussed above.

Indication of pattern of radiation
Again, this information is occasionally useful and can be obtained from a film but not from TLD. For example, if a filter casts a sharp shadow, this suggests a single exposure from one direction but a diffuse shadow suggests several smaller exposures from different directions. The presence of small, intense black spots on the film suggests contamination with unsealed radioactive material.

Table 11.7 Relative merits of film badges and thermoluminescent dosemeters as alternative methods of personnel monitoring

	Film Badge	TLD
1 Range of usefulness	0.2 mGy–6 Gy	0.1 mGy–10^4 Gy
2 Linearity of response	No	Yes
3 Calibration against radiation standards?	Yes	Yes
4 Response independent of radiation energy?	No	Yes (except at low kV)
5 Sensitive to temperature and humidity	Yes	No
6 Uniformity of response within batches	Yes	Yes (with care)
7 Maximum time of use	2 months	12 months
8 Compactness	Small	Very small
9 Permanent visual record?	Yes	No
10 Indication of type of radiation?	Yes	No
11 Indication of pattern of radiation exposure?	Sometimes	No
12 Cost to NHS	Expensive	Very expensive

Cost
Both systems are relatively expensive. Photographic film has now become expensive and is a consumable. TLD chips are re-usable but the capital

cost for their purchase and for the read-out and annealing equipment is high. A typical cost in the UK at 1986 prices might be £1.50 per worker per month.

All these factors are summarized in Table 11.7. There is no 'best buy'—each method is well suited to certain applications and both are used widely in the UK at the present time.

11.10 Relevant legislation

Most countries have now introduced legislation that relates either directly or indirectly to the use of X-rays and radioactive materials for medical purposes. This section presents briefly, in chronological order, the most important items of legislation in the United Kingdom. Generalization to other countries is made where appropriate.

The Radioactive Substances Act (1960)
The primary purpose of this Act was to exercise effective control over radioactive wastes. Control extends to the production and use of such materials only in so far as they eventually become waste. Hospitals in the UK are exempt from Registration as users of radioactive materials but are subject to those clauses dealing with the storage and disposal of radioactive waste.

The Radioactive Substances (Carriage by Road) Regulations (1974)
These were introduced to control the hazard to the public resulting from the transport of radioactive materials. When radioactive materials are transferred from one hospital to another, or indeed between two parts of the same hospital, arrangements must comply with the requirements of these regulations if it is necessary to use the public highway. A number of exemptions apply to the transport of low levels of activity for medical purposes by professional users, but packaging must always be adequate to withstand an accident and must always be labelled clearly with information on contents. It is recommended that the vehicle should carry a notice, displayed in a prominent position, giving instructions on the action to be taken in the event of an accident or emergency.

In the UK and many other countries the regulations relating to the transport of radioactive materials are based on IAEA Regulations (1979).

The Medicines (Administration of Radioactive Substances) Regulations (1978)

With respect to control over prescribing an investigation requiring exposure to radiation, there are, for historical reasons, significant differences between control over administration of radionuclides and administration of X-rays. Unless radiographs are being used for research, there is no limit in the UK to the number that a clinician or radiologist may take but the policies and recommendations of ICRP, as outlined earlier in this chapter, must be adhered to strictly. In some countries these recommendations have been or are being incorporated into legislation.

With radionuclides, however, there has always been some control over administration and many countries operate a licensing procedure. In general terms, the requirements for approval of an application for a medical licence are that

1 the proposed radiopharmaceuticals are ones that have been demonstrated to be safe for the patient and effective for diagnosis;

2 the proposed equipment, facilities and procedures are appropriate for effective use of the radiopharmaceuticals and are adequate to protect the health and safety of patient and public;

3 the radiopharmaceuticals will only be used by physicians whose training and experience in the basic principles of radiation and the clinical use of radionuclides are sufficient for their safe and effective use.

Current procedure under the Medicines (Administration of Radioactive Substances) Regulations (1978) is that all persons administering radioactive materials to patients must be licensed. The licence is specific as to the radionuclides that can be used, the tests that can be carried out and the premises where they can be conducted. The licence also covers doctors working under the supervision of the licensee and it is not envisaged that every clinician in a hospital would have a licence, or indeed would have the relevant experience to be granted a licence. Licences are issued by the Health Services Division of the Department of Health and Social Security.

Regulations vary markedly from one country to another and are sometimes much more complex, and perhaps more stringent, than those outlined above. For example in the United States, where the Food and Drug Administration (FDA) Agency, the Nuclear Regulatory Commission (NRC) of the Atomic Energy Commission and State Regulatory Agencies of Radiological Health all exercise control over the radiological health and safety aspects of radiopharmaceuticals, the NRC may issue several types of licence. These include, for example, a general licence to possess and use limited quantities of prepackaged individual doses of certain diagnostic radiopharmaceuticals, a private practice licence for physicians in private offices or an Institutional-specific licence of broad scope.

In general, physicians with the proper qualifications—appropriate training and experience, facilities, equipment and operating procedures—are readily licensed for well-established procedures where the pharmacological and radiological action of the drug, preferred route of administration, safety, side-effects, contra-indications, dose ranges and effectiveness are well known. However, the licence may specify the source from which the drugs are to be obtained and approval for a particular investigation will not necessarily extend to use of the same drug for a different investigation or administration by a different route. For newer radiopharmaceuticals, a full research protocol is required. The details of such a protocol are beyond the scope of this book.

The Ionizing Radiations Regulations (1985)

These have been introduced for a three-fold purpose:

1 to comply with the UK obligations under the Euratom Directive of September 1980, which lays down basic safety standards for radiation protection in the European Community;

2 to bring the existing provisions on radiological protection into line with the latest recommendations of the ICRP;

3 to replace the multiplicity of existing statutory and non-statutory provisions by a unified system of regulations and approved codes of practice applying to all sectors of the community as provided for in Section 1(2) of the Health and Safety at Work Act (1974).

It follows that, whereas the 1960 Radioactive Substances Act was primarily concerned with the environment, the 1986 Regulations are concerned with the safety of the individual. The Regulations are supplemented by an Approved Code of Practice which acts as an explanatory memorandum, and Guidance Notes for the protection of persons against radiations arising from medical and dental use are also planned. An important feature of these Regulations is the identification of 'local rules' as the formal means of setting down the procedures and work activities undertaken by the employer and the precautions relating to them. As the name implies, local rules may vary from one X-ray department to another and will, to a certain extent, depend on the procedures undertaken. However, their purpose is always the same, namely to safeguard patients, staff and visitors from unnecessary exposure. All staff must be made aware of their content.

11.11 Summary

The potentially harmful effects of ionizing radiation must be recognized and understood. Furthermore, it is important for radiologists to appreciate that increasingly sophisticated experiments have failed to provide evidence of a safe level of radiation for the two most important long term effects, carcinogenesis and mutagenesis. Nevertheless, with careful attention to detail in working procedures and compliance with approved codes of practice and local rules, radiation exposures to staff can be kept to an acceptable minimum and for patients the information to be gained can fully justify the small additional risk involved.

References and further reading

Alper T. (1979) *Cellular Radiobiology*. Cambridge University Press.

BEIR III (1980) *Report of the National Academy of Sciences on the Biological Effects of Ionizing Radiation*. Washington.

Bergonié J. & Tribondeau L. (1906) De quelques resultats de la radiotherapie et essai de fixation d'une technique rationnelle. *C. R. Seances Acad. Sci.* **143**, 983. English translation: Fletcher G. H. (1954) Interpretation of some results of radiotherapy and an attempt at determining a logical technique of treatment. *Radiat. Res.* **11**, 587.

Bithell J. F. and Stewart A. M. (1975) Pre-natal irradiation and childhood malignancy; a review of British data from the Oxford survey. *Br. J. Cancer,* **31**, 271.

Cember H. (1983) *Introduction to Health Physics*. Pergamon Press, Oxford.

Cohen B. L. (1980) The cancer risk from low level radiation. *Health Phys.* **39**, 659–678.

Coggle J. E. (1983) *Biological Effects of Radiation*. Taylor & Francis, London.

Dalrymple G. V., Gaulden M. E., Kollmorgan C. M. & Vogel H. H. eds (1973) *Medical Radiation Biology*. Saunders, Philadelphia.

Duncan W. & Nias A. H. W. (1977) *Clinical Radiobiology*. Churchill Livingstone, Edinburgh.

Gifford D. (1984) *A Handbook of Physics for Radiologists and Radiographers*. Wiley, Chichester.

Hall E. J. (1978) *Radiobiology for the Radiologist*. Harper & Row, New York.

Hughes D. (1982) *Notes on Ionizing Radiation, Quantities, Units, Biological Effects and Permissible Doses*. (Occupational Hygiene Monograph 5) Science Reviews Ltd.

Hughes J. S. & Roberts G. C. (1984) *The Radiation Exposure of the UK Population 1984 Review*. (Publication R173) National Radiological Protection Board.

IAEA (1979) *Regulations for the Safe Transport of Radioactive Materials—1973*, revised edition (as amended). (Safety Series No. 6) IAEA, Vienna.

International Commission on Radiation Units and Measurements (1980) *Radiation Quantities and Units*. (Publication 33) ICRU, Washington.

International Commission on Radiological Protection (1977) Recommendations (ICRP publication 26). *Ann. ICRP*, **1**(3).

International Commission on Radiological Protection (1979) Limits for intakes of radionuclides by workers (ICRP publication 30), part 1. *Ann. ICRP,* **2**(3/4).

International Commission on Radiological Protection (1980) Limits for intakes of radionuclides by workers (ICRP publication 30), part 2. *Ann. ICRP,* **4**(3/4).
International Commission on Radiological Protection (1981) Limits for intakes of radionuclides by workers (ICRP publication 30), part 3. *Ann. ICRP,* **4**(2/3).
International Commission on Radiological Protection (1982a) Protection against ionizing radiation from external sources used in medicine (ICRP publication 33). *Ann. ICRP,* **9**(1).
International Commission on Radiological Protection (1982b) Protection of the patient in diagnostic radiology (ICRP publication 34). *Ann. ICRP*, **9**(2/3).
International Commission on Radiological Protection (1982c) General principles of monitoring for radiation protection of workers (ICRP publication 35). *Ann. ICRP,* **9**(4).
International Commission on Radiological Protection (1983) Cost benefit analysis in the optimisation of radiation protection (ICRP publication 37). *Ann. ICRP,* **10**(2/3).
International Commission on Radiological Protection (1984a) Statement from the 1983 Washington meeting of the ICRP. *Ann. ICRP,* **14**(1).
International Commission on Radiological Protection (1984b) Non-stochastic effects of ionising radiation (ICRP publication 41). *Ann. ICRP,* **14**(3).
International Commission on Radiological Protection (1984c) A compilation of the major concepts and quantities in use by ICRP (ICRP publication 42). *Ann. ICRP*, **14**(4).
International Commission on Radiological Protection (1985) Principles of monitoring for the radiation protection of the population (ICRP publication 43). *Ann. ICRP,* **15**(1).
Ionizing Radiations Regulations (1985) Copies of the Regulations, Approved Codes of Practice, and notes for guidance referred to in this chapter are published by HMSO.
Medical Internal Radiation Dose Committee (1968) Pamphlets 1–3. Society of Nuclear Medicine, 404 Church Avenue, Suite 15, Maryville, TN 37801, USA.
Medical Internal Radiation Dose Committee (1975) Pamphlet 11. Society of Nuclear Medicine, 404 Church Avenue, Suite 15, Maryville, TN 37801, USA.
Otake M. & Schull V. J. (1984) In utero exposure to A-bomb radiation and mental retardation—a reassessment. *Br. J. Radiol.* **57**, 409.
Russell L. B. and Russell W. L. (1954) An analysis of the changing radiation response of the developing mouse embryo. *J. Cell Comp. Physiol.* **43**, Suppl. 1, 103.
Shapiro J. (1981) *Radiation Protection—a Guide for Scientists and Physicians*, 2nd edn. Harvard University Press.
Wakabayashi T., Kato H., Ikeda T. & Schull W. J. (1983) Studies of the mortality of A-bomb survivors. Report 7, Part III: Incidence of cancer in 1959–1978 based on the Tumour Registry, Nagasaki. *Radiat. Res.* **93**, 112.

Exercises

1 State the law of Bergonié & Tribondeau on cellular radiosensitivity. Name two cell types which obey the law and two which do not.
2 Explain why it is difficult to predict quantitatively the effects of very low doses of radiation. How and why might the prediction depend on the linear energy transfer (LET) of the radiation.
3 What is the difference between stochastic and non-stochastic radiation effects?

4 Sketch the most common forms of chromosome defect detectable after whole body irradiation.
5 What is the most sensitive period for the production of abnormalities in humans by irradiation *in utero*? Why does irradiation at other times, both later and earlier, effectively produce fewer abnormalities?
6 Review the evidence that ionizing radiation can cause harmful genetic effects.
7 List the organs and tissues of the body identified by ICRP publication 26 (1977) as most sensitive with regard to causing long-term death by ionizing radiation. Give an approximate risk of death for a whole body dose of 10 mSv.
8 What precautions should be taken regarding radiation protection in a paediatric X-ray clinic?
9 As consultant in charge of a department planning a new suite of rooms for neuroradiological investigations, what considerations would you have with regard to radiation protection when deciding on the structure and lay-out, and on the equipment installed?
10 List the precautions with regard to radiation protection, that must be taken when using a mobile X-ray image intensifier in an orthopaedic theatre.
11 Discuss the radiation protection requirements of a hospital laboratory suitable for preparing radiopharmaceuticals.
12 Summarize the radiological techniques that can reduce the dose to the patient.
13 Comment on the validity of the statement 'Any dose can be justified in diagnostic radiology'.
14 What are the requirements of an ideal personnel dosemeter?
15 What are the advantages and disadvantages of a film badge personnel dosimetry system when compared with a thermoluminescent system?

12

Ultrasound Imaging

12.1 Introduction

Because of the known harmful effects of ionizing radiation, there is a substantial role for any form of radiation that is capable of producing diagnostic quality images without causing ionization. An important example of a non-ionizing radiation is a sound wave, which is a longitudinal pressure wave that produces compressions and rarefactions of the 'particles' in the medium. In this context a 'particle' can be considered as a group of molecules in which the pressure and temperature is uniform and each particle is continuous with adjacent particles as shown in Fig. 12.1. A sound wave is non-ionizing because it cannot deposit enough energy in matter to disrupt chemical bonds.

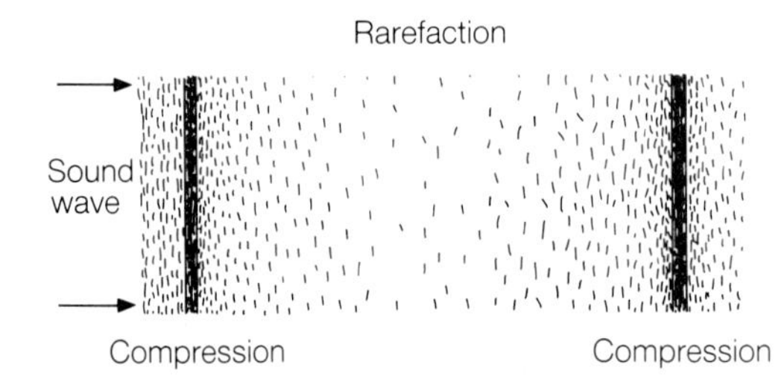

Fig. 12.1. Schematic representation of the bunching of particles at a compression in a sound wave.

The human ear can detect sounds up to a frequency of about 20 kHz. Ultrasound is any higher frequency wave that cannot be detected by the human ear and, as used in medicine, has a frequency between 1 and 15 MHz.

Because propagation of an ultrasound wave depends on moving particles in the medium, it is very dependent on the physical properties of the medium su ... us, for example, ultrasound travels app ... a solid than through a liquid because th ... n a solid. Ultrasound waves cannot trav ... ays and gamma rays of course can do so.

In gener ... ys and gamma rays can often be understo ... m as particles or photons, the most releva ... ose normally associated with visible light. Thus ultrasound travels at different velocities through different tissues, undergoes partial reflection at the boundary between two media, and suffers diffraction, scattering and attenuation. The information that may be obtained about the body by measuring, for example, reflections at boundaries and velocities of propagation, and the limitations placed on image quality by, say, diffraction and attenuation, are the major themes of this chapter.

12.2 Physical properties of sound waves

Insofar as sound, like light and X or gamma rays, is a form of wave motion, their properties are similar. Thus for example, sound waves can be represented as a sinusoidal wave motion as in Fig. 12.2. The wavelength is the shortest distance between two parts of the wave that are in phase, i.e. the distance XY or X_1Y_1, not the distance between two points of zero pressure departure which is only half a wavelength. The frequency f is the number of waves that would pass a given point in 1 s. The wave velocity v, the frequency f and wavelength λ are related by

$$v = f\lambda$$

As a particle within the medium will move back and forth about a mean position under the influence of the sound wave, the particle velocity can also be represented by a sine wave with the same wavelength and frequency as the sound wave. The peak of the velocity wave will be half a wavelength or 180° out of phase with the pressure wave. Note that this relationship applies only for a continuous wave where there is a single value of the wavelength and frequency along the wave. It does not apply to a pulsed wave, although the latter can often be considered as a mixture of continuous waves of different frequencies.

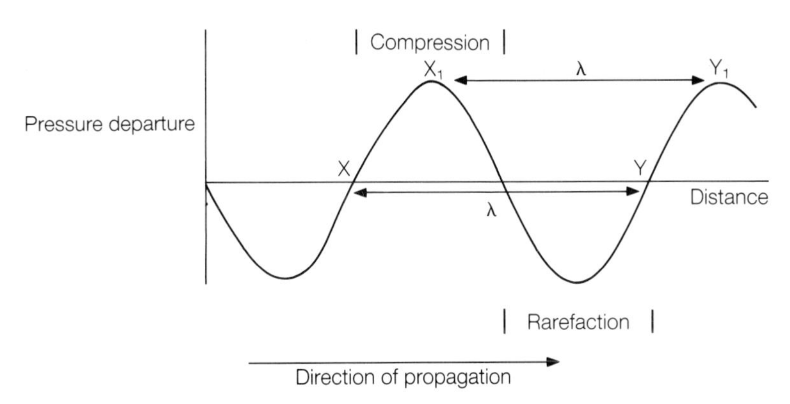

Fig. 12.2. Representation of a sound wave as a sinusoidal wave motion. The wave is propagating to the right at v m s^{-1}.

Ultrasound waves are distinguished on the basis of frequency because the frequency or frequency distribution in a sound wave stays constant as the wave passes from one medium to another while the velocity and wavelength change. Each medium is identified ultrasonically by its characteristic impedance Z where $Z = \varrho v$, ϱ being the density of the medium and v the velocity of sound in the medium. Some typical values for ϱ, v and Z are shown in Table 12.1. Note that since the speed of sound varies very little with frequency, it is not necessary to specify the frequency at which measurements are made.

Table 12.1. Values for the density, speed of sound and characteristic impedance for different materials

Material	Density ($kg\ m^{-3}$)	Speed of sound ($m\ s^{-1}$)	Characteristic impedance ($kg\ m^{-2}\ s^{-1}$)
Air	1.2	3.3×10^2	4×10^2
Water	1×10^3	1.48×10^3	1.48×10^6
Soft Tissue	1.1×10^3	1.54×10^3	1.63×10^6
Bone	1.9×10^3	4.08×10^3	7.8×10^6

12.3 Physical interactions of sound waves with matter

As the ultrasound beam passes through a homogeneous medium, some energy is lost from the beam and is converted to heat. Tissues in the body are not however homogeneous and in passing through a tissue further energy is lost by scatter (termed Rayleigh scatter) from small regions in the tissue with a different acoustic impedance from the remainder of the tissue. These discontinuities are less than a wavelength in size and are thus too small for specular reflection to take place. Losses due to scatter and heat production are collectively called attenuation. It is this diffuse reflection which gives different tissues their characteristic echo patterns.

The unit in which beam attenuation is expressed is the decibel (dB). Thus if the intensity incident on a medium is I_0 and the intensity emerging is I_1, the intensity of the emerging beam is said to be 10 $\log_{10} I_1/I_0$ dB lower than the intensity of the incident beam.

The decibel is not a unit of power or intensity but is simply a way of expressing a ratio. The use of a $\log_{10}$ ratio allows a very wide variation of intensities to be accommodated and also allows the total attenuation of several adjacent volumes to be calculated easily by simply adding their individual attenuations. Consider for example the situation shown in Fig. 12.3. If the intensity leaving the first slab is $I_0/2$, the attenuation across this slab is

$$10 \log_{10} \frac{I_0/2}{I_0} \text{ dB} = 10 \log_{10} 0.5 \text{ dB}$$

$$= 10 \times (-0.301) \text{ dB} = -3 \text{ dB or a 3 dB reduction.}$$

If we add a second, thicker, slab of the same material, across which the intensity drops from $I_0/2$ to $I_0/10$, then the attenuation across the second slab is

$$10 \log_{10} \frac{I_0/10}{I_0/2} \text{ dB} = 10 \log_{10} 0.2 \text{ dB} = -7 \text{ dB}$$

The attenuation from the front face of the composite slab to the back face is

$$10 \log_{10} \frac{I_0/10}{I_0} \text{ dB} = 10 \log_{10} 0.1 \text{ dB} = -10 \text{ dB}$$

This is the sum of the attenuations across the individual slabs.

The attenuation of a material is often quoted as dB per unit length of material and this is termed the attenuation coefficient α.

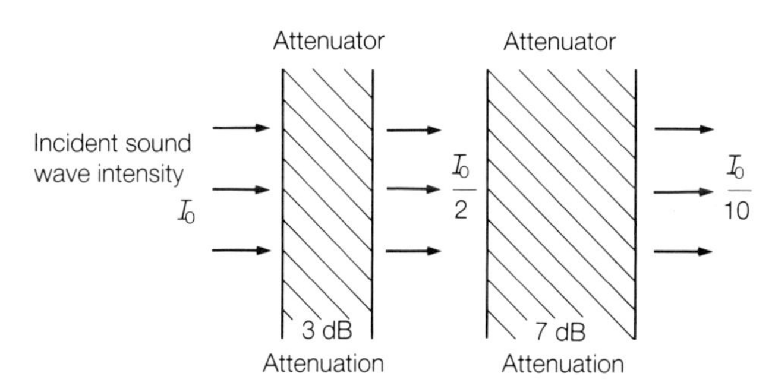

Fig. 12.3. Diagram showing that the effect of two attenuators is additive when expressed in decibels.

Unlike the speed of sound, which is frequency-independent, the attenuation coefficient is very dependent upon the frequency of the ultrasound wave. For many soft tissues, α at 1 MHz is quite close to 0.1 dB mm^{-1}, that is to say the intensity will fall to a half in about 3 cm of tissue. As the frequency f increases, α increases such that α/f remains approximately constant. For bone, the value of α at 1 MHz is much higher (approximately 1.3 dB mm^{-1}) and in the frequency range 1–2 MHz increases more nearly in proportion to f^2 than to f.

12.4 Interaction at material interfaces

12.4.1 Stationary boundary

If an ultrasound beam strikes a boundary between two media with different acoustic impedances, then the beam is split into three components as shown in Fig. 12.4. The effect is very similar to that of a light beam hitting a glass block and the various components are described as in optics.

Part of the beam is reflected according to the laws of reflection. The angle of incidence equals the angle of reflection whatever the nature of the materials forming the boundary. The transmitted beam is deviated (refracted) from the path of the incident beam either away from the normal when $v_1 < v_2$ or towards the normal when $v_1 > v_2$ according to Snell's law.

$$\frac{\text{The sine of the angle of incidence}}{\text{The sine of the angle of refraction}} = \frac{v_1}{v_2}$$

If $v_1 = v_2$ or if the beam strikes the boundary at right angles (regardless of the values of v_1 and v_2), no refraction takes place. In practice, because variations in the speed of sound in different solid materials are small, the beam is only deviated slightly from its original path. The transmitted ultrasound beam will pass through the second medium to interact at the next boundary.

The third component is termed scattered ultrasound and is produced at any angle in an unpredictable manner. It occurs because biological interfaces act to ultrasound as a scratched mirror does to light, i.e. as an imperfect reflector, and the reflected beam is reduced in intensity.

If scattering is ignored, the relationship between the incident beam intensity and the reflected beam intensity can be predicted. For simplicity, only ultrasound waves incident normally on a boundary will be considered. Since in most ultrasound investigations the same transducer is used to generate and detect the ultrasound beam, i.e. the incident and reflected beams must be normal or very nearly normal to the boundary, this simplification is not unreasonable. It will be further assumed that the interface is perfectly flat and that both media are homogeneous. In this case, when a

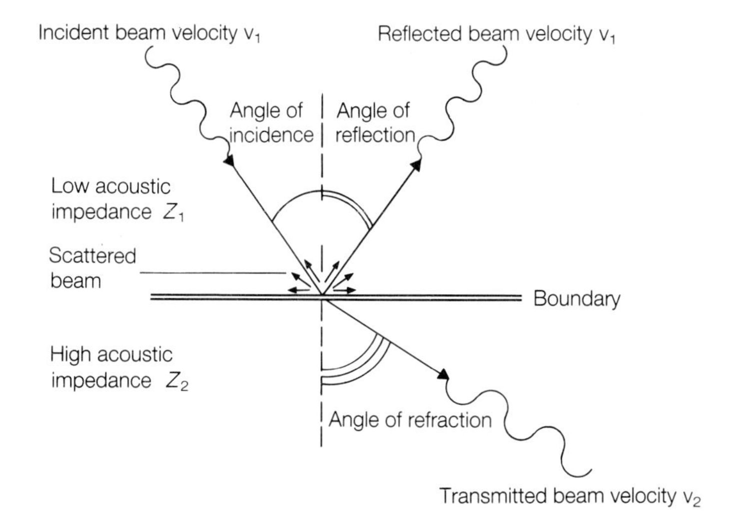

Fig. 12.4. Partial reflection, partial transmission and scatter when an ultrasound beam is incident at an angle on the boundary between two media of different characteristic impedance.

beam travelling in medium 1, impedance Z_1, is incident on a boundary with medium 2, impedance Z_2, the fraction of the ultrasound beam intensity reflected back along its incident path, R is given by

$$R = \left(\frac{Z_2 - Z_1}{Z_2 + Z_1}\right)^2$$

Assuming no losses, the transmitted intensity

$$T = (1 - R) = \frac{4Z_1Z_2}{(Z_2 + Z_1)^2}$$

If $Z_1 = Z_2$ there is no reflection. Furthermore, substitution of a few numbers quickly demonstrates that when Z_1 and Z_2 are very different, the amount of energy reflected is high, irrespective of whether Z_1 or Z_2 is the larger. Therefore big differences in impedance must be avoided if one wishes to transmit appreciable intensities of ultrasound into the soft tissues of the body.

12.4.2 Moving boundary

In establishing the relationships in the previous section, the boundary between the two media was assumed to be stationary. If, as in Fig. 12.5a

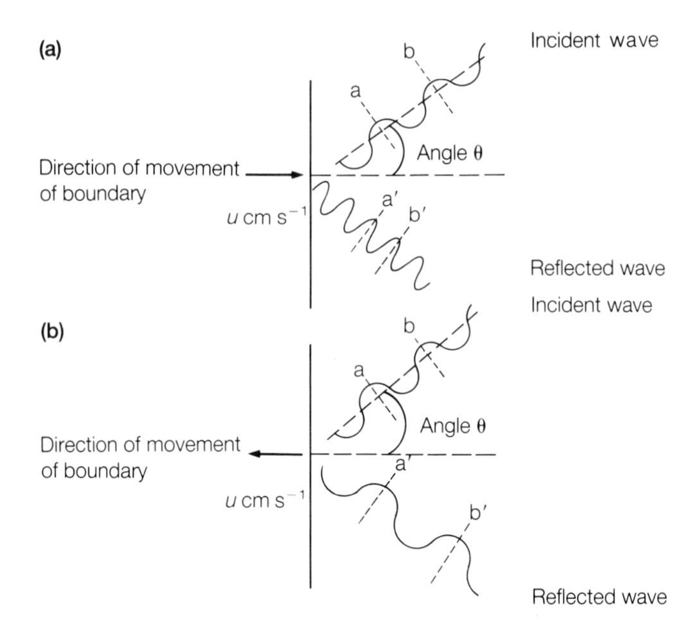

Fig. 12.5. Effect of a moving boundary on the wavelength of a reflected sound wave (the Doppler effect, (a) when the boundary is approaching the wave, the wavelength is shortened, (b) when the boundary is receding, the wavelength is longer.

the reflecting boundary is moving towards the incident wave, each part of the beam is reflected earlier than it would have been if the reflector had been stationary. The reflected beam still has the same form but the distance a′b′ over which one wave occurs has been reduced by twice the distance the reflecting surface moved forward during the time interval between points a and b hitting it, i.e. the wave length has been reduced. It will be recalled that the velocity of a sound wave is related to the frequency f and wavelength λ of the beam by $v = f\lambda$. If v remains constant, as it must since the incident and reflected beams are in the same medium, then

$$f_1\lambda_1 = f_2\lambda_2$$

where f_1, λ_1 are the frequency and wavelength before reflection, and f_2, λ_2 are the frequency and wavelength after reflection. If λ_2 is less than λ_1, then f_2 must be greater than f_1, i.e. the frequency of the reflected beam has risen.

When the reflecting boundary is moving away from the sound wave as in Fig. 12.5b, the opposite effect takes place. Here the wavelength of the reflected beam is greater than that of the incident beam. Following the same argument as before, the frequency of the reflected beam must, in this case, fall. This change in frequency f_D when a beam is reflected from a boundary moving with velocity u is called the Doppler effect. u is always much less than v.

If the angle between the direction of movement of the boundary and the ultrasound beam direction is θ, and u is much less than v

$$f_D = \frac{2f \cos \theta \ u}{v}$$

At normal incidence $f_D = 2fu/v$. The change in frequency is thus proportional to the velocity of the moving boundary and hence may be used to measure this velocity.

12.5 Generation of ultrasound and ultrasound transducers

A transducer is a device which, in general terms, converts a physical effect into a representative electrical signal and vice versa. An ultrasound transducer is a device that is capable both of converting an electrical signal into ultrasound waves and converting ultrasound waves back into electrical signals.

Ultrasound waves are produced by an electric signal which causes the dimensions of the transducer to change rapidly, an effect called the piezo electric effect. In piezo electric crystals, the electric charges bound within the lattice can be considered to be in the form of dipoles as shown in Fig. 12.6a. If an electric field is applied across the faces of the crystal as in Fig. 12.6b, the charges are repelled and the dipoles are compressed. This reduces the distance between the two faces of the crystal. If the polarity of the electric field is reversed as in Fig. 12.6c, the dipoles are expanded and

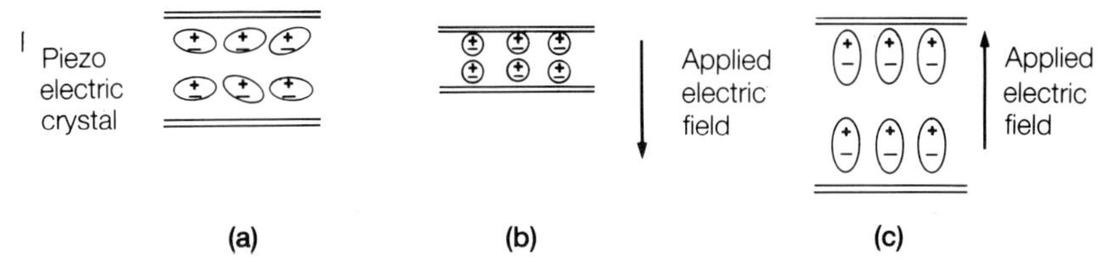

Fig. 12.6. Schematic representation of (a) the dipoles in a piezo electric crystal, (b) the compression effect of an applied electric field, (c) the expansion effect of a field applied in the opposite direction.

the distance between the two faces is greater. If the polarity of the electric field changes in a sinusoidal manner, then a similar motion is produced in the crystal faces. This is transmitted to the air in contact with the face of the crystal and hence an ultrasound wave is generated.

To understand the reverse effect, consider the situation shown in Fig. 12.7a where, in the normal state of the crystal, the charge distribution in the dipoles is such that all the charges cancel and there is no charge difference detectable across the faces of the crystal. If the crystal is mechanically deformed by compression as in Fig. 12.7b or by expansion as in Fig. 12.7c, the charge distribution within the crystal is altered. This distortion of the position of the dipoles induces a charge on the surface of the crystal which changes as the crystal is compressed or expanded. Although many natural crystals, such as quartz, exhibit the phenomenon of piezo electricity, the material used most commonly for ultrasound transducers is lead zirconate titanate (PZT), which is a synthetic ceramic. Lead zirconate titanate is a ferro-electric material, the crystals of which contain domains of atoms in which quite large dipole moments can be induced. During manufacture the lead zirconate titanate is heated above its Curie point to destroy any existing charge polarization. While the material is cooling, a strong electric field is applied across it. This causes the polarization of the charge in each domain to line up with the electric field and it remains in that direction when the field is removed after the material has cooled. The precise properties of the material can vary depending on the impurities present.

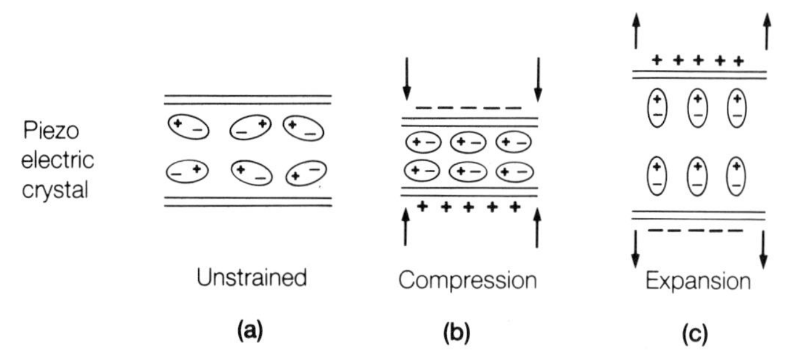

Fig. 12.7. Schematic representation of the dipoles in a piezo electric cyrstal (a) when unstrained, (b) when compressed, (c) when expanded.

Being a ceramic, the transducer can be moulded and fired into any shape and can also be polarized in any direction. Although many different shapes can be produced, the most common are either a disc with flat faces or part of the surface of a sphere. In both cases the thickness of PZT must be the same over the transducer surface. A thin layer of silver is applied to each of the faces so that electric connections can be made.

12.5.1 Probe construction

In the majority of ultrasound investigations the transducer used to transmit the sound waves is also used to detect them after reflection. This means that the ultrasound waves must be sent out in very short pulses with the transducer spending most of its time in the listening mode to detect the reflected waves between pulse transmission.

The transducer has a resonant frequency which is determined by its thickness. At this frequency the opposite sides of the crystal move in and out at the same time and a wave is produced of wavelength equal to twice the thickness of the transducer, e.g. assuming $v = 4000 \text{ m s}^{-1}$ then for a crystal 1 mm thick $\lambda = 2$ mm and $f = 2$ MHz.

As the faces of the transducer go in and out, they radiate energy into the media that are adjacent to them. A fraction of this energy, determined by the relative characteristic impedances of the media and the transducer, is reflected back into the transducer. At the resonant frequency, the reflected wave is exactly in phase with the piezo electric wave and reinforces it. The movement of the sides of the transducer is greatest at this frequency and thus the transducer produces the largest output signals when driven by reflected ultrasound waves at the resonant frequency.

12.5.2 Beam shape

Ultrasound of wavelength λ emitted from a flat transducer of radius r will travel in an approximately parallel beam for a distance d (Fig. 12.8a), where $d = r^2/\lambda$. This part of the beam is called the near field and is the part normally used for imaging. Beyond the near field the beam diverges at an angle θ where $\sin \theta = 1.22\lambda/2r$.

Shaping the transducer alters the profile of the ultrasound beam as shown in Fig. 12.8b. If the radius of curvature is d (where d is obtained as above) then the minimum beam width is at approximately $0.6d$. If the radius of curvature is much smaller, $d/4$ (Fig. 12.8c), then the minimum beam width is at $0.24d$. These curved probes are said to focus the ultrasound beam. Although the beam is now much narrower, which improves resolution (see section 12.9), the depth in tissue over which it can be

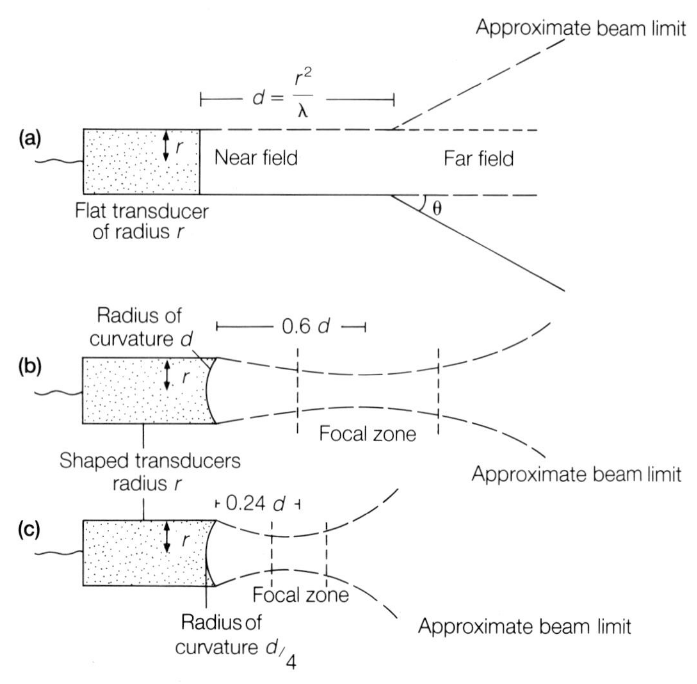

Fig. 12.8. Beam shapes produced by (a) a flat transducer of radius r, (b) a focused transducer with radius of curvature d, (c) a focused transducer with radius of curvature $d/4$.

usefully used is restricted to the focal zone. The greater the focusing, the shorter the focal zone.

12.5.3 Transducer arrays

The above transducers have fixed focal distances and if areas of interest in a patient are at substantially different depths then a change in probe may be necessary. Varying focus probes can be produced by using transducer arrays, which are essentially of two types. The first (Fig. 12.9a) operates by sub-dividing the circular transducer element into a number of concentric rings or annuli. Each annulus operates independently of the others. The initiating pulse is common to all annuli but undergoes different time delays in its transit to the annuli. By arranging these time delays carefully, the individual annuli can be made to transmit at slightly different times. The effect of this is to focus the beam. Normally the transmission focal distance is kept constant, usually at approximately half the penetration depth. By altering the time delays during the period the reflected pulses are received however, the focal spot can be effectively moved. It is set close to the

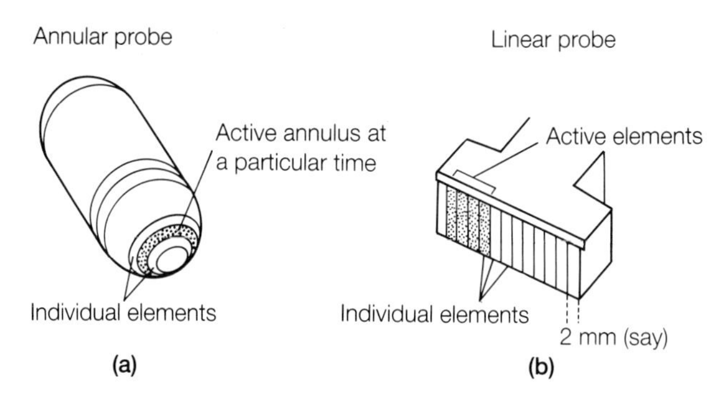

Fig. 12.9. Variable focus transducer probes produced by either (a) sub-dividing the circular transducer element into a number of concentric rings or annuli, or (b) using a linear array of small elements that can be fired individually or in groups.

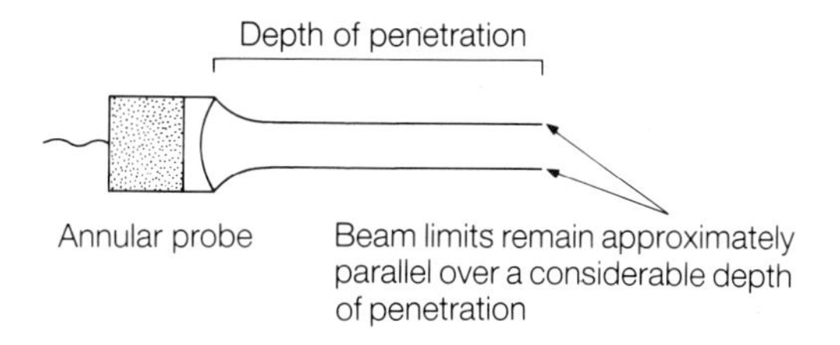

Fig. 12.10. Schematic representation of the beam profile that can be achieved with a circular transducer array.

transducer when the signals from surfaces close to the transducer are received, i.e. the first ones, and is then progressively moved away from the transducer face as pulses from greater depth return. The result is that the transducer is ideally focused over the full depth of penetration giving the effective beam shape in Fig. 12.10. The delay lines are then reset to the mid-way position for the next pulse.

The linear array shown in Fig. 12.9b consists of many very small rectangular elements typically 1–4 mm wide. The transducer is constructed either by assembling individual elements or by taking a slab and cutting grooves in it. The elements are normally fired in groups. If all are fired at the same time a plane beam is produced (Fig. 12.11a). However, if time delays are introduced, as in the annular probe, not only can the beam be focused (Fig. 12.11b) but also its direction can be changed (Fig. 12.11c, d). It is now possible to scan the beam across the patient by progressively moving the group of elements that is activated (Fig. 12.12). Additionally, if

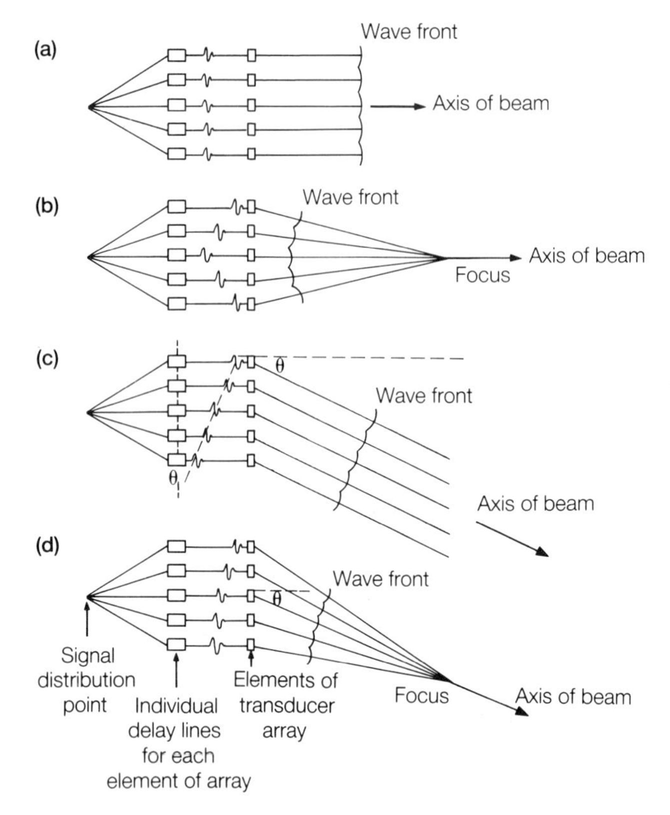

Fig. 12.11. Demonstration of electronic steering of the output from an array of transducers. (a) If all transducers are fired in phase (illustrated by the pulse positions) a plane wave normal to the array is generated. (b) If delays are introduced to simulate a curved transducer surface, the wave front is focused. (c) Delays introduced to simulate a transducer surface at angle θ. (d) Effects (b) and (c) combined to produce a focused beam at angle θ.

required, the beam direction can be changed between pulses by electronically changing the delay on the pulses so that the beam is made to sweep across the patient.

12.6 Pulse length

The length of pulse has a critical effect on axial resolution (resolution in the direction of the ultrasound beam). Figure 12.13 shows the effect if pulses are too long. The reflected beam from the first interface is still being detected by the transducer when the reflected beam from the second interface starts to arrive. The transducer cannot distinguish between the

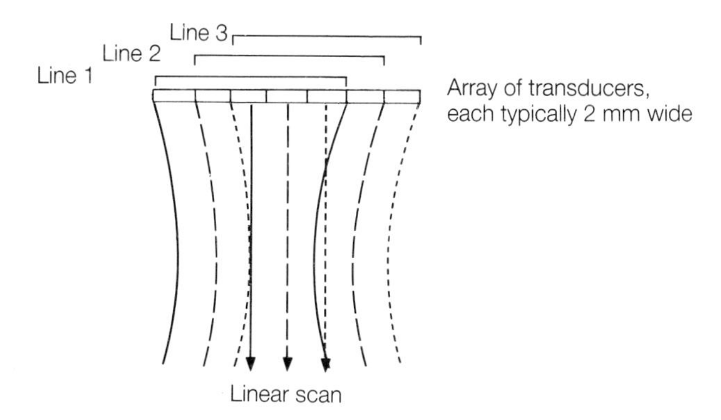

Fig. 12.12. Production of a linear scan by progressive activation of different groups of elements. *Solid lines*: beam profile when line 1 of transducers is activated. *Dashed lines*: beam profile when line 2 of transducers is activated. *Dotted lines*: beam profile when line 3 of transducers is activated.

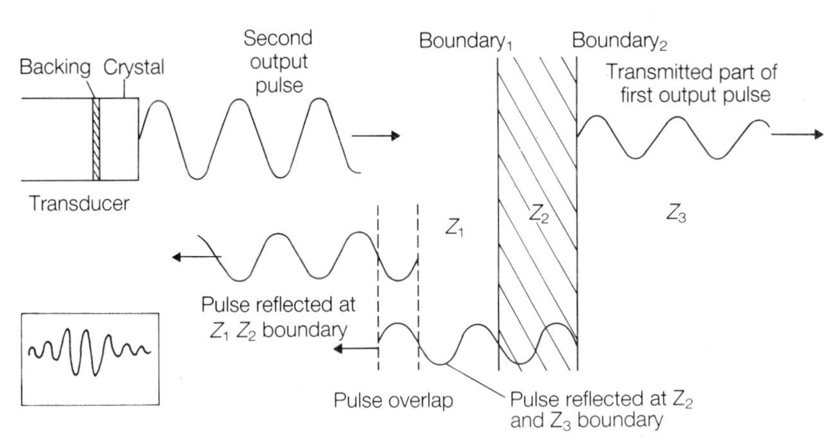

Fig. 12.13. Demonstration that axial resolution depends on pulse length. If the pulse is too long the reflected signal from the Z_2Z_3 boundary will overlap with the reflection from the Z_1Z_2 boundary and the boundaries will not be separately distinguishable. Note that the actual pulse shape will be more similar to that shown in the inset. It is shown here as a simple sine wave for clarity.

signals and thus cannot resolve the two interfaces. The shorter the transmitted pulse the better the axial resolution.

However, if a very short electrical pulse is applied across the faces of a free transducer, the transducer will continue to oscillate at its resonant frequency for some time after the electrical pulse has stopped, with the amplitude of the oscillation slowly decreasing (Fig. 12.14). This is called 'ringing' and is due to that part of the energy wave within the transducer

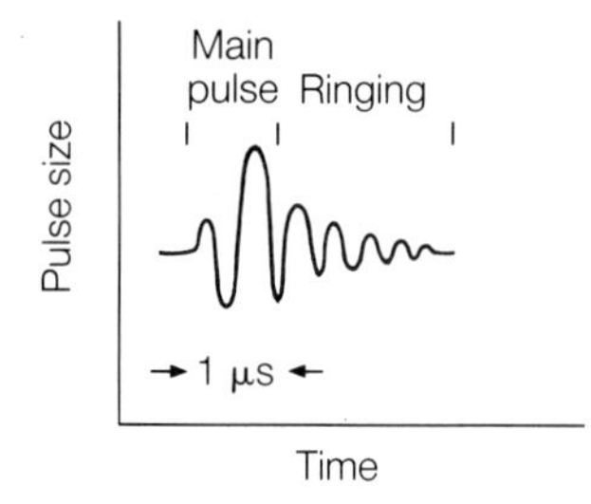

Fig. 12.14. 'Ringing' (the production of oscillations of decreasing amplitude) after the main pulse has been generated.

that is continually being reflected at the transducer faces. With air at the back of the transducer this ringing could continue for 20 or 30 wave cycles and would seriously limit axial resolution. This is because the characteristic impedance of air is several hundred times less than that of the PZT so the interface would reflect virtually all of the beam.

To eliminate ringing, the air on the back of the transducer must be replaced by a material of similar acoustic impedance to the transducer so that the energy passes rapidly from the transducer. This material, called the backing, must absorb the energy and not eventually reflect it back into the transducer. The most widely used backing material is epoxy resin loaded with tungsten powder. The tungsten helps to scatter the beam and also raises the density and acoustic impedance of the epoxy resin to values close to those of the transducer. With a backing of this kind, the pulse amplitude can be reduced to less than one third in two wave cycles. With backing, the system no longer vibrates at its resonant frequency and the very rapid reduction in pulse amplitude widens the frequency spectrum in the pulse as shown in Fig. 12.15.

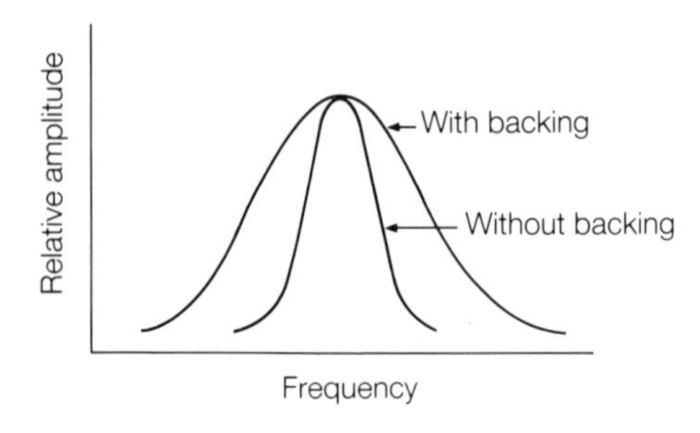

Fig. 12.15. The frequency spectrum in an ultrasound pulse either without backing or with backing to provide effective damping of the pulse.

12.7 Impedance matching

The face of the transducer used to transmit and receive ultrasound must be coupled to the body in such a way as to achieve maximum transmission.

The characteristic impedances of the three materials involved, air, tissue and PZT, were given in Table 12.1. Because the impedance of air is very different from that of the other two materials, even a thin layer of air between transducer and tissue will result in very little of the ultrasound beam being passed into the tissue and all the energy will be reflected back into the transducer. A coupling oil or gel, with a suitable characteristic impedance therefore has to be used to ensure that there are no air gaps between the transducer and tissue.

Transmission is further improved by inserting between the face of the transducer and the tissue an intermediate layer of material with a characteristic impedance approximately mid-way between tissue and PZT. The ideal value of Z_{layer} is $(Z_{\text{PZT}} \times Z_{\text{tissue}})^{\frac{1}{2}}$. This material must be $\lambda/4$ thick to ensure maximum transmission as shown in Fig. 12.16. This matching layer will also act as a filter because the short pulse of ultrasound has a wide frequency spectrum and the $\lambda/4$ thickness is only strictly optimized for one of those frequencies. The efficiency of transmission falls off as the frequency moves away from this particular value.

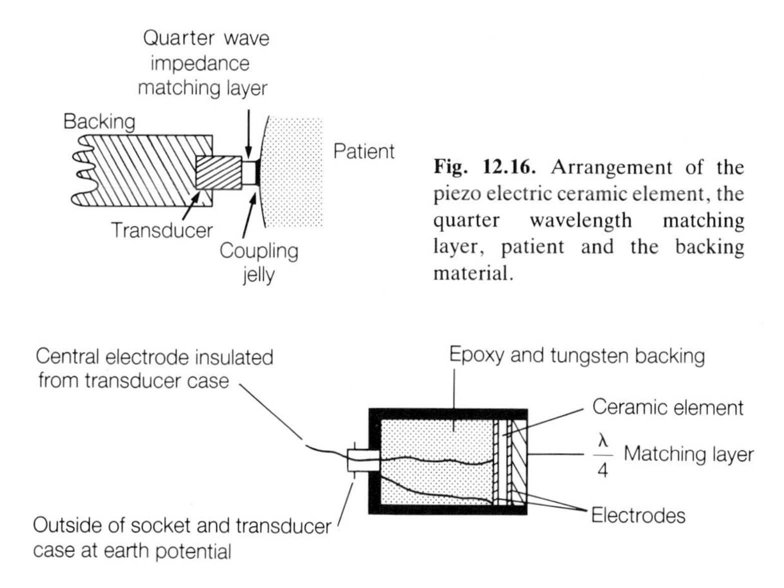

Fig. 12.16. Arrangement of the piezo electric ceramic element, the quarter wavelength matching layer, patient and the backing material.

Fig. 12.17. Simplified cross-section through a transducer probe.

In addition to the acoustical properties of the matching layer, consideration must also be given to its mechanical strength, its resistance to solvents and its ease of manufacture. Also the transducer, with its associated backing and coupling frontal layers, has to be incorporated into a holder before it can be used. The transducer must be mechanically insulated from this holder, otherwise mechanical 'ringing' induced in the holder will lengthen the pulse. A simplified cross-section through a probe is shown in Fig. 12.17. Note that for electrical safety both the front face and the case of the transducer must be at earth potential. This is necessary as the voltage pulse supplied is several hundred volts.

12.8 Size of pulse echo

As discussed in sections 12.3 and 12.4, when an ultrasound beam passes through tissue it is attenuated by about 1 dB per MHz of frequency every centimetre. This results in the signals from deep within the body being reduced by much more than would be expected from the reflection coefficients. In addition to the echo pulse there is also background noise which appears as small random signals from the transducer as shown in Fig. 12.18a. Because the intensity of the reflection is used to indicate the characteristics of an interface, this reduction in intensity due to attenuation makes interpretation very difficult. Furthermore, simple amplification of the smaller signals would make the more intense signals so strong they would saturate the electronics and become indistinguishable. To overcome this problem, a varying gain is applied to the incoming signals. If the tissues through which the beam is passing are acoustically fairly uniform, the gain is increased uniformly with time (Fig. 12.18b) so that the strong reflected signals arising from interfaces close to the surface are amplified by a small

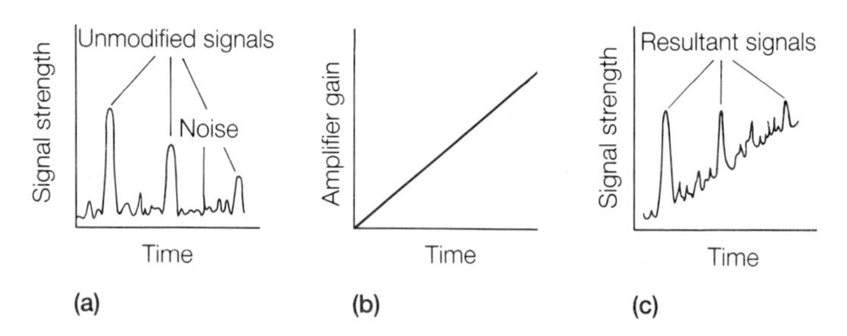

Fig. 12.18. Recorded signals from interfaces in the presence of noise and heavy attenuation. (a) Unmodified signals from identical boundaries at different depths. (b) Linear amplification with time of the returning signal strengths. (c) In the resultant signals, reflections from similar boundaries have similar amplitudes irrespective of depth.

amount, and the signals from distant surfaces are amplified by a large amount. This is termed swept gain and the result is shown in Fig. 12.18c. Other gain versus time functions may be necessary if the ultrasound passes through regions in the body that are acoustically very different. If the swept gain is correctly set, signals from interfaces with the same reflectivity will have the same strength.

12.9 Beam penetration and resolution

At high frequencies the ultrasound beam is absorbed very rapidly and frequencies of around 10 MHz can only be used for superficial investigations such as ophthalmic examinations. At 5 MHz the useful penetration depth is approximately 8 cm, and at 2.5 MHz it is approximately 15 cm. The design of the probe also has a significant effect on the useful penetration. As designs improve, particularly for array probes (section 12.5.3), the useful depth of the probe at a particular frequency is increasing.

Resolution of the ultrasound beam must be considered in two directions, axial resolution along the direction of the beam and lateral resolution at right angles to the beam. As discussed in section 12.6, the length of the ultrasound pulse limits axial resolution with as short a pulse as possible being required. Increasing the frequency of the pulse also improves the resolution. The improvement in transducer design, referred to above, allowing greater useful penetration at higher frequencies, is making significant contributions to improvement in axial resolution. Although it is difficult to give precise information on axial resolution attainable, note that for a 1 MHz generator emitting a pulse 1 μs long, the pulse length in water will be about 1.5 mm (one wavelength) and the theoretical axial resolution 3 mm.

Lateral resolution is limited by the beam width at the point of reflection. For single transducers this is a compromise between the ultrasound frequency and the size of the transducer. From section 12.5.2, increasing the diameter of the probe will increase the length of the near field for a flat transducer and increase the length of the focal zone for a focused transducer. Resolution will thus be improved over a greater depth. However the maximum resolving power of the system will have been reduced because the minimum beam width has been increased. The converse is true as the transducer size is reduced. Focusing obviously improves the lateral resolution. For many applications, lateral resolution using the new transducer designs may approach, within a factor of perhaps 3–5, the theoretical limit

Table 12.2. Typical quoted resolutions for modern diagnostic imaging ultrasound probes

Use	Frequency (MHz)	Wavelength in water/tissue (mm)	Lateral resolution (mm)	Axial resolution (mm)
General purpose	3.5	0.4	2.0	0.4
Cardiac	5.0	0.3	1.0	0.3
Linear array	5.0	0.3	1.2	0.4
Neonatal	7.5	0.2	0.75	0.2

set by the wavelength of the ultrasound. Some quoted resolution values are shown in Table 12.2.

Increasing the frequency increases the depth of the near field and focal zone, although until the recent advances in transducer design this option had only limited application. At higher frequencies the physical limit on depth penetration must also be considered.

12.10 A scans (amplitude scans)

The velocity of sound in all tissues is approximately constant. Thus it is possible to establish a relationship between the depth of an interface beneath an ultrasound transducer and the length of time it takes for a pulse to travel to the interface and back. In an **A** scan, the magnitude of the reflected beam is used to drive the *y* axis of an oscilloscope as shown in Fig. 12.19. When a return pulse is detected, a peak is produced on the screen.

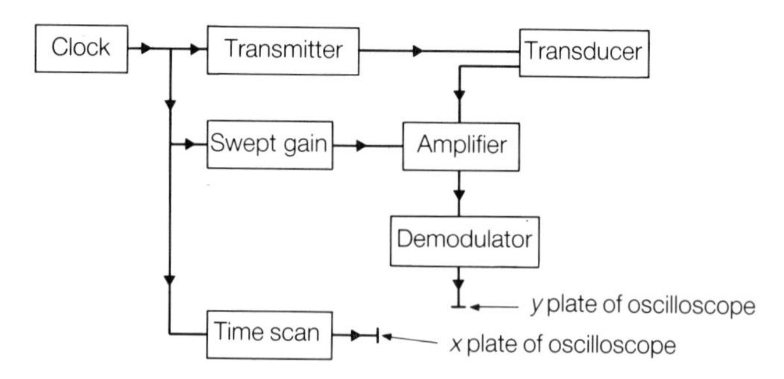

Fig. 12.19. Block diagram showing how in an **A** scan the reflected beam is used to drive the *y* plate of an oscilloscope.

The rate of movement of the spot across the x axis on the oscilloscope is adjusted by the time-base so that by measuring the distance between peaks, and assuming a value for the speed of sound in tissue, the distance between the reflecting interfaces in the tissue can be calculated.

The transducer is held in a fixed position and measurements are made along one line only. In spite of the use of swept gain, the range of sizes of the reflected pulses is so large that a cathode ray oscilloscope cannot display them in a meaningful manner. Thus the pulses are generally electronically processed before being displayed.

Sufficient pulses must be transmitted for the image to remain constant on the screen, the actual rate depending on the time base of the oscilloscope. Care must be taken not to employ too high a pulse rate otherwise interference between the emitted and reflected pulses will take place.

A scans are easy to understand but are not normally used in modern diagnostic ultrasound machines because they contain only limited information.

12.11 B scans (brightness modulated scans)

In a **B** scan, the amplitude of the reflected signal is displayed on the cathode ray screen as a spot of variable brightness. A large signal produces a bright spot, a small signal a dim spot. The transducer is moved round the circumference of the plane in the patient from which information is required and the angle and position of the probe are recorded. This is undertaken either by using an electronic detecting system or mechanically, using a system of pulleys and potentiometers.

Each scan line is then displayed on the cathode ray screen in the correct geometrical location relative to other scan lines. By taking successive scan lines as the transducer is moved to different positions on the patient or tilted at different angles, a two-dimensional image of the plane is built up. If all mechanical alignments are adjusted correctly, the reflection from a surface at one position of the transducer will register in exactly the same position on the screen irrespective of the position from which it is detected (see Fig. 12.20).

The two-dimensional picture was originally recorded using a camera with an open shutter, effectively using the film to store each scan line and integrate the total. This technique has now been replaced by electronic methods which allow the image to be viewed on a cathode ray tube as it is built up.

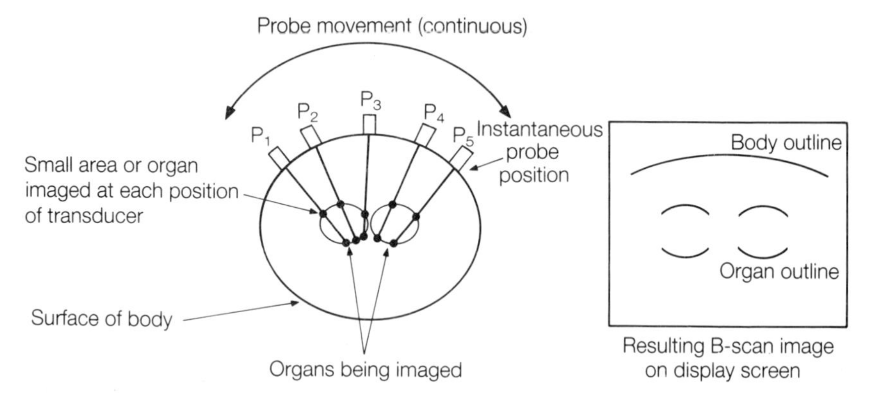

Fig. 12.20. Schematic representation of a **B** scan. (a) The organ is imaged from many different directions using a pencil beam of ultrasound. (b) With correct registration the reflections form the reflecting surface.

12.12 M mode (motion mode scanning)

This technique can show movement of structures in relation to time and is used primarily in cardiac work. Having established the area of interest, a single line of a **B** scan is used, directed as closely as possible at right angles to the moving interfaces. The light spots, of different brightness according to the strength of the reflected pulse, are generally imaged onto light-sensitive paper. This paper is moved at a constant speed and as the position of the reflecting interface changes, a trace of the movement is produced. As the velocity of the paper is known, the time interval between different parts of the trace can be obtained. It is also possible to superimpose on the trace other physiologial measurements such as an ECG or pressure measurements.

12.13 Real-time scanning

In **A**, **B** and **M** mode scanning, the image is produced by a single ultrasound transducer or an array of transducers electronically constrained to act as a single transducer. The image gives no indication of movement except in the **M** scan where information on movement is obtained by the external addition of a time-base to a **B** scan image.

In real-time scanning it is possible to study movement of organs and vessels on a brightness modulated scan (**B** scan) by scanning the ultrasound pulses rapidly across the section of the body under investigation. This is executed in one of two ways (Fig. 12.21). The first uses a linear array of transducers. Groups of transducers are activated electronically in turn so that the pulses are scanned across the section (see section 12.5.3). These pulses can produce either a rectangular shaped field with the linear array along the top edge, or a triangular shaped field with the array as the apex. The second method uses a rotating head with a number of transducers on it. As each transducer is scanned through the section, a triangular shaped field is produced, again with the transducer at the apex. When the next transducer scans across the field, the picture is updated with any interfaces that have moved since the last scan being shown in their new position. The ultrasound pulses are scanned across the plane so quickly that a flicker-free image is produced which is updated so rapidly that movements can be measured.

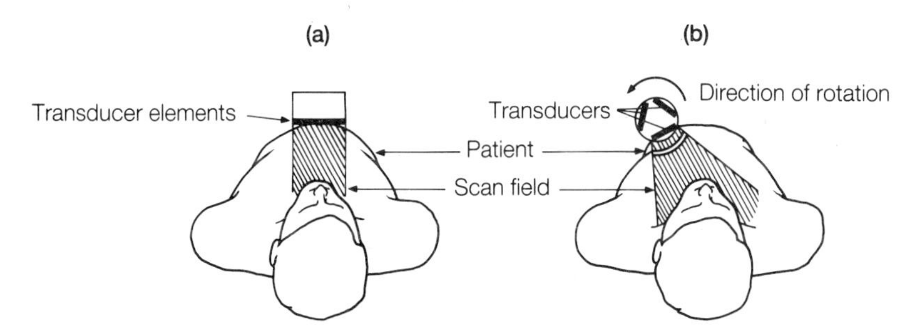

Fig. 12.21. The scan field produced with two different types of real-time scanning. (a) Linear array producing a rectangular-shaped field. (b) A rotating head produces a triangular-shaped field.

The introduction of linear array probes and rotating probes was not in itself sufficient to produce real-time images. This was also dependent on the utilization of scan converters which are described in section 12.15. However, the quality of pictures on modern real-time scanners is now so good that information other than just boundary interfaces can be observed. The amount and quality of isotropic scatter referred to in section 12.3 varies from one tissue type to another and this information can be used by an experienced operator to help diagnosis.

12.14 Doppler imaging

The rate of flow of blood along a vessel or the rate of opening or closing of a valve in the heart can often provide highly significant diagnostic information. In section 12.4 the effect of a moving surface on the frequency of a reflected ultrasound beam was described. This change in frequency may be used to measure the rate at which the reflecting surface is moving. The first Doppler systems used a continuous wave ultrasound beam which is shown diagramatically in Fig. 12.22. The probe consists of two transducers, one for transmitting the other for receiving. As the name implies, the probe is continuously emitting an ultrasound beam so the signal detected by the receiving transducer is a mixture of the pure ultrasound beam frequency reflected from stationary interfaces and the Doppler shifted frequencies produced by moving interfaces. This gives a beat frequency equal to the difference between the pure and shifted frequencies (see Fig. 12.23). The Doppler shifted frequencies are extracted electronically from the signal.

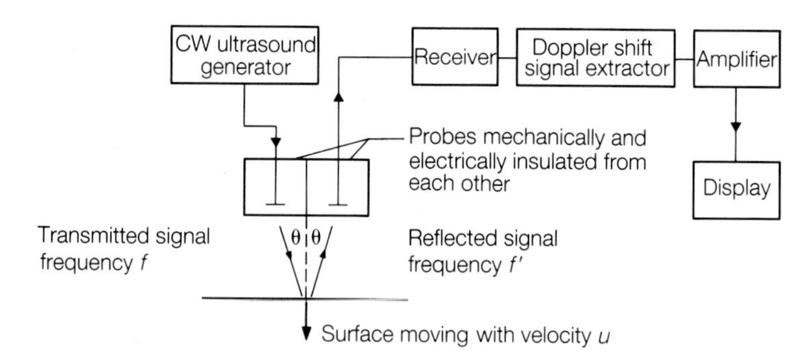

Fig. 12.22. Use of continuous wave (CW) ultrasound to record Doppler-shifted frequencies.

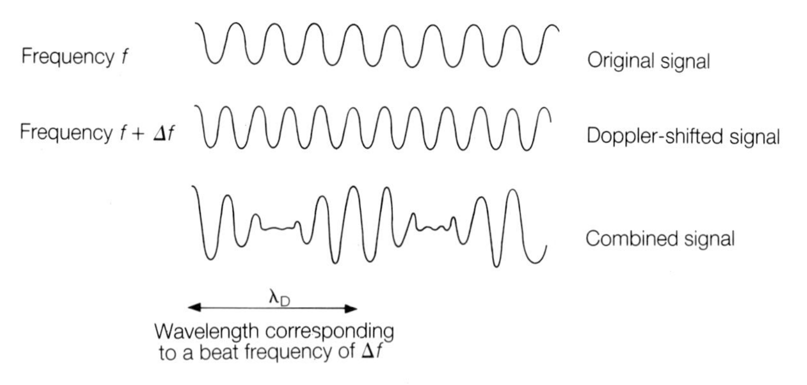

Fig. 12.23. The original signal and Doppler-shifted signal shown separately and combined. The combined signal is not to the same scale as the other two. In practice the value of Δf would be very small and the wavelength between beats correspondingly long.

Many real time scanners now include the facility to carry out 'pulsed Doppler' measurements with the same probe that is used to produce the real-time scan. If a very short pulse (about 1 μs) is transmitted, the receiver may be set only to register Doppler shifted signals for a short period at a specific time. These signals are associated with tissue at a very specific depth. Since both the conventional scan and the Doppler information are displayed on the CRT at the same time, it is possible to see exactly which surface is giving the Doppler signals and this allows very precise measurements to be taken.

Longer pulses are used for the Doppler measurements than for imaging but because the beam is still pulsed, electronic extraction of information on the change of frequency is more complicated than in the continuous wave measurements. Note that by recording the two sets of scan data simultaneously, it is not necessary to know the velocity of sound in tissue very precisely (since the same value is assumed for each observation). However, it is necessary to ensure that the transmitted pulses are very evenly spaced in time, usually at about 100 μs intervals.

12.15 Display equipment

The return echoes obtained from a **B** scan, as well as giving positional information, give information on the strength of the echo and hence on the type of tissue reflecting the beam. The early displays did not utilize this information.

In order to gain information on the strength of the reflected signals, it is necessary to be able to show a range of light densities. Thus, for example, it has been found that, generally, one shade of grey can be distinguished from another when it is twice as bright. This is described as one shade of grey, and is the basis of the Grey Scale used in many display systems. Adopting this definition, approximately 10 shades of grey can be visualized on a standard monitor. Some equipment utilizes colour bands but this display method has not competed as strongly with grey scales in ultrasound as it has in nuclear medicine.

When one grey shade is twice as bright as another, there is a 3 dB signal change, so 10 shades give a monitor a 30 dB range. However, the intensities of the reflected pulses received by the transducer can cover a range of 100 dB. To accommodate this range on the monitor, it is either compressed to 30 dB by selectively amplifying some signals and not others, or all the signals above or below preselected upper and lower levels are shown as white or black respectively.

The signals from the transducer are converted into a signal suitable for a monitor by either a scan converter tube or a digital scan converter. The scan converter tube is shown in simplified form in Fig. 12.24. It consists of an evacuated tube assembly similar to a cathode ray tube. Instead of having a fluorescent screen, the face of the tube is a matrix of approximately 1000 × 1000 insulated elements. These elements can store electrical charge. As the signals are received from the transducer, the electron gun deposits charge on the matrix elements that represent the line along which the ultrasound scan is being made. The electron beam is moved by the plates and the amount of charge deposited is proportional to the signal from the transducer. The **B** scan is thus created, not as a brightness distribution, but as a charge distribution across the matrix. This process is called the 'write process'. The charge distribution is converted to a TV monitor picture by using a second electron beam to 'read' the charge distribution. This beam scans the matrix in synchrony with the raster of the TV monitor. Where there is charge deposited on a matrix element, it is repelled and a current flows in the grid. This current signal is used to drive the electron gun of the television monitor to produce the scan picture. The charge pattern is stored until updated following another sweep of the transducer. Alternatively it can be frozen to enable measurements to be made once the required image is obtained.

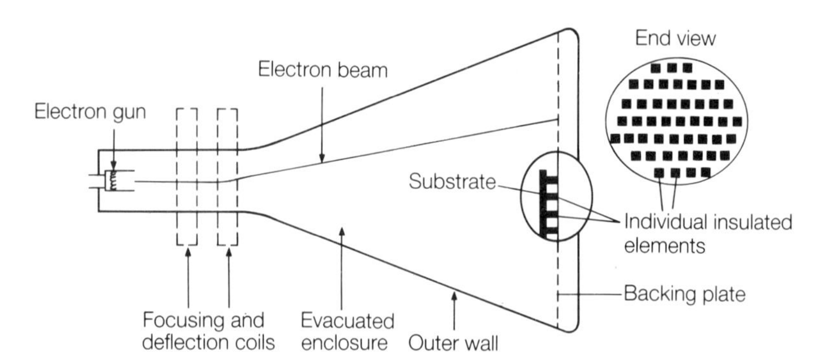

Fig. 12.24. Essential components of an analogue scan converter tube. The inset shows an end-on view of the array of elements.

In modern units the scan converter tube has been almost completely replaced by a digital system in which each matrix element has been replaced by a location in a microcomputer memory. The magnitude of the reflected signal is converted into a digital signal which represents a shade of grey on the grey scale and is stored at its correct location until updated. The stored digital signals are read electronically and used to drive the electron gun of a TV monitor.

12.16 Artefacts

Artefacts are apparently real reflections which do not in fact correspond to tissue boundaries in the plane of the scan. Some common causes of artefacts will now be mentioned.

Multiple reflections

The pulse which returns to the transducer from a boundary is not completely absorbed within the probe because there is a characteristic impedance discontinuity between the transducer and the tissue surface. The portion not absorbed is reflected back into the tissue and behaves like a transmitted pulse, delayed in time. The beam can thus oscillate back and forth between the transducer and the boundary with reduced intensity each time it is reflected until it becomes too small to be measured (Fig. 12.25).

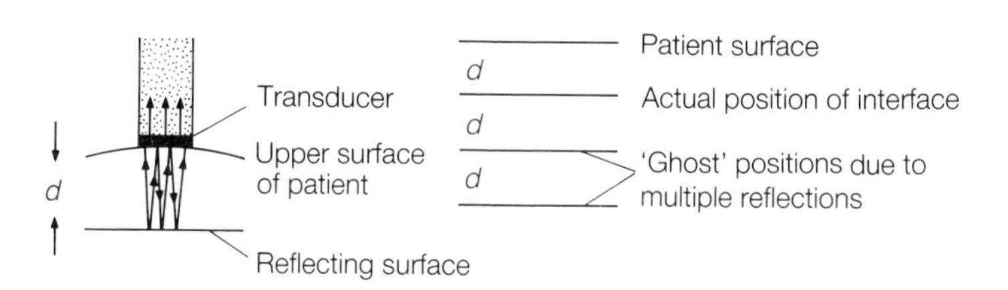

Fig. 12.25. The creation of artefactual boundaries as a result of multiple reflections. Each time the ultrasound pulse reaches the tissue-transducer boundary part of it is reflected back into the tissue. The multiple reflections generate the same effect as weakly reflecting boundaries at $2d$, $3d$, ...

The registration electronics assumes that each return pulse has come from an interface a further distance d into the tissue. This produces a series of boundary lines each d apart, gradually getting weaker as they apparently progress into the tissue. This artefact is normally only a problem with a strongly reflecting boundary near to the surface, but it can also occur within large vessels such as the aorta.

Double reflection

The ultrasound beam may be reflected by a boundary and then reflected back along the same path by a second object. This second object then appears as a virtual image on the far side of the boundary (Fig. 12.26).

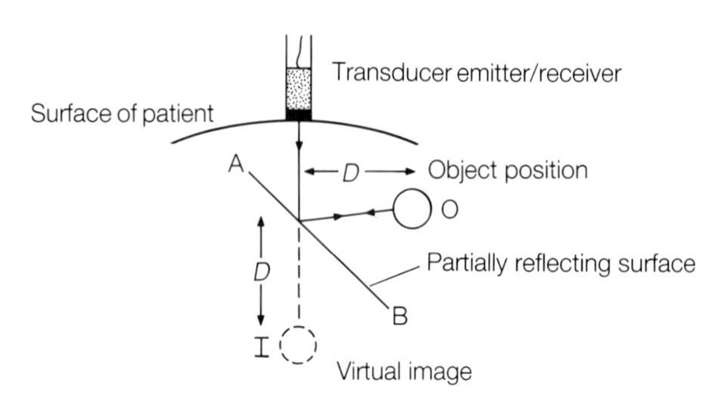

Fig. 12.26. Creation of an artefact by double reflection. An impedance mismatch along the line AB will create a partially reflecting surface. As a result of double reflection an object at O will appear to be at I.

Errors in registration

In a **B** scan the speed of sound through different tissues is assumed to be the same. If the same vessel, when viewed from different angles, is effectively covered by different tissue, then this assumption can produce images of the vessel in slightly different positions (Fig. 12.27). This artefact is not apparent in real-time imaging.

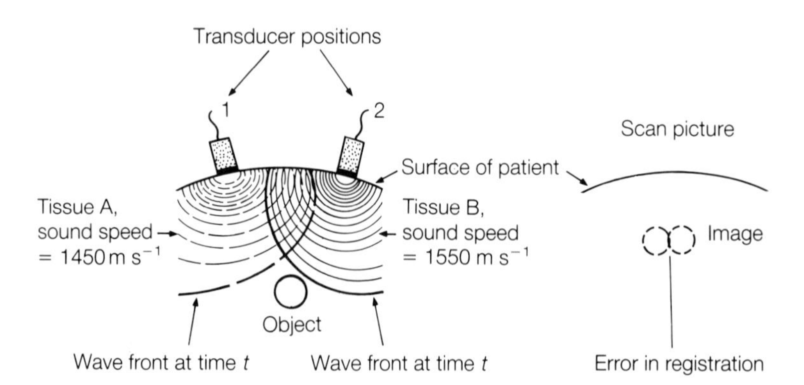

Fig. 12.27. Schematic representation of error of registration. In this example the error arises from assuming, incorrectly, that the speed of sound in overlying tissue is the same when the transducer is in position (1) as when it is in position (2). In fact the wave front will be delayed from position (1) relative to position (2) and reflection will take place later. The final outcome is that different 'views' of the object do not superimpose.

Shadowing
A structure such as bone reflects virtually all the ultrasound beam and this precludes investigation of the area behind it unless the transducer can be moved to another position from where the sound wave can avoid the bone.

Refraction
Because the ultrasound beam changes direction slightly at the interface between two tissues of different sound velocity, the position of the reflected beam can be displaced from its real position.

Experience, and a good knowledge of anatomy, allow most artefacts to be ignored without difficulty.

12.17 Biological effects of ultrasound

The possible harmful biological effects of ultrasound fall into three major categories, heat absorption, microstreaming and cavitation.
1 *Tissue heating* as a result of the absorption of ultrasound energy is the major cause of damage in cancer ablation techniques. However in diagnostic investigations it will not be a source of hazard because the average ultrasonic power will only be about 0.1 mW mm^{-2} and the heat generated will be dissipated, mainly by conduction and convection, including blood flow, with a scarcely perceptible temperature rise.
2 *Streaming* is the movement of particles or liquid molecules under a sonic torque. The particles move along the axis of the beam away from the transducer and return along a path towards the perimeter of the field. This effect would appear to affect cell membrane permeability and is proportional to the intensity of the ultrasound beam.
3 *Cavitation* is the interaction of the sound wave with microbubbles present in the liquid or liquid-like medium. A continuous sound wave at low intensity causes the bubble to oscillate stably but as the sound wave intensity is increased an unstable state is reached in which the bubble increases to several times its original value and then contracts violently. During the collapse, which occurs so quickly that it is adiabatic, the bubble temperature may rise to several thousand degrees. Visible light is emitted and water vapour in the bubble is dissociated into free radicals. Intermittent ultrasonic radiation is less conducive to cavitation induction and it is possible that exposures of the type used for diagnostic imaging, e.g. 1 μs

pulses every 1 ms, cannot cause unstable cavitation at practically attainable energy levels.

4 *Miscellaneous effects* include structural changes due to the direct mechanical effects of acceleration, pressure or shear stress on the membranes of cells causing them to vibrate and change their permeability, and agglomeration of particles in a fluid at pressure peaks in the ultrasound wave, an effect that has been seen to cause blood cell stasis.

The possibility of harmful effects occurring must be a function of intensity levels of sound being used. For typical values see Table 12.3. Note that great care is required in the interpretation of quoted ultrasound intensities. First, the peak spatial intensity must be distinguished from the average intensity across the whole sound field. The latter will be at least three times less and with focused beams used in some high resolution diagnostic equipment may be 100 times less than the former. Secondly the time averaged value of the intensity will be less than the peak temporal

Table 12.3. Typical approximate intensities (spatial peak-time averaged) for ultrasound in medicine

Use	Intensity	Mode
Diagnostic	0.01–1 mW mm^{-2}	Pulsed
	0.01–0.3 mW mm^{-2}	Doppler continuous wave
Physiotherapy	5–20 mW mm^{-2}	Continuous wave
Surgical	0.5–200 W mm^{-2}	Focused multiple beams

value by the ratio of off/on times which may be as high as 1000 for pulse echo equipment. Thus, in an extreme case, the spatial peak – temporal peak value could be 10^5 times greater than the spatial average – temporal average value.

On the basis of an extensive review of the reported biological effects of ultrasound on both cells and organisms, the American Institute of Ultrasound in Medicine has concluded that

> In the low megahertz frequency range there have been, as of this date, no independently confirmed significant biological effects in mammalian tissues exposed to intensities (spatial peak temporal average as measured in a free field in water) below 1 mW mm^{-2}. Furthermore, for ultrasonic exposure times (total time; this includes off-time as well

> as on-time for a repeated pulse regime) less than 500 s and more than 1 s, such effects have not been demonstrated at higher intensities when the product of intensity and exposure time is less than 0.5 J mm^{-2}.

However, largely on the basis of reports of unstable cavitation in ultrasound beams, the National Council on Radiation Protection and Measurements in the United States (1983) has recommended that

> Research should be carried out to investigate the possibility that biologically significant cavitation or bubble activity occurs in human tissue under conditions of diagnostic and therapeutic medical ultrasound. Such studies should include (a) physical studies of the response of bubbles (cavitation 'nuclei') to short, high intensity repeated pulses characteristic of pulse echo techniques; (b) investigations of the potential of mammalian tissue for containing bubbles or nuclei and (c) the biological significance of any bubble activity produced by medical ultrasound in tissue.

12.18 Summary

Ultrasound differs from X and gamma rays in two important ways. First, it is not an ionizing radiation so it does not carry the risks associated with such radiations. Secondly, it is propagated as a result of longitudinal compressions and rarefactions of the medium through which it travels. Hence the behaviour of the sound wave provides information about the properties of the medium through which it is travelling.

Ultrasound is a wave motion and exhibits many of the properties of waves such as reflection and diffraction. A property analogous to the refractive index of a medium for light waves is the characteristic impedance of a medium. This is equal to the product of density and sound wave velocity in the medium and determines the fraction of the incident power in a sound wave that will be reflected at the boundary. The waves are heavily attenuated in tissue, attenuation being expressed in decibels. In soft tissues the attenuation coefficient increases with frequency and this factor limits high frequency investigations to relatively superficial structures.

Transducers for the generation of ultrasound are constructed from piezo electric crystals, frequently of lead zirconate titanate. Important features of the construction include the creation of a large dipole moment to facilitate conversion of electrical energy into sound, the appropriate thickness to ensure resonance at the required frequency, and, when used to generate pulses, appropriate backing to prevent multiple reflections. The same crystal normally acts as transducer and receiver. Careful impedance matching is required to achieve maximum energy transfer to the patient.

The design of the tranducer and the frequency selected determine both the axial and lateral resolution of the system according to the laws of diffraction. With the relatively recent development of transducer arrays, lateral resolution may approach, within a factor of 3–5, the theoretical limit imposed by the ultrasound wavelength.

The older amplitude modulated scans (**A** scans) and brightness modulated scans (**B** scans) have now been largely replaced by real-time scanning using either a linear array of transducers or several transducers on a rotating head. Movement of organs may be studied and image quality is now so good that information may be obtained from the ultrasound scattered within tissues as well as from the ultrasound reflected at boundaries.

The use of the Doppler effect to study movement and flow in the body now finds quite widespread application and when Doppler techniques are used in conjunction with real-time imaging very precise information is available on the origin of the Doppler signal.

Ultrasound does not carry the risk of ionizing radiation and to date there is no confirmed evidence of damage to patient or foetus from diagnostic ultrasound. However, harmful biological effects are known to occur at higher sound intensities and until the results of current research programmes are known, some attempt should be made to measure output and to ensure that ultrasound intensities are well below maximum recommended levels.

References and further reading

Atkinson P. & Woodcock J. P. (1982) *Doppler Ultrasound and its Use in Clinical Measurement*. Academic Press, London.

Hill C. R. & ter Haar G. (1982) Ultrasound. In *Non Ionizing Radiation Protection*, ed. M. J. Suess, pp. 199–228. World Heatlh Organization.

Lunt R. M. (1978) *Handbook of Ultrasonic B-scanning in Medicine*. Cambridge University Press.

McDicken W. N. (1976) *Diagnostic Ultrasonics—Principles and Use of Instruments*. Crosby Lockwood Staples, St Albans.

National Council on Radiation Protection and Measurement (1983) *Biological Effects of Ultrasound: Mechanisms and Clinical Implications*. (Report 74) NCRP, Bethesda, Maryland.

Shirley I. M., Blackwell R. J., Cusick G., Farman D. J. & Vicary F. R. (1978) *A User's Guide to Diagnostic Ultrasound*. Pitman, Tunbridge Wells.
Wells P. N. T. ed. (1977) *Ultrasonics in Clinical Diagnosis*, 2nd edn. Churchill Livingstone, Edinburgh.
Wells P. N. T. (1977) *Biomedical Ultrasonics*. Academic Press, London.
Wells P. N. T. (1982) Ultrasonic imaging. In *Scientific Basis of Medical Imaging*, ed. P. N. T. Wells, pp. 138–193. Churchill Livingstone, Edinburgh.

Exercises

1 Why is the low megahertz range used for ultrasonic imaging? Give, with reasons, the transducer frequencies and focal distances you would choose to make ultrasound examinations of
(a) an eye,
(b) a liver.

2 What factors determine the reflection coefficients of an ultrasound pulse at tissue interface? What is meant by specular and diffuse reflection?

3 A source of pulsed ultrasound and a target are separated by tissue with ultrasonic properties characteristic of normal human soft tissues. Discuss the effect of each of the following on the amplitude of the ultrasonic pulse reflected back to the source:
(a) the size and shape of the target,
(b) the densities of the target substance and intervening tissue,
(c) the speed of sound in the target substance and intervening tissue,
(d) the distance between source and target,
(e) the frequency of the ultrasound.

4 Find the position of the last axial maximum in a continuous wave ultrasound beam from a vibrating circular piston transducer operating at 2 MHz. The diameter of the transducer is 20 mm and the velocity of sound 1500 m s^{-1}. Calculate also the angle of divergence in the far field.

5 Describe the construction of an ultrasonic probe for diagnostic imaging, explaining how the various components affect the performance of the transducer and the quality of the image. What factors limit the resolution of the probe?

6 Explain briefly how a focused beam of ultrasound can be obtained from the excitation of a number of transducer elements in a linear array. What are the advantages over a linear pulse?

7 List the features seen in an ultrasound **B** scan image that could be used to distinguish between a solid tumour and a cyst.

8 Describe the operation of the swept gain control on an ultrasound

B scanning machine. Show two different forms of swcpt gain and suggest examinations for which each might be appropriate.

9 Discuss the use of **M** (or time-position) scans and real-time ultrasound images in the investigation of heart disease. Indicate the advantages and disadvantages of each.

10 Outline the main features of a real time ultrasound scanning machine and describe different modes of transducer operation that can be used to produce an ultrasound image in real time. Discuss the advantages of real time systems in different areas of clinical application.

11 Explain the Doppler effect and outline the principles of the continuous wave Doppler ultrasound technique for obtaining information on blood flow velocities. Describe the main features of a Doppler wave form depicting blood flow in an artery, and discuss how changes in these features are used to obtain information on the state of the artery.

12 Discuss the development of pulsed Doppler ultrasound devices and show how they can extend the application of the Doppler technique in the diagnosis of arterial disease.

13 Discuss the advantages of a grey scale system for displaying ultrasound images, and indicate the sort of diagnostic information that can be obtained from them.

14 Explain how artefacts may arise in ultrasonic images. How are the artefacts due to multiple echoes avoided in practice?

13

Magnetic Resonance Imaging

13.1 Introduction

The possibility of exploiting the magnetic properties of materials to explore the chemical structure of matter has been recognized and used for many years in magnetic resonance (MR) spectroscopy but extension of such ideas to *in vivo* imaging is a relatively recent development.

The basis of both spectroscopy and imaging is that spinning charges in the atom, i.e. protons and, somewhat surprisingly, neutrons in the nucleus, or outer orbital electrons, convey to the atom and to any material containing the atom certain magnetic properties. In general, particles are paired in such a way that their spins cancel and materials which have an even number of particles exhibit very weak magnetic properties. Ferromagnetics are a special class of materials in which the atoms contain several unpaired particles. Whole domains of atoms align in the same direction and these materials have permanent magnetization.

The atoms of greatest interest for MR imaging are of low atomic number, i.e. they are biologically important, and have unpaired spinning particles in the nucleus. Each nucleus behaves as a small magnet and in a static magnetic field each magnet takes up one of a number of preferred orientations such that the material as a whole has a resultant magnetization, or magnetic moment, along the line of the applied field. Under carefully defined conditions the nuclei can draw energy from an applied magnetic field. The magnetic effects associated with nuclear particles are very much smaller than those associated with spinning or orbiting electrons, but unlike the latter effects which die away very rapidly once the applied field is removed, they persist long enough for measurements to be made.

The amount of energy taken from the field is a measure of the number of nuclei present and the rate at which the nuclei lose that energy gives information about their environment. Such measurements form the basis of MR imaging, and although a full mathematical treatment is beyond the scope of this book, by considering the basic principles it should be possible to understand the origins of the measured signals and the mode of operation of simple imaging devices.

13.2 Basic principles of electromagnetism

Two fundamental principles of electromagnetism may be used as a starting point. First, when a charge moves it creates a magnetic field around it (Fig.

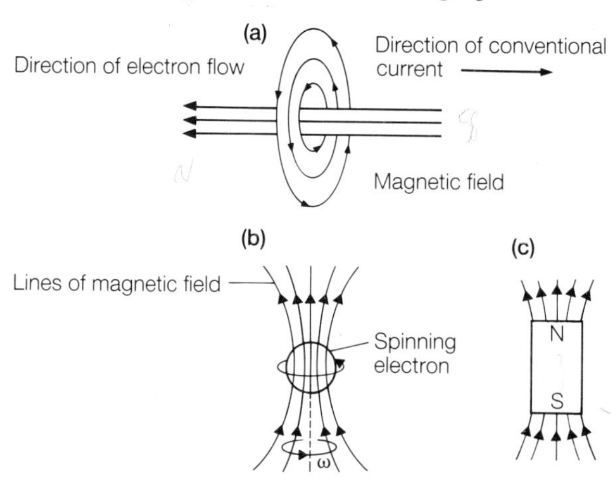

Fig. 13.1. Simple diagrams showing that moving charges and spinning charges have magnetic fields associated with them. (a) There are circular lines of magnetic field round a wire carrying current. (b) and (c) continue as text. The spinning electron behaves like a small bar magnet of magnetic moment ω.

13.1). For example, the movement of electrons along a wire in an electric motor creates a magnetic field which turns the motor. Also shown in Fig. 13.1 is the effect of a rotating or spinning charge. Both these motions can be regarded as creating a small bar magnet with a magnetic moment that can be represented by the vector μ which is aligned along the axis of spin. (Recall that a vector quantity is one which has both magnitude and direction.)

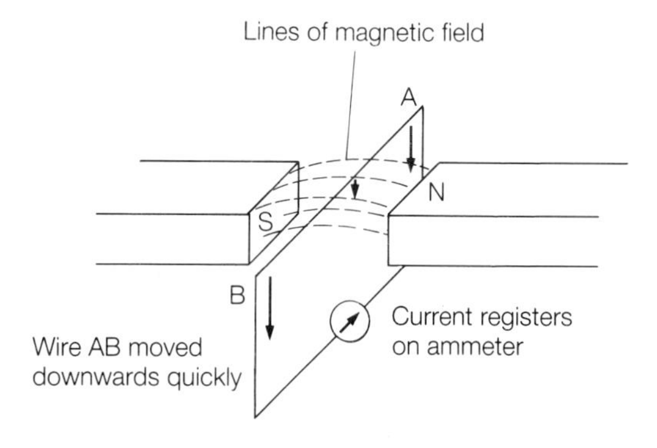

Fig. 13.2. When a conductor moves through a magnetic field, an emf is induced across its ends. If the wire is part of a circuit containing an ammeter, a current will flow through the ammeter.

Second, if a conductor moves through a magnetic field as in Fig. 13.2, then an electromotive force (emf) is produced across its ends. The converse also produces an emf, i.e. a stationary conductor in a moving magnetic field. These emfs may be readily detected by the potential difference they create across a measuring device.

13.3 Magnetic fields around atomic particles

All atomic particles spin and have associated with them a spin quantum number which is either zero, an integer or a half integer. Electrons have spin $\frac{1}{2}$ and an associated magnetic moment, and this is utilized in the more widely known electron spin resonance investigation techniques. Protons have spin $\frac{1}{2}$ and a corresponding magnetic moment. Neutrons also have a magnetic moment associated with them. Although neutrons have a net zero charge, they appear to behave as if positive and negative charges were distributed within them in an uneven manner. This creates a magnetic moment because the charges are rotating at different rates and magnetic moments are determined by electrical currents or moving charges rather than by the charges themselves.

Table 13.1. Nuclear properties of some elements of potential interest to MR imaging

Nucleus	Spin	Elemental abundance in man (% wt)	Relative atomic abundance in man	Gyromagnetic ratio (rad s^{-1} T^{-1})
^{1}H	½	10.0	1.0	2.68×10^{8}
^{12}C	0	22.9	0.19	—
^{14}N	1	2.6	0.018	1.93×10^{7}
^{16}O	0	61.4	0.39	—
^{19}F	½	4×10^{-3}	2×10^{-5}	2.52×10^{8}
^{31}P	½	1.1	3.6×10^{-3}	1.08×10^{8}

MR imaging is essentially concerned with the magnetic moments associated with spinning nuclear particles, but if an even number of protons and an even number of neutrons is present then, although the magnetic moments align with each other, the spins pair in opposite directions and the magnetic fields cancel. MR investigations can thus only be carried out on nuclei with an odd number of protons or for example in the case of carbon-13, an odd number of neutrons. The nuclear properties of various potentially interesting elements are given in Table 13.1. As can be seen, of the elements that are most abundant in man, H-1, C-12 and O-16, only hydrogen can be used for MR studies. The other three elements in this table with non-zero nuclear spin will be mentioned again later.

13.4 Effect of an external magnetic field

When a nucleus spins, the strength of the equivalent bar magnet is related to the spin property by the gyromagnetic ratio γ. Thus, in terms of the angular momentum $\mathbf{p}$ the magnetic moment $\mu = \gamma\mathbf{p}$. Since the angular momentum is related to spin $\mathbf{I}$ by $\mathbf{p} = \mathrm{h}\mathbf{I}/2\pi$, where h is the Planck constant,

$$\mu = \frac{\gamma \mathrm{h}\mathbf{I}}{2\pi} \qquad \text{(equation 13.1)}$$

When placed in an external magnetic field, the magnetic vector μ can only take well-defined orientations relative to the field. For elements of spin $\frac{1}{2}$, there are two possible orientations as shown in Fig. 13.3a. These two directions are in fact predicted by quantum theory and the resultant magnetic vector $\mathbf{M}_0$ cannot line up in any other position unless influenced by a second external magnetic field. If coordinate axes are arbitrarily chosen so that the magnetic field is parallel to the z axis and has an associated magnetic flux density $\mathbf{B}_0$ say, the vector μ can be considered to have two components, a component μ_0 which is parallel to $\mathbf{B}_0$ and a component in the xy plane. In equilibrium nuclear magnets point in many directions as shown in Fig. 13.3b and 13.3c and the resultant in the xy plane is zero.

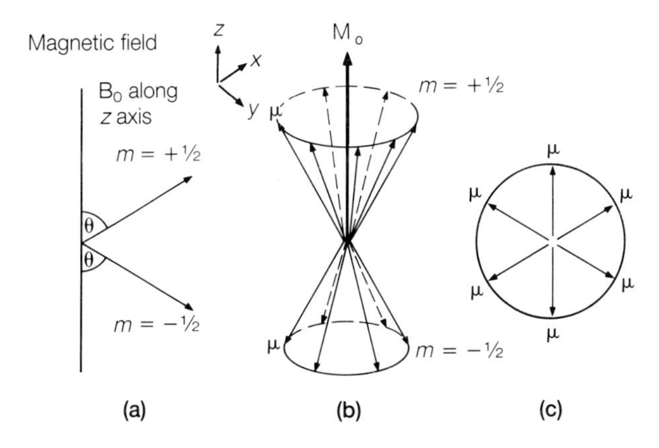

Fig. 13.3. Orientations of individual proton magnetic moments in a uniform magnetic field. In a uniform magnetic field nuclei with spin $\frac{1}{2}$ can only take one of two fixed orientations such that their magnetic moments are inclined at $\pm\theta$ to the field (Fig. 13.3a). All possible orientations of the magnetic moments describe two cones as shown in Fig. 13.3b. Projected into the xy plane, the μ vectors are randomly orientated with no net resultant (Fig. 13.3c).

Concentrating for the moment on the μ_0 components, those nuclei for which μ_0 is antiparallel to $\mathbf{B}_0$ have more energy than those for which μ_0 is parallel to $\mathbf{B}_0$. The difference in energy between the two positions is given by

$$\Delta E = 2\mu \cdot \mathbf{B}_0$$

(In a non-vector notation $\Delta E = 2\mu_0 B_0 = 2\mu B \cos\theta$.) Nuclei can be stimulated to transfer from one energy state to the other but since the energy levels are quantized, that is to say precisely defined, only stimuli that provide exactly the right amount of energy cause transitions. Since, also by quantum theory $E = hf$, ΔE is associated with a characteristic frequency given by

$$f = \Delta E/h = \gamma B_0/2\pi \text{ (since } I = \tfrac{1}{2} \text{ for protons)}$$

Thus an alternating magnetic field of frequency $\gamma B_0/2\pi$, or angular frequency

$$\omega_0 = \gamma B_0 \qquad \text{(equation 13.2)}$$

will stimulate changes, but slightly different frequencies will not. For protons in an external magnetic field of flux density $B_0 = 0.1$ tesla, ΔE is about 10^{-8} eV and f is about 4 MHz, i.e. in the radiowave frequency band of the electromagnetic spectrum. Note that in the remainder of this chapter the symbol B and the unit tesla (T) will always be used to represent the flux density associated with a magnetic field although this may not always be spelled out.

13.5 Net magnetic moment in a magnetic field

Although it has been indicated that the μ_0 component of the individual magnetic vector associated with a given nucleus can be aligned in one of two ways, no indication has been given of the relative number pointing in each direction when many nuclei are subject to an external magnetic field. In fact there is an almost equal division between the number pointing parallel to the magnetic field and the number pointing anti-parallel. For every million hydrogen nuclei, on average only 1.4 more point with the field than against it at body temperatures. Nevertheless, for a sample containing 10^{24} protons, a very typical figure, more than 10^{18} protons contribute to the MR signal.

The number of active nuclei N_A is given by

$$N_A = \frac{Nh}{2\pi} \cdot \frac{\omega}{2kT}$$

where N is the total number of spins, T the temperature in K, and k the Boltzmann constant.

In the macroscopic situation, such as exists in a patient, the resultant magnetic vector of all the spins must be considered. For reasons explained in Fig. 13.3, in equilibrium this resultant is parallel to $\mathbf{B}_0$ and may be denoted by $\mathbf{M}_0$. Hence, substituting from equations 13.1 and 13.2 and setting $I = \frac{1}{2}$ for protons,

$$\mathbf{M}_0 = N_A \cdot \mu \simeq N\mu^2 \cdot \frac{\mathbf{B}_0}{kT}$$

The strength of MR signals will depend on the magnitude of $\mathbf{M}_0$ and to maximize $\mathbf{M}_0$ requires:

1 low T, but for *in vivo* work this is fixed;
2 large μ and hence large γ;
3 large N, hence the advantage of working with an element that is abundant in the body;
4 large B_0.

Referring back to Table 13.1, H-1 is clearly the most promising nucleus for reasons of spin, sensitivity and abundance. Note that fluorine-19 and phosphorus-31 have low relative atomic abundance in man. Nitrogen-14 is reasonably abundant but has a low gyromagnetic ratio.

Increasing B_0 improves the signal to noise ratio and this advantage outweighs other considerations, other than cost, up to flux densities of about 0.5 T for which superconducting magnets are required. However, at even higher values of B_0 other factors have to be considered. Eventually, a point is reached where, because the attenuation of radio waves in the body increases with frequency, the system becomes less efficient with increasing B_0 for imaging protons. Higher fields are necessary for *in vivo* spectroscopy of other elements such as phosphorus-31. The highest field strength at which top quality, high resolution proton imaging will be possible remains to be determined.

13.6 Precession

Up to this point, the alignment of magnetic vectors, relative to the external magnetic field, has been described as if it were a stable, fixed position. In fact the magnetic vectors do not remain fixed in space. Because the protons are spinning and have angular momentum, the individual magnetic vectors behave like toy spinning tops in the Earth's gravitational field. They wobble, or precess, around the axis of the main field (Fig. 13.4). The frequency at which each one precesses is predicted by Larmor's theorem which states that, ω_L, the angular frequency of precession is related to the external magnetic field and the gyromagnetic ratio by

$$\omega_L = \gamma \mathbf{B}_0$$

Note that this is the same frequency as that which stimulated transitions between the energy levels so $\omega_L = \omega_0$. Vectors are used here to show that both rotations must be in the plane to which B_0 is the normal.

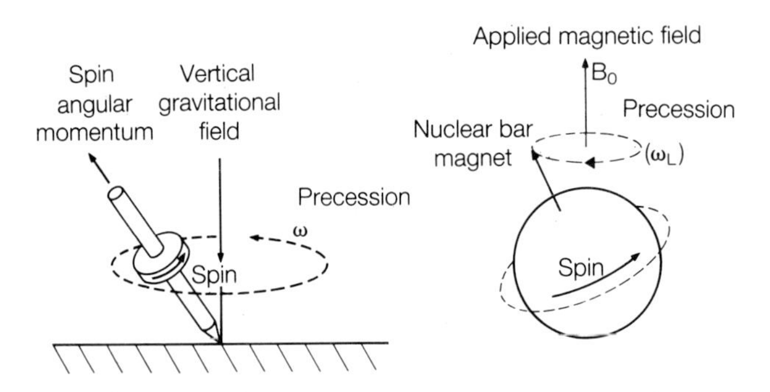

Fig. 13.4 The analogy between a magnetic nucleus precessing in a magnetic field (shown on the right) and a spinning top (shown on the left).

13.7 Production of MR signals

13.7.1 Measurement of proton density

At equilibrium, the resultant magnetic vector $\mathbf{M}_0$ is parallel to $\mathbf{B}_0$. Before $\mathbf{M}_0$ can be used to supply information, it must be moved away from equilibrium, thus acquiring energy. This can be achieved by applying another field $\mathbf{B}_1$, perpendicular to $\mathbf{B}_0$. Since $\mathbf{M}$ is precessing round the line of the fixed magnetic field, $\mathbf{B}_1$ will only apply a steady force to $\mathbf{M}_0$ if it also rotates at angular frequency ω_L. As already stated, ω_L corresponds to a radiofrequency wave of a few MHz.

Now $\mathbf{M}$ not only precesses about $\mathbf{B}_0$ but also precesses about $\mathbf{B}_1$ so that the resultant magnetic vector traces out a spiral motion (Fig. 13.5).

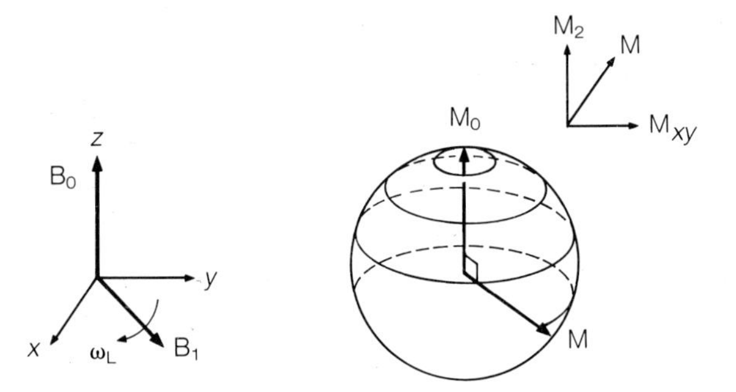

Fig. 13.5. The effect of applying a field $\mathbf{B}_1$ rotating at frequency ω_L on the net magnetic vector is shown. On the right **M** tilts away from its equilibrium position and precesses about the z axis. At any instant **M** is the resultant of two components M_z and M_{xy}. When **M** has tilted through 90° it only has a component in the xy plane.

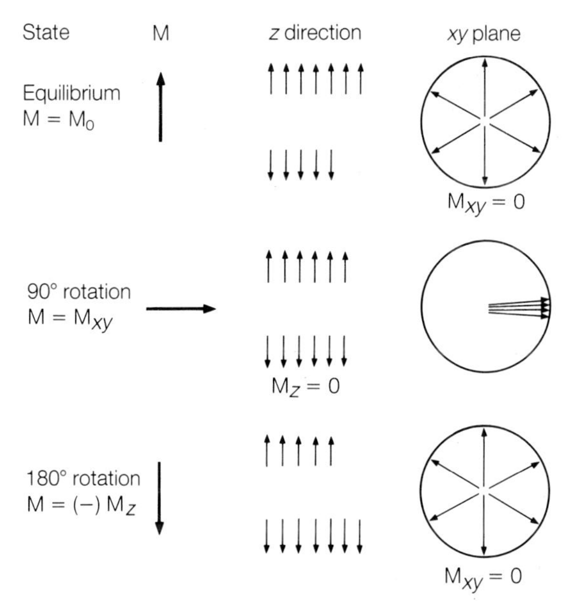

Fig. 13.6. Orientation of individual magnets in the z direction and in the xy plane for different states of the net magnetic vector. $\mathbf{B}_0$ is vertically upwards (positive z direction) throughout.

A simplified explanation of this behaviour in terms of the individual magnets is shown in Fig. 13.6. Application of $\mathbf{B}_1$ at the characteristic angular frequency ω_L (= ω_0) causes some nuclei to acquire energy and change to the antiparallel direction, thereby making M_z smaller or even negative. Note that M_z, the z component of a non-equilibrium value of **M**, must be distinguished from $\mathbf{M}_o$, the equilibrium value of **M** which also acts along the z axis. The radiofrequency pulse also causes some magnets to rotate a little faster and some a little slower. The magnets are said to have been 'phased', either partially if there is a preferred orientation, or completely if they all point in the same direction. Their directions in the xy plane are no longer random and **M** acquires a component M_{xy}.

When **M** is not parallel to $\mathbf{B}_0$, it cuts the lines of force associated with $\mathbf{B}_0$ during rotation and if a coil is wrapped round the body, an emf is induced in it. If the $\mathbf{B}_1$ field is now switched off the emf decreases sinusoidally as **M** both rotates around $\mathbf{B}_0$ and gradually returns to the vertical, and this process is known as free induction decay. This emf will cause a current to flow which can be amplified and used to create a signal (Fig. 13.7). The amplitude of the emf is proportional to **M** and if values of $\mathbf{B}_0$ and ω_L have been chosen such that $\omega_L = \gamma \mathbf{B}_0$ for hydrogen, the emf is directly proportional to the number of protons that have been excited and gives a measure of the number of protons in the volume interrogated. It is maximum when **M** has rotated through 90°.

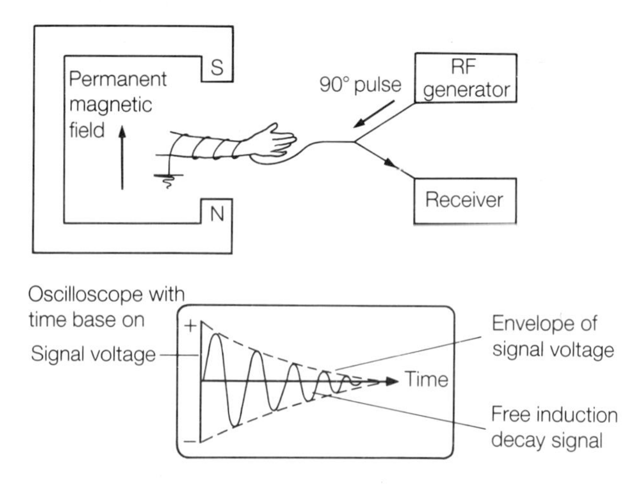

Fig. 13.7. Arrangement of a coil round the specimen or patient and the resulting free induction decay signal.

Note that further rotation of **M** results in a decrease of signal not an increase. When **M** has rotated through 180°, the emf is again zero even though this is not an equilibrium position.

13.7.2 T_1 and T_2 time constants

The net magnetic vector **M** is the resultant of two components M_z and M_{xy}, which are frequently called the longitudinal and transverse magnetization vectors. When the oscillating field $\mathbf{B}_1$ is removed, M_z and M_{xy} change in an exponential way with respect to time such that **M** returns gradually to its equilibrium value. However M_z and M_{xy} do not change at the same rate and are characterized by two time constants, T_1 the longitudinal magnetic relaxation time constant and T_2 the transverse magnetic relaxation time constant respectively. T_1 is always longer than T_2 in biological materials.

T_1 is governed by the rate at which energy can be transferred from spinning protons to the rest of the lattice thereby allowing the protons to 'flip back' into the parallel orientation. It is sometimes called the spin-lattice relaxation time. There is a strong inverse correlation between the degree of structure in the lattice and the T_1 value since the greater the degree of structure, the more readily energy is transferred and the shorter T_1.

T_2 depends on how long the protons in the sample can be held precessing in phase after removal of the radio frequency pulse. At the end of a 90° pulse they are all in phase but when the pulse is removed, they rapidly lose coherence (Fig. 13.8). This is partly due to spin–spin interactions and T_2 is called the spin–spin relaxation time that one wishes to measure. However, other factors, notably local variations in magnetic field, also contribute to loss of spin coherence, the observed rate constant being denoted by $T_2{}^*$.

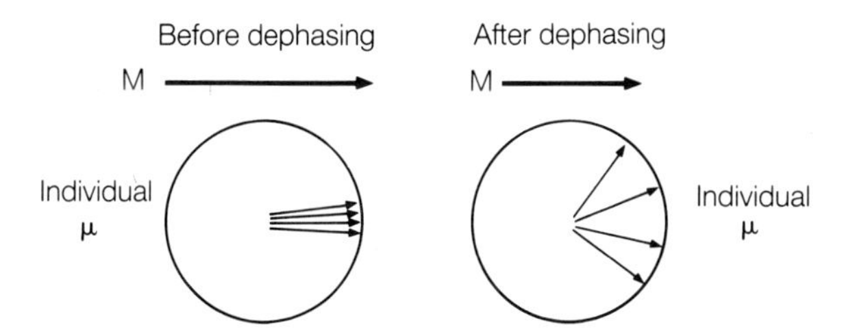

Fig. 13.8. Orientation of magnetic vectors in the *xy* plane before and after dephasing as a result of spin–spin relaxation and field inhomogeneities. Note that the resultant **M** continues to act in the *xy* plane but is smaller in magnitude after dephasing.

Note that since T_1, T_2 and $T_2{}^*$ are rate constants describing exponential decays, the mathematics of section 1.6 can be applied. For example the *xy* component of **M** will decay at a rate given by

$$M_{xy}(t) = M_{xy}(0)\exp(-t/T_2{}^*)$$

The time for M_{xy} to decay to half is equal to $\ln 2 \cdot T_2{}^*$.

13.7.3 Measurement of T_1

The rate of decay of M_z cannot be measured directly since M_z is parallel to the external magnetic field. However, the instantaneous value of M_z can be found at any time by applying a 90° pulse and measuring immediately the numerically equal value of M_{xy}. Hence the sequence: 180° pulse—variable time decay; 90° pulse known as the inversion-recovery sequence, will give the signals shown in Fig. 13.9 from which the time for the signal to decay to half and hence T_1 may be found.

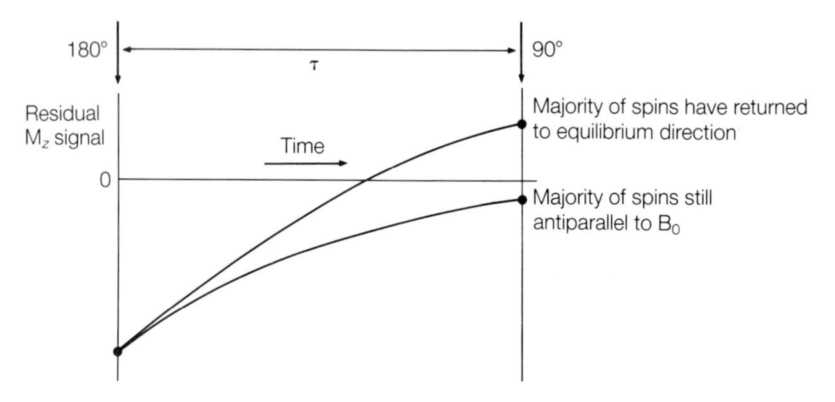

Fig. 13.9. Curves showing that the 180°, τ, 90° sequence can produce either a positive or negative signal depending on whether the relaxation rate is fast or slow. In either case, provided exponential decay is assumed, T_1 may be calculated.

If the decay of M_z is assumed to be exponential, just two measurements, one after a 90° pulse, the other after a 180° pulse; $\tau = \tau'$; 90° sequence are required. τ' should be chosen to be of the same order as T_1 to give good discrimination from materials with different T_1 values.

This pulse sequence for the measurement of T_1 is easy to understand. One disadvantage is that a time interval of three to four times T_1 should be allowed to elapse before further measurements are made to allow the system to return to equilibrium. In practice alternative sequences are used to permit faster imaging.

13.7.4 Measurement of T_2

At first sight T_2 should be easy to measure simply by applying a 90° pulse and then observing the rate of decay of the M_{xy} magnetization.

In practice the time constant of this decay $T_2{}^*$ is much less than T_2 because dephasing of the spins occurs not only as a result of spin–spin interactions but also because of inhomogeneities in the magnetic field. If the field on a particular nucleus is slightly greater than $\mathbf{B}_0$, then since $\omega = \gamma \mathbf{B}$, this nucleus will precess a little faster than the others and hence get out of phase. Fortunately this dephasing mechanism is systematic, whereas spin–spin interaction is random, and the two can be separated by a process known as spin-echo. This process is summarized in Fig. 13.10. The end result is that the measured M_{xy} value represents spin–spin decay only, during the time interval.

As for T_1, pulse sequences used in practice are somewhat more elaborate. For details see for example Lerski (1985).

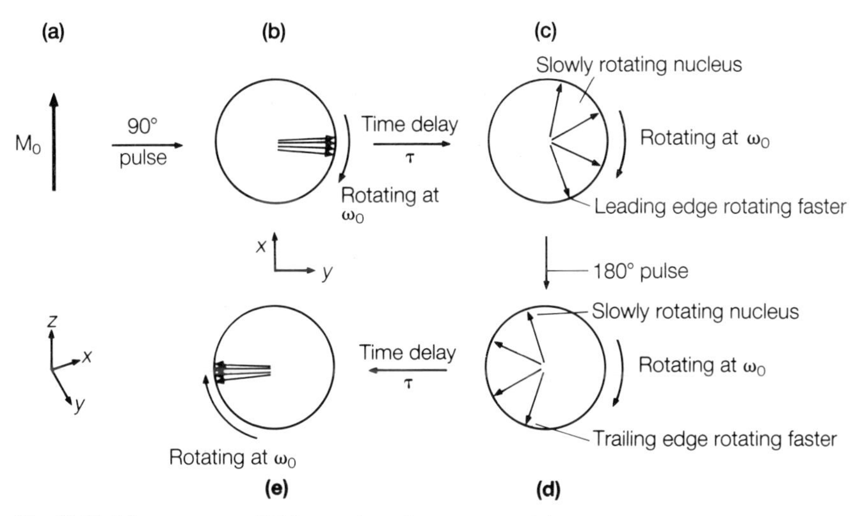

Fig. 13.10. Measurement of T_2 by a spin -echo sequence. (a) Equilibrium state. (b) 90° pulse, all vectors in phase. (c) Nuclear vectors dephase due to field inhomogeneities and due to spin–spin interactions (not shown). (d) 180° pulse applied about the x' axis in the rotating frame. After the 180° pulse, the nucleus that was rotating fastest continues to rotate fastest but is now trailing behind the other nuclei. (e) After time τ the 'fast' nucleus catches up, the 'slow' nucleus is caught by the pack and dephasing due to field inhomogeneities is, for a moment eliminated. The M_{xy} value is measured at this instant.

13.8 Imaging techniques

To localize a signal to a known small volume of the body, information in at least three separate directions is required. There are several ways in which this can be achieved and the subsequent sections simply outline some of those used most frequently.

13.8.1 Selective excitation of an image plane

If, in addition to the uniform field B_0 acting conventionally in the vertical z direction, a static field B_z (y) also acting in the z direction but varying with position along the long axis of the patient, conventionally the y axis, is applied, the total field will vary as shown in Fig. 13.11a. If the body is now subjected to a single frequency, excitation will occur only at the precise point where $\omega = \gamma[B_0 + B_z (y)]$, γ being the appropriate value for protons.

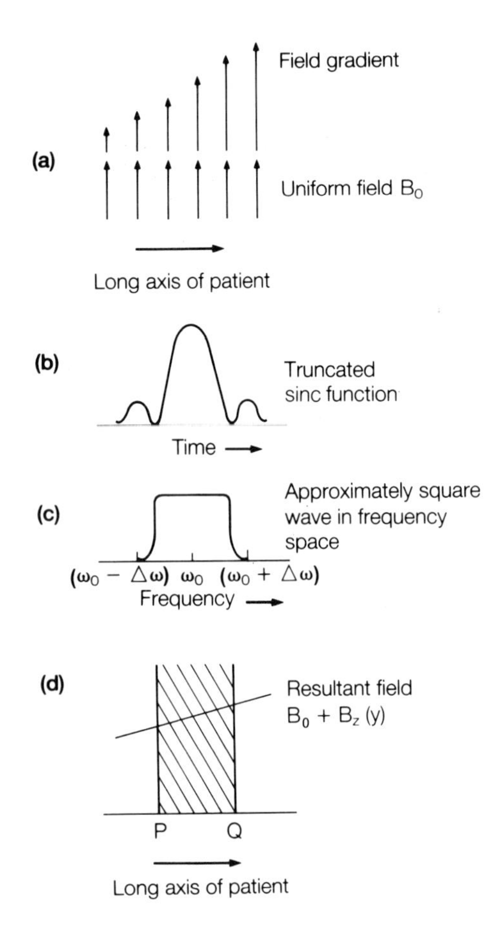

Fig. 13.11. Use of field gradients and a pulse of EM radiation to excite a slice in the patient. (a) A field gradient is superimposed on the uniform field (not to scale). (b) A pulse is applied for a short time. (c) The frequencies making up this pulse range from $\omega - \Delta\omega$ to $\omega + \Delta\omega$. (d) The protons in slice PQ are excited. At P, $(\omega - \Delta\omega) = \gamma B_P$, and at Q,$(\omega + \Delta\omega) = \gamma B_Q$.

If a narrow range of frequencies, extending from $\omega - \Delta\omega$ to $\omega + \Delta\omega$ is used, protons in the slice of tissue which extends from

$$(\omega - \Delta\omega) = \gamma[B_0 + B_z(y_1)]$$

to

$$(\omega + \Delta\omega) = \gamma[B_0 + B_z(y_2)]$$

will be excited (Figs 13.11b,c,d). According to Fourier analysis, the shorter the pulse the greater the number of frequencies it contains. Thus slice thickness is determined by the pulse duration and the steepness of the gradient of the variable field $B_z(y)$.

Note that if a second field gradient were applied, varying say along the x axis, the excited protons would all lie along a line. If yet a third gradient were applied, only a small volume element would have the resonant value of B.

13.8.2 Localization within a slice

One dimension may be coded to frequency using the inverse of the selective excitation principle. If, during read out, an additional field $B_z(x)$ is applied, again in the z direction but now varying in the x direction, the field at a given point x_1 on the x axis will be $B_0 + B_z(x_1)$ and the frequency of the signal emitted from that point will be $\gamma[B_0 + B_z(x_1)]$. The frequency of the signal emitted from a different point x_2 will be $\gamma[B_0 + B_z(x_2)]$. Hence analysis of the frequency spectrum of the output signal gives a projection of the selected plane onto the x axis.

The third dimension can be obtained by repeating this process at different angles and using the technique of filtered back projection as in X-ray computed tomography.

13.8.3 'Spin warp' method

This is an alternative method of reconstruction using the fact that each component of the signal has both a frequency and a phase. If a magnetic field gradient is applied in such a way and for such a time that the phases of the nuclei in different parts of the slice can be altered in a regular manner, producing a 'spin warp', it is possible to obtain a direct reading in the third dimension without using tomographic reconstruction techniques.

An important difference between the two reconstruction procedures is in the effect of non-uniformities in the magnetic flux density B_0. In the spin warp method, the image will be shifted slightly causing minor distortion but the more serious reconstruction artefacts resulting from filtered back projection do not occur.

13.9 Choice of imaging system

The most important component of the imaging system, and the only one there is room to consider in detail here, is the magnet. Early MR imagers contained resistive magnets in which the magnetic field is obtained by passing a current through a soft iron coil of magnetic material. A substantial amount of heat is generated because of the finite resistance of the coil. Such magnets provide all the features required to perform basic imaging for both head and whole body studies and certainly played an important role in launching the subject. Furthermore at low field strengths, of the order of 0.04–0.08 T, they are relatively cheap to operate.

However, there are three key requirements as regards magnet performance, and resistive magnets have proved inadequate in all of them. First, the field must be very homogeneous—otherwise the unique relationship between a signal of a particular frequency and a particular part of the body will be lost. Second, the field must be very stable for the same reason. Finally, for reasons related to signal strength and signal:noise ratios (see section 13.5), field strengths greater than the maximum attainable with resistive magnets (perhaps 0.2–0.3 T) are desirable.

A superconductor is a material that becomes resistanceless when cooled to liquid helium temperatures (about 4 K) and in this condition, when wound as a magnet, it is capable of creating very high magnetic fields with virtually no power dissipation. Furthermore, the fields are much more uniform and stable than those produced by a resistive magnet. As a direct consequence of the introduction of such magnets, image quality has improved dramatically and over 90% of the magnets now being produced for imagers are of the superconducting type.

Some problems remain and important new applications depend on further magnet development. For example, superconducting magnets are expensive to manufacture. The major cost is the superconducting wire so opportunities for savings are limited. These magnets are also expensive to run since they must be replenished with cryogens, both liquid helium, and liquid nitrogen to cool the helium, at regular intervals. These costs may be reducible by the incorporation of small reliquifying plants. Finally, because of the high field strength, there are significant stray magnetic fields out to quite large distances. These stray fields currently cause major installation problems and need to be reduced.

Research is continuing into the development of magnets capable of providing the higher field strengths (2.0 T and above) and even greater uniformity required for *in vivo* spectroscopy. Work is also in progress to try to reduce the time required to change the magnetic field strength. Because a very large amount of magnetic field energy is stored, this is not a trivial problem but experience to date suggests that different magnetic field strengths may be optimum for different processes. Head scans may be best at 0.5 T, body scans at 1.5 T. Field strengths of 2.0 T and above are certainly desirable for spectroscopy. The prospect of being able to change the field strength sufficiently quickly to permit different groups of investigations to be made under optimum conditions is clearly very attractive.

Two other essential components of the imager are the radiofrequency coils that control the perturbing magnetic field, B_1, and the field gradient coils. Finally electronic devices are required to perform a number of functions. They must apply a radio frequency voltage in pulsed sequence across the radiofrequency coil and must also control the field gradients. They must detect the magnetic resonance signal—either the free induction decay or, if a pulse sequence is used, the spin echo. Finally, they must reconstruct the image from the data and permit it to be manipulated, much as for any other form of digitized image, so that it may be displayed in the required form.

13.10 Biological significance of imaged protons and contrast

The simplest measure to make, using the procedure of section 13.7.1, is proton density. This will give quite good contrast between bone and soft tissue, or between soft tissue and gas in the lung because their proton densities are dissimilar. However, discrimination between soft tissues is poor because their proton densities are too similar.

The range of T_1 values observed in biological tissues is much wider. Even within soft tissues the range can be very great especially at low field strength (see Table 13.2). Pathological tissues generally have longer T_1 values than their normal counterparts and very aqueous structures, such as fluid cysts, have the longest relaxation times. T_1 values are frequency dependent and the relative values for different tissues also vary with frequency.

Several commercial machines use a combination of the T_1 value and the proton density value to enhance contrast between neighbouring tissues in the image.

Table 13.2. Typical T_1 values for different normal and pathological human soft tissues measured on MR images obtained at 1.7 MHz (0.04 T)

Tissue	T_1 (ms)
Liver	150
Fat	170
Cerebrum (white matter)	225
Cerebrum (grey matter)	275
Brain infarction	350
Liver tumour	400
Cerebro-spinal fluid	450
Simple Serous cyst	900
Urine	1000

Note that T_1 values are frequency dependent being longer at higher resonant frequencies (higher values of B_0).

13.11 Hazards of MR imaging

Three potential sources of hazard exist; the static magnetic field, the varying magnetic field (i.e. switched gradient field) and the radio frequency fields (for a review see Saunders 1982).

Possible contra-indications are heart disease, epilepsy and pregnancy. Some pacemakers are significantly affected by magnetic fields and great care should be taken with patients who use a pacemaker. The study should probably not be undertaken without full knowledge of its effect on that particular type of pacemaker.

Indirect risks include suffocation by gas from a quenching superconducting magnet, mechanical or electrical failure and, perhaps the most likely mechanical injury, the result of metal objects being attracted into the magnet.

13.11.1 Static magnetic fields

These vary from about 0.02 T to 2.0 T. The National Radiological Protection Board (NRPB) (1984) has recommended an upper limit of 2.5 T for imaging studies in the UK, although no lasting effect following exposure to fields of this magnitude has been measured. It is further recommended that staff operating the equipment should not be exposed for prolonged periods to more than 0.02 T to the whole body or more than 0.2 T to the arms or hands.

The possible harmful effect is depolarization of blood cells following induction of a potential difference across a blood vessel with blood flowing at right angles to the magnetic field. This depolarization has a threshold of 40 mV across a cell. However the maximum potential difference created across the aorta is only 15 mV T^{-1} so the potential across an individual cell is only a fraction of 40 mV even with the highest magnetic fields in current use.

13.11.2 Switched magnetic fields

The NRPB suggested limit for the maximum rate of change of magnetic field is 20 T s^{-1} for pulses longer than 10 s. Although the rate of change of magnetic field, either when switching on or off the main magnet, or in the gradient coils, is much less than this figure, little work has actually been performed to study effects *in vivo* and this should be done.

13.11.3 Radio frequency fields

Radio frequency fields produce a rise in temperature in the exposed tissue. At particular risk are tissues such as the testes and eyes which have a poor vascular supply and thus remove heat inefficiently. The NRPB recommend that skin and rectal temperature measurements should not show a rise in body temperature of more than one centigrade degree and the specific absorption rate in any mass of tissue not exceeding 1 g should be limited to 4 W kg^{-1}. The radio frequency power output varies significantly between different MR imagers and the amount of energy absorbed increases significantly with increasing frequency. Each machine must therefore be considered separately.

13.12 Conclusions and future prospects

In this chapter, the physical principles of magnetic resonance imaging have been discussed. The fundamental concept is that a nucleus spinning in a static magnetic field is able to draw energy from an applied alternating magnetic field but this can only happen if the frequency of the alternating field (ω_L) is related to the flux density of the static magnetic field B_0 by the relationship $\omega_L = \gamma B_0$ where γ is the gyromagnetic ratio. When nuclei have been excited in this way their resultant magnetic vector precesses around the magnetic field and induces an EMF in a coil wrapped around the patient. The magnitude of this EMF provides information about the number of nuclei in the excited state.

By using field gradients to vary the value of the static magnetic field in space, and the resonance condition $\omega_L = \gamma B_0$, it is possible to identify the region from which a particular signal is being received by its frequency. If the field gradient is applied in one dimension, the signal is localized to a slice, if applied in two dimensions it originates from a line and if applied in three dimensions it identifies a small volume element. In practice several different techniques are available for generating spatial information and hence MR images.

Elements of biological importance with non-zero nuclear spins include hydrogen, nitrogen and phosphorus. However, only hydrogen is present in sufficient relative atomic abundance in man to produce high resolution images *in vivo*, the size of the resolution element being determined primarily by the signal:noise ratio.

MR imaging has already demonstrated a number of important advantages. First, it is able to measure several different properties of the system, notably proton density, T_1 and T_2 values. Second, with particular reference to T_1 measurements, it provides a unique opportunity to study the aqueous environment of soft tissues *in vivo*. The more loosely water is bound to the structure, the longer T_1, a good example being the elevated T_1 values for malignant tissues. It has been known for many years, from work on experimentally induced tumours in animals, that the water content of malignant tissue increases at an early stage, but the phenomenon could not previously be studied *in vivo*. The ubiquitous presence of water in the body makes this aspect of MR imaging a valuable tool. Third, and again with particular reference to T_1, Table 13.2 shows a range of values of some 400%—contrast the differences in attenuation, sometimes as small as 1–2%, that have to be distinguished in conventional radiology. Thus if T_1 values of normal tissue can be measured with a high degree of reproducibility, this wide range provides a valuable discriminatory technique. Finally, MR imaging does not involve the use of ionizing radiation so there is no reason to anticipate the long-term harmful effects of carcinogenesis and mutagenesis associated with X-rays. Indeed there are no known deleterious effects attached to MR imaging at present, although the National Radiological Protection Board has issued guidelines limiting the strength of static and alternating fields as a precautionary measure. Note that it is a misconception to state that MR imaging is 'non-invasive'. It is not possible to obtain information about the body by any physical imaging technique unless the body is first 'invaded' by some form of physical signal.

In this chapter, only the basics of MR imaging have been presented but the technique has tremendous potential for further development. For example by changing the sequence in which magnetic field gradients are applied, it is possible to obtain directly sagittal and coronal sections. Contrast CT imaging where such views can only be obtained by taking a series of transverse sections and reconstructing other views by mathematical techniques. There are also several ways in which T_1 values can be modified, notably by the use of paramagnetic contrast agents. All compounds containing transition elements, for example manganese and cobalt, are paramagnetic but they tend to be toxic. Free radicals are also paramagnetic and intensive research can be expected into the development of non-toxic, stable, free radicals. Techniques are also being developed for quantitation of flow by MR imaging (Bryant *et al.* 1984, O'Donnell 1985). Finally there will be important developments in low resolution imaging and *in vivo* spectroscopy using elements other than hydrogen, notably phosphorus.

It is difficult to anticipate the ultimate role of MR imaging, which will have to compete with developments in other imaging technologies. For example, digital radiology will become more powerful as methods of data capture, storage, display, processing, quantitative analysis and image restoration, and communication are all steadily improved. In ultrasonic imaging, a relatively recent development has been its use in conjunction with Doppler flow studies. Combining these two techniques is beginning to produce some very promising results especially in the assessment of carotid arterial disease. MR imaging itself has made dramatic progress during its short lifetime and reference to X-ray CT suggests that a full 10 years may elapse before the quality of MR images approaches its full potential.

Irrespective of further improvements in image quality, the future roles of MR and other imaging techniques will be heavily influenced by considerations of simplicity of procedures, patient throughput, facility of data communication and storage, and cost. In the latter context, one must question whether the days of photographic film in the X-ray department are numbered, and whether the year 2000 will see a conversion to information collection and retrieval systems based on digital storage media and digital to analogue conversion for video display on the radiologist's personal console.

The physics of radiology has undergone a period of rapid change during the past two decades. Although such a rate of change is unlikely to be sustained, it is clear that new developments will require continuing updating of the radiologist's knowledge and understanding of the physics and technology of his subject.

References and further reading

Andrew E. R. (1982) Nuclear magnetic resonance imaging. In *Scientific Basis of Medical Imaging*, ed. P. N. T. Wells, pp. 212–236. Churchill Livingstone, Edinburgh.

Bryant D. J., Payne J. A., Firmin D. N. & Longmore D. B. (1984) Measurement of flow with NMR imaging using a gradient pulse and phase difference technique. *J. Comp. Assist. Tomog.* **8**, 588–593.

Gadian D. G. (1982) *Nuclear Magnetic Resonance and its Application to Living Systems*. Clarendon Press, Oxford.

Lerksi R. A., ed. (1985) *Physical Principles and Clinical Applications of Nuclear Magnetic Resonance*. (IPSM 2) Hospital Physicists' Association, London.

Mansfield P. & Morris P. G. (1983) *NMR Imaging in Biomedicine*. Supplement 2, Advances in Magnetic Resonance. Academic Press, New York.

National Radiological Protection Board (UK) (1984) *Advice on Acceptable Limits of Exposure in Nuclear Magnetic Resonance Clinical Imaging*. HMSO, London.

O'Donnell M. (1985) NMR blood flow imaging using multiecho, phase contrast sequences. *Med. Phys.* **12**, 59–64.

Saunders R. D. (1982) Biological hazards of NMR. In *NMR Imaging: Proceedings of the International Symposium on NMR Imaging, Winston, Salem, October 1981*, ed. R. L. Witcofski, N. Karstaedt & C. L. Partain, pp. 65–67.

Valk J., MacLean C. & Algra P. R. (1985) *Basic Principles of Nuclear Magnetic Resonance Imaging*. Elsevier, Amsterdam.

Young S. W. (1984) *Nuclear Magnetic Resonance Imaging. Basic Principles.* Raven Press, New York.

Exercises

1 Explain what is meant by proton magnetic resonance.

2 Discuss the factors that make protons the most suitable nuclei for *in vivo* MR imaging.

3 Indicate briefly the type of information that proton magnetic resonance can give about the body.

4 Water is vital to life. Why is it becoming important in diagnosis?

5 What is the function of the radiofrequency coil in an MR imaging machine?

6 Describe the behaviour of
(a) the individual nuclear magnetic moments
(b) the macroscopic magnetization
when (i) a 90° pulse (ii) a 180° pulse is applied.

7 Explain what is meant by a free induction decay signal. Why is it not possible to obtain T_2 from measurements of this signal?

8 Describe two pulse sequences that will permit calculation of T_1 values. Indicate any advantages and disadvantages of each method for *in vivo* imaging.

9 Why does a magnetic field gradient make it possible to tell whether water protons are on one side of a sample placed in an MR spectrometer or the other?

10 What factors determine slice thickness when the technique of selective activation is used?

11 Compare and contrast the production of a tomogram by a CT scanner and an MR imager.

Index